Formal Methods for Nonmonotonic and Related Logics

Karl Schlechta

Formal Methods for Nonmonotonic and Related Logics

Vol I: Preference and Size

Springer

Karl Schlechta
CNRS, LIF UMR 7279
Aix-Marseille Université
Marseille, France
and Frammersbach, Germany

ISBN 978-3-319-89652-6 ISBN 978-3-319-89653-3 (eBook)
https://doi.org/10.1007/978-3-319-89653-3

Library of Congress Control Number: 2018960248

© Springer Nature Switzerland AG 2018

This work is subject to copyright. All rights are reserved by the Publisher, whether the whole or part of the material is concerned, specifically the rights of translation, reprinting, reuse of illustrations, recitation, broadcasting, reproduction on microfilms or in any other physical way, and transmission or information storage and retrieval, electronic adaptation, computer software, or by similar or dissimilar methodology now known or hereafter developed.

The use of general descriptive names, registered names, trademarks, service marks, etc. in this publication does not imply, even in the absence of a specific statement, that such names are exempt from the relevant protective laws and regulations and therefore free for general use.

The publisher, the authors and the editors are safe to assume that the advice and information in this book are believed to be true and accurate at the date of publication. Neither the publisher nor the authors or the editors give a warranty, express or implied, with respect to the material contained herein or for any errors or omissions that may have been made. The publisher remains neutral with regard to jurisdictional claims in published maps and institutional affiliations.

This Springer imprint is published by the registered company Springer Nature Switzerland AG
The registered company address is: Gewerbestrasse 11, 6330 Cham, Switzerland

Foreword

The study of nonmonotonic consequence relations began in the 1980s, and has been developing vigorously ever since. The literature is now so vast that it has become difficult for graduate students and investigators alike to obtain a clear and up-to-date picture of what has exactly has been done and how it all fits together.

This book, by one of the most eminent researchers in the field, fills much of the gap. The first volume shows systematically how concepts and facts about many nonmonotonic logics reflect the behavior of underlying semantic structures - sets of classical models, equipped with relations of preference, operations of choice, and related devices. The second volume traces the ways in which closely related techniques have been developed to support neighboring areas – belief change, defeasible inheritance, counterfactual conditionals, and more.

The author's magisterial exposition is accompanied by intuitive and heuristic asides, and includes many solved exercises that will be of great assistance to the reader. Every mathematical logician working in the area, or thinking of getting into it, should have the double volume on hand for consultation, and any library with a section on mathematical logic should have a copy for reference.

London, December 2017 David Makinson
(London School of Economics)

Preface

Introduction

This book is a textbook, mainly discussing formal results and techniques for nonmonotonic and related logics. Therefore, motivation and description of the general context are quite limited, and given only when needed to understand the formal material, or to illustrate the path from intuition to formalisation.

The text presents results and proof methods, but also translations of motivational and philosophical considerations to formal constructions. This is true, in particular, for Section 5.7 which contains hardly any formal results, but shows how to reconcile various ideas in one formal construction. Section 7.3 is mostly about translating intuitive requirements for deontic logics into formal properties, too. It is a personal book, in the sense that it treats methods and results either used, or, more frequently, developed by the author, sometimes in cooperation with collegues.

For more introductory and motivational material, the reader is referred to other publications by the author, and in particular to [Sch04]. It is strongly advised to have a copy of this text at hand.

The intended audience is advanced students, but also researchers in the domain, who want to have a compendium of methods developed by a colleague. The book contains many exercises, the solutions are either in the appendix, or in [Sch04].

Many chapters are rather independent from each other, the numbers of the chapters give already a suggestion for reading. Chapter 3 however, contains many abstract ideas which may serve as intuitive guidelines. (It is put after the chapter on preferential structures, as these structures give an example from which the size concept is abstracted.)

Previously Published Material and Acknowledgements

This text contains no new material - except the short Section 5.8. It is based on previously published material, the following list will give the main sources, more information is given locally in the text.

In particular, significant parts in Section 1.4, Section 1.6, Section 1.7, and Section 4.4 were published in [Sch04].

- Section 1.2 has evolved over time, mainly: [Sch92], [Sch96-1], [Sch00-1], [Sch00-2], [GS08c], [GS09f].
- Section 1.3: [Sch92], [Sch96-1], [Sch00-2].
- Section 1.4: [Sch96-1], [Sch04], [GS08d], [GS10].
- Section 1.5: [Sch96-1], [GS08a], [GS09f].
- Section 1.6 and Section 1.7: [Sch04].
- Section 1.8: [Sch92], [Sch95-3], [Sch97-2], [BLS99], [Sch99], [SGMRT00].
- Chapter 2: [GS08b], [GS09f].
- Section 3.1: [Sch97-4].
- Section 3.2 has evolved over time, mainly: [GS08c], [GS09f].
- Section 3.3: [Sch95-1], [Sch97-2].
- Section 4.2 has evolved over time, mainly: [SLM96], [LMS01], [Sch04], [GS09f].
- Section 4.3: [Sch91-1], [SLM96], [LMS01], [Sch04], [GS08f], [GS08h].
- Section 4.4: [DS99], [SD01], [Sch04].
- Section 5.2 and Section 5.3: [Sch97-2].
- Section 5.4: [Sch93].
- Section 5.5: [GS08f].
- Section 5.6: [Sch90].
- Section 5.7: [GS16].
- Chapter 6: [GS09c], [GS10], [GS16].
- Chapter 7: [GS10].
- Chapter 8: [GS16].
- Section 9.1: [SM94].
- Section 9.2: [Sch91-2], [Sch97-2].
- Section 9.3: [Sch95-2], [Sch97-2].
- Section 9.4: [GS16].

I would like to thank my co-authors, and also Ronan Nugent of Springer, for his very patient and helpful comments.

Frammersbach 2017   Karl Schlechta

Contents (Volume I)

Volume I Preference and Size

1 Preferential Structures 3
  1.1 Introduction 4
  1.2 Basic Definitions and Overview 5
      1.2.1 Introduction 5
      1.2.2 General Properties 5
            1.2.2.1 Algebraic Properties 5
            1.2.2.2 Logical Properties 11
      1.2.3 General Nonmonotonic Logic 19
            1.2.3.1 Algebraic Properties 20
            1.2.3.2 Logical Properties 24
            1.2.3.3 Connections Between Algebraic and Logical Properties 26
      1.2.4 Preferential Structures 32
            1.2.4.1 Algebraic Properties 33
            1.2.4.2 Logical Properties 41
      1.2.5 Algebraic and Structural Semantics 42
            1.2.5.1 Abstract or Algebraic Semantics 42
            1.2.5.2 Structural Semantics 45
      1.2.6 Tables for Logical and Semantical Rules 45
      1.2.7 Tables for Preferential Structures 45
  1.3 Basic Cases 54

		1.3.1	Introduction 54
		1.3.2	General Preferential Structures 55
			1.3.2.1 General Minimal Preferential Structures 55
			1.3.2.2 Transitive Minimal Preferential Structures 58
		1.3.3	Smooth Minimal Preferential Structures with Arbitrarily Many Copies 60
			1.3.3.1 Discussion 60
			1.3.3.2 The Constructions 61
			1.3.3.3 Smooth and Transitive Minimal Preferential Structures 66
		1.3.4	The logical characterization 70
	1.4	Ranked Case 71
		1.4.1	Ranked Preferential Structures 71
			1.4.1.1 Introduction 71
			1.4.1.2 The Details 72
		1.4.2	$\mathcal{A}$-Ranked Structures 80
			1.4.2.1 Introduction 80
			1.4.2.2 Representation Results for $\mathcal{A}$-Ranked Structures .. 80
	1.5	The Smooth Case Without Domain Closure 88
		1.5.1	Introduction 88
		1.5.2	Problems without closure under finite union 88
			1.5.2.1 Introduction 88
			1.5.2.2 Introduction to Plausibility Logic 88
			1.5.2.3 Completeness and Incompleteness Results for Plausibility Logic 91
			1.5.2.4 A Comment on the Work by Arieli and Avron 96
		1.5.3	Smooth Preferential Structures Without Domain Closure Conditions 97
			1.5.3.1 Introduction 97
			1.5.3.2 Detailed Discussion 100
	1.6	The Limit Variant 116
		1.6.1	Introduction 116
		1.6.2	The Algebraic Limit 118
		1.6.3	The Logical Limit 120

		1.6.3.1	Translation Between the Minimal and the Limit Variant 120
		1.6.3.2	Logical Properties of the Limit Variant 124
	1.6.4	Simplifications of the General Transitive Limit Case 126
	1.6.5	Ranked Structures Without Copies 128
		1.6.5.1	Introduction 128
		1.6.5.2	Representation 128
		1.6.5.3	Partial Equivalence of Limit and Minimal Ranked Structures 130
1.7	Preferential Structures Without Definability Preservation 134
	1.7.1	Introduction 134
		1.7.1.1	The Problem 135
	1.7.2	Characterisations Without Definability Preservation 137
		1.7.2.1	Introduction 137
		1.7.2.2	General and Smooth Structures Without Definability Preservation 140
		1.7.2.3	Ranked Structures 147
		1.7.2.4	The Logical Results 149
	1.7.3	The General Case and the Limit Version Cannot Be Characterized 151
		1.7.3.1	Introduction 151
		1.7.3.2	The Details 152
1.8	Various Results and Approaches 157
	1.8.1	Introduction 157
	1.8.2	The Role of Copies in Preferential Structures 158
		1.8.2.1	The Infinite Case 158
		1.8.2.2	One Copy Version 160
	1.8.3	A Counterexample to the KLM-System 161
		1.8.3.1	Introduction 161
		1.8.3.2	The Formal Results 162
	1.8.4	A Nonsmooth Model of Cumulativity 164
		1.8.4.1	Introduction 164
		1.8.4.2	The Formal Results 165
	1.8.5	A New Approach to Preferential Structures 169

		1.8.5.1	Introduction 169
		1.8.5.2	Validity in Traditional and in Our Preferential Structures 172
		1.8.5.3	The Disjoint Union of Models and the Problem of Multiple Copies 174
		1.8.5.4	Representation in the Finite Case 177
	1.8.6	Preferred History Semantics for Iterated Updates 181
		1.8.6.1	Introduction.................................. 181
		1.8.6.2	Some Important Logical Properties of Updates ... 188
		1.8.6.3	A Representation Theorem 191
	1.8.7	Orderings on $\mathcal{L}$ and Completeness Results 201
		1.8.7.1	Introduction 201
		1.8.7.2	A Natural Ordering 201
		1.8.7.3	Comparison to Orders in [KLM90] and [LM92] .. 206
		1.8.7.4	The Results of [GM94] 208
		1.8.7.5	Completeness Results 209
		1.8.7.6	The Rank 215
	1.8.8	Preferential Choice Representation Theorems for Branching Time Structures 216
		1.8.8.1	Overview 216
		1.8.8.2	A Ranked and Smooth Preferential Representation for a Deontic Choice Function ... 217
		1.8.8.3	An Extension of the Katsuno/Mendelzon Update Semantics 223

2 Higher Preferential Structures 233

2.1 Introduction 233
2.2 IBRS 234
2.2.1 Definition and Comments 234
2.2.2 The Power of IBRS 238
2.2.3 Abstract Semantics for IBRS and Its Engineering Realization 239
2.2.3.1 Introduction 239
2.2.3.2 A Circuit Semantics for Simple IBRS Without Labels 242

Contents (Volume I)

- 2.3 Higher Preferential Structures 244
  - 2.3.1 Introduction 244
  - 2.3.2 The General Case 252
  - 2.3.3 Discussion of the Totally Smooth Case 257
  - 2.3.4 The Essentially Smooth Case 260
  - 2.3.5 Translation to Logic 264

3 Abstract Size 267

- 3.1 Introduction 267
  - 3.1.1 Comparison of Three Abstract Coherent Size Systems 268
- 3.2 Basic Definitions and Overview 269
  - 3.2.1 Introduction 269
    - 3.2.1.1 Additive and Multiplicative Laws About Size 270
  - 3.2.2 Additive Properties 270
    - 3.2.2.1 Discussion of the Tables 3.2 and 3.3 272
    - 3.2.2.2 A Partial Order View 276
    - 3.2.2.3 Discussion of Other, Related, Rules 277
  - 3.2.3 Coherent Systems 278
    - 3.2.3.1 Definition and Basic Facts 278
    - 3.2.3.2 Implications Between the Finite Versions 280
    - 3.2.3.3 Implications Between the ω Versions 281
    - 3.2.3.4 Rational Monotony 284
    - 3.2.3.5 Size and Principal Filter Logic 284
  - 3.2.4 Multiplicative Properties 287
    - 3.2.4.1 Multiplication of Size for Subsets 288
    - 3.2.4.2 Multiplication of Size for Subspaces 290
    - 3.2.4.3 Conditions for Abstract Multiplication and Generating Relations 296
  - 3.2.5 Modular Relations and Multiplication of Size 298
    - 3.2.5.1 Hamming Distances 305
    - 3.2.5.2 Some Examples 307
  - 3.2.6 Tables for Abstract Size 309
- 3.3 Defaults as Generalized Quantifiers 316
  - 3.3.1 Introduction 316

		3.3.1.1	In More Detail 317
	3.3.2	Semantics and Proof Theory 318
		3.3.2.1	Overview of This Section 318
		3.3.2.2	Semantics 320
		3.3.2.3	Proof Theory 322
		3.3.2.4	Soundness and Completeness 322
		3.3.2.5	Extension to Normal Defaults with Prerequisites . 324
		3.3.2.6	Extension to $\mathcal{N}$-Families 325
	3.3.3	Strengthening the Axioms 327
		3.3.3.1	Overview of This Section 327
		3.3.3.2	The Details 328
		3.3.3.3	An Alternative Semantics for a Predicate Logic Version of P and R 329
	3.3.4	Sceptical Revision of Partially Ordered Defaults 330
		3.3.4.1	Overview of This Section 330
		3.3.4.2	Introduction 331
		3.3.4.3	Basic Definitions and Approaches 331

Contents (Volumes I and II)

Volume I Preference and Size

1 Preferential Structures 3
   1.1 Introduction 4
   1.2 Basic Definitions and Overview 5
        1.2.1 Introduction 5
        1.2.2 General Properties 5
                1.2.2.1 Algebraic Properties 5
                1.2.2.2 Logical Properties 11
        1.2.3 General Nonmonotonic Logic 19
                1.2.3.1 Algebraic Properties 20
                1.2.3.2 Logical Properties 24
                1.2.3.3 Connections Between Algebraic and Logical Properties 26
        1.2.4 Preferential Structures 32
                1.2.4.1 Algebraic Properties........................ 33
                1.2.4.2 Logical Properties 41
        1.2.5 Algebraic and Structural Semantics 42
                1.2.5.1 Abstract or Algebraic Semantics 42
                1.2.5.2 Structural Semantics 45
        1.2.6 Tables for Logical and Semantical Rules 45
        1.2.7 Tables for Preferential Structures 45
   1.3 Basic Cases 54

XVII

	1.3.1	Introduction 54
	1.3.2	General Preferential Structures 55
		1.3.2.1 General Minimal Preferential Structures 55
		1.3.2.2 Transitive Minimal Preferential Structures 58
	1.3.3	Smooth Minimal Preferential Structures with Arbitrarily Many Copies 60
		1.3.3.1 Discussion 60
		1.3.3.2 The Constructions 61
		1.3.3.3 Smooth and Transitive Minimal Preferential Structures 66
	1.3.4	The logical characterization 70
1.4	Ranked Case 71
	1.4.1	Ranked Preferential Structures 71
		1.4.1.1 Introduction 71
		1.4.1.2 The Details 72
	1.4.2	$\mathcal{A}$-Ranked Structures 80
		1.4.2.1 Introduction 80
		1.4.2.2 Representation Results for $\mathcal{A}$-Ranked Structures .. 80
1.5	The Smooth Case Without Domain Closure 88
	1.5.1	Introduction 88
	1.5.2	Problems without closure under finite union 88
		1.5.2.1 Introduction 88
		1.5.2.2 Introduction to Plausibility Logic 88
		1.5.2.3 Completeness and Incompleteness Results for Plausibility Logic 91
		1.5.2.4 A Comment on the Work by Arieli and Avron 96
	1.5.3	Smooth Preferential Structures Without Domain Closure Conditions 97
		1.5.3.1 Introduction 97
		1.5.3.2 Detailed Discussion 100
1.6	The Limit Variant 116
	1.6.1	Introduction 116
	1.6.2	The Algebraic Limit 118
	1.6.3	The Logical Limit 120

		1.6.3.1	Translation Between the Minimal and the Limit Variant 120
		1.6.3.2	Logical Properties of the Limit Variant 124
	1.6.4	Simplifications of the General Transitive Limit Case 126
	1.6.5	Ranked Structures Without Copies 128
		1.6.5.1	Introduction 128
		1.6.5.2	Representation 128
		1.6.5.3	Partial Equivalence of Limit and Minimal Ranked Structures 130
1.7	Preferential Structures Without Definability Preservation 134
	1.7.1	Introduction 134
		1.7.1.1	The Problem 135
	1.7.2	Characterisations Without Definability Preservation 137
		1.7.2.1	Introduction 137
		1.7.2.2	General and Smooth Structures Without Definability Preservation 140
		1.7.2.3	Ranked Structures 147
		1.7.2.4	The Logical Results 149
	1.7.3	The General Case and the Limit Version Cannot Be Characterized 151
		1.7.3.1	Introduction 151
		1.7.3.2	The Details 152
1.8	Various Results and Approaches 157
	1.8.1	Introduction 157
	1.8.2	The Role of Copies in Preferential Structures 158
		1.8.2.1	The Infinite Case 158
		1.8.2.2	One Copy Version 160
	1.8.3	A Counterexample to the KLM-System 161
		1.8.3.1	Introduction 161
		1.8.3.2	The Formal Results 162
	1.8.4	A Nonsmooth Model of Cumulativity 164
		1.8.4.1	Introduction 164
		1.8.4.2	The Formal Results 165
	1.8.5	A New Approach to Preferential Structures 169

		1.8.5.1	Introduction 169
		1.8.5.2	Validity in Traditional and in Our Preferential Structures 172
		1.8.5.3	The Disjoint Union of Models and the Problem of Multiple Copies 174
		1.8.5.4	Representation in the Finite Case 177
	1.8.6	Preferred History Semantics for Iterated Updates 181
		1.8.6.1	Introduction 181
		1.8.6.2	Some Important Logical Properties of Updates ... 188
		1.8.6.3	A Representation Theorem 191
	1.8.7	Orderings on $\mathcal{L}$ and Completeness Results 201
		1.8.7.1	Introduction 201
		1.8.7.2	A Natural Ordering 201
		1.8.7.3	Comparison to Orders in [KLM90] and [LM92] .. 206
		1.8.7.4	The Results of [GM94] 208
		1.8.7.5	Completeness Results 209
		1.8.7.6	The Rank 215
	1.8.8	Preferential Choice Representation Theorems for Branching Time Structures 216
		1.8.8.1	Overview 216
		1.8.8.2	A Ranked and Smooth Preferential Representation for a Deontic Choice Function ... 217
		1.8.8.3	An Extension of the Katsuno/Mendelzon Update Semantics 223

2 Higher Preferential Structures 233

2.1 Introduction 233

2.2 IBRS 234

    2.2.1 Definition and Comments 234

    2.2.2 The Power of IBRS 238

    2.2.3 Abstract Semantics for IBRS and Its Engineering Realization 239

		2.2.3.1	Introduction 239
		2.2.3.2	A Circuit Semantics for Simple IBRS Without Labels 242

	2.3	Higher Preferential Structures 244	
		2.3.1 Introduction 244	
		2.3.2 The General Case 252	
		2.3.3 Discussion of the Totally Smooth Case 257	
		2.3.4 The Essentially Smooth Case 260	
		2.3.5 Translation to Logic 264	
3	**Abstract Size** 267		
	3.1	Introduction 267	
		3.1.1 Comparison of Three Abstract Coherent Size Systems 268	
	3.2	Basic Definitions and Overview 269	
		3.2.1 Introduction 269	
		3.2.1.1 Additive and Multiplicative Laws About Size 270	
		3.2.2 Additive Properties 270	
		3.2.2.1 Discussion of the Tables 3.2 and 3.3 272	
		3.2.2.2 A Partial Order View 276	
		3.2.2.3 Discussion of Other, Related, Rules 277	
		3.2.3 Coherent Systems 278	
		3.2.3.1 Definition and Basic Facts 278	
		3.2.3.2 Implications Between the Finite Versions 280	
		3.2.3.3 Implications Between the ω Versions 281	
		3.2.3.4 Rational Monotony 284	
		3.2.3.5 Size and Principal Filter Logic 284	
		3.2.4 Multiplicative Properties 287	
		3.2.4.1 Multiplication of Size for Subsets 288	
		3.2.4.2 Multiplication of Size for Subspaces 290	
		3.2.4.3 Conditions for Abstract Multiplication and Generating Relations 296	
		3.2.5 Modular Relations and Multiplication of Size 298	
		3.2.5.1 Hamming Distances 305	
		3.2.5.2 Some Examples 307	
		3.2.6 Tables for Abstract Size 309	
	3.3	Defaults as Generalized Quantifiers 316	
		3.3.1 Introduction 316	

		3.3.1.1	In More Detail 317
	3.3.2	Semantics and Proof Theory 318
		3.3.2.1	Overview of This Section 318
		3.3.2.2	Semantics 320
		3.3.2.3	Proof Theory................................ 322
		3.3.2.4	Soundness and Completeness 322
		3.3.2.5	Extension to Normal Defaults with Prerequisites . 324
		3.3.2.6	Extension to $\mathcal{N}$-Families 325
	3.3.3	Strengthening the Axioms 327
		3.3.3.1	Overview of This Section 327
		3.3.3.2	The Details 328
		3.3.3.3	An Alternative Semantics for a Predicate Logic Version of P and R 329
	3.3.4	Sceptical Revision of Partially Ordered Defaults 330
		3.3.4.1	Overview of This Section 330
		3.3.4.2	Introduction 331
		3.3.4.3	Basic Definitions and Approaches 331

Volume II Theory Revision, Inheritance, and Various Abstract Properties

4 Theory Revision and Sums 339
4.1	Introduction 339
4.2	Basic Definitions and Overview 340
	4.2.1 Introduction 340
	4.2.2 The AGM Approach 341
	4.2.3 Algebraic Properties 344
	4.2.4 Logical Properties 345
	4.2.5 Connections Between Algebraic and Logical Properties ... 346
	4.2.6 Tables for Theory Revision 348
4.3	Theory Revision 350
	4.3.1 Introduction 350
	4.3.2 Revision by Distance 351
	4.3.2.1 Introduction 351
	4.3.2.2 The Algebraic Results 351

Contents (Volumes I and II)   XXIII

- 4.3.2.3 The Logical Results 363
- 4.3.2.4 There Is No Finite Characterization 367
- 4.3.3 The Limit Case 370
- 4.3.4 Revision and Definability 371
  - 4.3.4.1 "Soft Characterisation" – The Algebraic Result .. 372
  - 4.3.4.2 The Logical Result 374
- 4.3.5 Update 375
  - 4.3.5.1 Introduction 375
  - 4.3.5.2 Hidden Dimensions 375
- 4.3.6 Theory Revision and Probability 384
  - 4.3.6.1 Introduction 384
  - 4.3.6.2 Epistemic Preference Relations 386
  - 4.3.6.3 Measuring Theories, and an Outlook for a Different Treatment of Theory Revision 392
- 4.3.7 Revision and Independence 394
  - 4.3.7.1 Problem and Background 394
  - 4.3.7.2 Factorisation 396
  - 4.3.7.3 Factorisation and Hamming Distance 401
- 4.3.8 Extension of the Multiple Relations Idea of [BCMG04] to the infinite case 402
  - 4.3.8.1 Introduction 402
  - 4.3.8.2 The Framework of [BCMG04] 405
  - 4.3.8.3 Construction and Proof 407
- 4.4 Sums and the Farkas Algorithm 417
  - 4.4.1 Introduction 417
    - 4.4.1.1 The General Situation and the Farkas Algorithm .. 417
    - 4.4.1.2 Update by Minimal Sums 418
    - 4.4.1.3 "Between" and "Behind" 420
  - 4.4.2 The Farkas Algorithm 420
  - 4.4.3 Representation for Update by Minimal Sums 422
    - 4.4.3.1 Introduction 422
    - 4.4.3.2 An Abstract Result 423
    - 4.4.3.3 Representation 425

		4.4.3.4	There Is No Finite Representation for Our Type of Update Possible 428
	4.4.4	"Between" and "Behind" 430	
		4.4.4.1	There Is No Finite Representation for "Between" and "Behind" 431

5 Defeasible Inheritance Theory 435

5.1 Introduction 435
    5.1.1 Overview 435

5.2 A Detailed Survey of Inheritance Theory a la Thomason et al. 437
    5.2.1 Introduction 437
        5.2.1.1 Basic Discussion 438
    5.2.2 Directly Sceptical Split Validity Upward Chaining Off-Path Inheritance 443
        5.2.2.1 The Definition of $\models$ (Validity of Paths) 446
        5.2.2.2 Properties of $\models$ 452

5.3 Review of Other Approaches and Problems 454
    5.3.1 Introduction 454
    5.3.2 Fundamental Differences 455
        5.3.2.1 Extension-Based Versus Directly Skeptical Definitions 455
        5.3.2.2 Upward Versus Downward Chaining 456
        5.3.2.3 On-Path Versus Off-Path Preclusion 457
        5.3.2.4 Split-Validity Versus Total-Validity Preclusion ... 457
        5.3.2.5 Intersection of Extensions Versus the Intersection of Their Conclusion Sets 457
    5.3.3 Problems Specific to Certain Approaches 459
        5.3.3.1 Discussion of the [HTT87] Approach, the Problem of Positive Support 459
        5.3.3.2 The Extensions Approach – Coherence Properties 463

5.4 Directly Sceptical Inheritance Cannot Capture the Intersection of Extensions 467
    5.4.1 Introduction 467
        5.4.1.1 History and Motivation 467

Contents (Volumes I and II) XXV

		5.4.1.2	Relevance of the Question for Inheritance Theory and Beyond 471
	5.4.2	Definitions, Statement, and Proof of Theorem 471
		5.4.2.1	Basic Cell (Figure 5.22) 474
		5.4.2.2	Combining Basic Cells (Figure 5.24) 476
		5.4.2.3	Final Construction of Γ 476
5.5	Detailed Translation of Inheritance to Modified Systems of Small Sets 479
	5.5.1	Normality 479
	5.5.2	Small Sets 480
5.6	A Semantics for Defeasible Inheritance 490
	5.6.1	Outline 490
	5.6.2	The Construction 490
	5.6.3	Discussion 491
	5.6.4	A Model from the Axioms 492
5.7	A Unified Structure 494
	5.7.1	Summary 494
	5.7.2	Desiderata 497
		5.7.2.1	Overall Aim 497
		5.7.2.2	Situation 498
		5.7.2.3	Rare Influence Changes and Its Consequences ... 500
	5.7.3	The Solution 502
		5.7.3.1	The Construction 505
	5.7.4	Discussion 509
		5.7.4.1	General Remarks 509
		5.7.4.2	Rarity and Its Coding by Inheritance 510
		5.7.4.3	Modularity 513
		5.7.4.4	Graceful Degradation and Coherence 513
		5.7.4.5	Core and Extensions 514
		5.7.4.6	Contradictions 514
		5.7.4.7	Philosophy of Science 514
		5.7.4.8	The Different Aspects of Our Construction 515
		5.7.4.9	Modifications 515
5.8	Influence Change and Inheritance 517

		5.8.1	Introduction 517
		5.8.2	Size 517
			5.8.2.1 Definitions and Connection to Inheritance 518
			5.8.2.2 Generalization 527
		5.8.3	Influence Change and Inheritance 528
			5.8.3.1 Intersection of Extensions Versus Direct Scepticism 528
			5.8.3.2 Specificity 529
			5.8.3.3 Upward or Downward Chaining? 531
		5.8.4	Influence Change and the Construction in Section 5.7 533
6	**Interpolation** 535
	6.1	Introduction 536
		6.1.1	Problem and Method 536
		6.1.2	Monotone and Antitone Semantic and Syntactic Interpolation 537
			6.1.2.1 Semantic Interpolation 537
			6.1.2.2 The Interval of Interpolants 539
			6.1.2.3 Syntactic Interpolation 539
			6.1.2.4 Finite Goedel Logics 540
		6.1.3	Introduction to Section 6.5 541
			6.1.3.1 Interpolation and Size 542
			6.1.3.2 Equilibrium Logic 545
			6.1.3.3 Interpolation for Revision and Argumentation 545
			6.1.3.4 Language Change to Obtain Products 546
	6.2	Monotone and Antitone Semantic Interpolation 548
		6.2.1	The Two-Valued Case 548
		6.2.2	The Many-Valued Case 550
	6.3	The Interval of Interpolants in Monotonic or Antitonic Logics 554
		6.3.1	Introduction 554
		6.3.2	Examples and a Simple Fact 555
		6.3.3	$+$ and $-$ (in f^+ and f^-) as New Semantic and Syntactic Operators 556
			6.3.3.1 Motivation 557

Contents (Volumes I and II)   XXVII

		6.3.3.2	Formal Definition and Results 557
		6.3.3.3	The Special Case of Classical Logic 558
		6.3.3.4	General Results on the New Operators 559
6.4	Monotone and Antitone Syntactic Interpolation 565
	6.4.1	Introduction 565
	6.4.2	The Classical Propositional Case 566
	6.4.3	Finite (Intuitionistic) Goedel Logics 567
		6.4.3.1	The Definitions 567
		6.4.3.2	Normal Forms and f^+ 570
		6.4.3.3	An Important Example for Non-existence of Interpolation 576
		6.4.3.4	The Additional Operators J, A, F, Z 579
		6.4.3.5	Special Finite Goedel Logics 583
6.5	Semantic Interpolation for Non-monotonic Logic 584
	6.5.1	Discussion 584
	6.5.2	Interpolation of the Form $\phi \mathrel{	\!\sim} \alpha \vdash \psi$ 585
	6.5.3	Interpolation of the Form $\phi \vdash \alpha \mathrel{	\!\sim} \psi$ 587
	6.5.4	Interpolation of the Form $\phi \mathrel{|\!\sim} \alpha \mathrel{|\!\sim} \psi$ 589
		6.5.4.1	Introduction 589
		6.5.4.2	Some Examples 590
		6.5.4.3	Interpolation and $(\mu * 1)$ 592
		6.5.4.4	Interpolation and $(\mu * 4)$ 596
		6.5.4.5	Interpolation for Equivalent Formulas 597
	6.5.5	Interpolation for Distance-Based Revision 599
		6.5.5.1	Hamming Distances and Revision 599
		6.5.5.2	Discussion of Representation 600
	6.5.6	The Equilibrium Logic EQ 601
		6.5.6.1	Introduction and Outline 601
		6.5.6.2	Basic Definition and Definability of Chosen Models 601
		6.5.6.3	The Approach with Models of Value 2 603
		6.5.6.4	The Refined Approach 605
6.6	Context and Structure 607
6.7	Interpolation for Argumentation 608

7	**Neighbourhood Semantics and Deontic Logic** 611				
	7.1	Introduction 611			
		7.1.1	Some Details 613		
			7.1.1.1	Motivation 613	
			7.1.1.2	Tools to Define Neighbourhoods 614	
			7.1.1.3	Additional Requirements 615	
			7.1.1.4	Interpretation of the Neighbourhoods 617	
			7.1.1.5	Extensions 617	
	7.2	Tools and Requirements for Neighbourhoods and How to Obtain Them 618			
		7.2.1	Tools to Define Neighbourhoods 618		
			7.2.1.1	Background 618	
			7.2.1.2	Algebraic Tools 619	
			7.2.1.3	Relations 621	
			7.2.1.4	Distances 622	
		7.2.2	Additional Requirements for Neighbourhoods 624		
		7.2.3	Connections Between the Various Concepts 626		
			7.2.3.1	The Not Necessarily Independent Case 629	
			7.2.3.2	The Independent Case 631	
			7.2.3.3	Remarks on the Counting Case 633	
		7.2.4	Neighbourhoods in Deontic and Default Logic 634		
	7.3	Abstract Semantics of Deontic Logic 635			
		7.3.1	Introductory Remarks 635		
			7.3.1.1	Context 635	
			7.3.1.2	Central Idea 636	
			7.3.1.3	A Common Property of Facts and Obligations ... 636	
			7.3.1.4	Derivations of Obligations 636	
			7.3.1.5	Orderings and Obligations 637	
			7.3.1.6	Derivation Revisited 638	
			7.3.1.7	Relativization 638	
			7.3.1.8	Numerous Possibilities 638	
			7.3.1.9	Overview 639	
		7.3.2	Philosophical Discussion of Obligations 639		

		7.3.2.1	A Fundamental Difference Between Facts and Obligations: Asymmetry and Negation 639
		7.3.2.2	"And" and "or" for Obligations 640
		7.3.2.3	Ceteris Paribus – A Local Property 642
		7.3.2.4	Global and Mixed Global/Local Properties of Obligations 643
		7.3.2.5	Soft Obligations 644
		7.3.2.6	Overview of Different Types of Obligations 644
		7.3.2.7	Summary of the Philosophical Remarks 646
	7.3.3	What Is an Obligation? 646
	7.3.4	Conclusion 648
7.4	A Comment on Work by Aqvist 648
	7.4.1	Introduction 648
	7.4.2	There Are (at Least) Two Solutions 649
	7.4.3	Outline 651
	7.4.4	$Gm \vdash A$ Implies $G \vdash A$ (Outline) 655
7.5	Hierarchical Conditionals 656
	7.5.1	Introduction 656
		7.5.1.1	Description of the Problem 656
		7.5.1.2	Outline of the Solution 658
	7.5.2	Formal Modelling and Summary of Results 660
	7.5.3	Overview 663
	7.5.4	Connections with Other Concepts 665
		7.5.4.1	Hierarchical Conditionals and Programs 665
		7.5.4.2	Connection with Theory Revision 666
	7.5.5	Formal Results and Representation for Hierarchical Conditionals 666

8 Abstract Independence 671
8.1 Introduction, Basic Definitions, and Notation 672
8.1.1 Probabilistic Independence 672
8.1.2 Set and Function Independence 674
8.2 Discussion of Some Simple Examples and Connections 676
8.2.1 $X \times Z$ 676
8.2.2 $X \times Z \times W$ 677

		8.2.3	$X \times Y \times Z$ 678
		8.2.4	$X \times Y \times Z \times W$ 679
	8.3	Basic Results for Set and Function Independence 680
	8.4	New Rules, Examples, and Discussion for Function Independence . 683
		8.4.1	Example of a Rule Derived from the Basic Rules 684
		8.4.2	More New Rules 686
	8.5	There Is No Finite Characterization for Function Independence 690
		8.5.1	Discussion 690
		8.5.2	Composition of Layers 691
		8.5.3	Systematic Construction 691
		8.5.4	The Cases to Consider 693
		8.5.5	Solution of the Cases 694
		8.5.6	Final Argument 696
			8.5.6.1 Comment 696
	8.6	Systematic Construction of New Rules for Function Independence . 696
		8.6.1	Consequences of a Single Triple 696
		8.6.2	Construction of Function Trees 697
		8.6.3	Examples 698

9 Various Aspects of Nonmonotonic and Other Logics 705

	9.1	Local and Global Metrics for Counterfactuals 705
		9.1.1	Introduction 705
		9.1.2	Basic Definitions 706
		9.1.3	Results 707
		9.1.4	Outline of the Construction for Theorem 9.1.4 710
		9.1.5	Detailed Proof of Theorem 9.1.4 711
	9.2	Extensions by Approximation from Below 715
		9.2.1	Introduction 715
		9.2.2	Cautious Monotony Does Not Extend 717
			9.2.2.1 Idea 717
			9.2.2.2 Construction of $\overline{\overline{A}}$ 717
			9.2.2.3 The Extension $\hat{A}$ (as in [FLM90]) 718
		9.2.3	Weak Distributivity Entails Partial Distributivity 719
		9.2.4	On Different Infinite Extensions of $\mathrel|\!\sim$ 719

		9.2.5	Extension by Unbounded Subsets 721
		9.2.6	A Final Example 722
	9.3	Logic and Analysis 724
		9.3.1	Overview, Motivation, and Basic Definitions 724
			9.3.1.1 Overview 724
			9.3.1.2 Motivation to Consider Continous Logics, the Intuition Behind Our Definition 726
			9.3.1.3 Average Difference Between Two Logics 728
			9.3.1.4 Relation Between the Motivational and the Technical Part 728
		9.3.2	Technical Development 729
			9.3.2.1 Outline of the Technical Part 729
			9.3.2.2 The Topological Construction 731
			9.3.2.3 We Turn to Logics 736
			9.3.2.4 A Measure on $Th_{\mathcal{L}}$, Integration of the Difference Between two Logics 740
	9.4	The Talmudic KAL Vachomer Rule 742
		9.4.1	Summary 742
		9.4.2	Introduction 743
			9.4.2.1 The Problem 743
			9.4.2.2 Historical Origin 744
		9.4.3	The AGS Approach 745
			9.4.3.1 Description 745
			9.4.3.2 A Problem with the Original AGS Algorithm 747
		9.4.4	There Is No Straightforward Inductive Algorithm for the AGS Approach 748
			9.4.4.1 Even the Case with Simple (Not Multi) Sets Is Quite Complicated 748
			9.4.4.2 The Multiset Case 749
		9.4.5	The Arrow Counting Approach 753
			9.4.5.1 Definition and Discussion 753
			9.4.5.2 Comparison of the AGS and the Arrow Counting Approach 754
A	Solutions to Exercises in Vol. I 761

		A.1	Exercises in Chapter 1.1	761
		A.1.1	Exercises in Section 1.2	761
		A.1.2	Exercises in Section 1.3	763
		A.1.3	Exercises in Section 1.4	764
		A.1.4	Exercises in Section 1.5	765
		A.1.5	Exercises in Section 1.6	765
		A.1.6	Exercises in Section 1.7	766
		A.1.7	Exercises in Section 1.8	766
		A.1.8	Exercises in Chapter 2	772
		A.1.9	Exercises in Chapter 3	773
B	**Solutions to Exercises in Vol. II**			777
	B.1	Exercises in Chapter 4		777
		B.1.1	Exercises in Section 4.3	777
	B.2	Exercises in Chapter 5		779
	B.3	Exercises in Chapter 6		784
	B.4	Exercises in Chapter 7		788
	B.5	Exercises in Chapter 8		790
	B.6	Exercises in Chapter 9		792

References 797

Index 805

Volume I
Preference and Size

Chapter 1
Preferential Structures

Abstract The basic property of "normal" preferential structures is that minimization is upward absolute. If $x, y \in X$, $x \prec y$, i.e., y is a non-minimal element in X, and $X \subseteq Y$, then y will be a non-minimal element in Y, too - as $x \in Y$. This results in the fundamental "algebraic" property that for $X \subseteq Y$ $\mu(Y) \cap X \subseteq \mu(X)$, where $\mu(X)$ is the set of minimal elements of X. A second (trivial), elementary algebraic property is that $\mu(X) \subseteq X$. If X is the set of models of some theory T, and we define a logic $\mathrel|\!\sim$ by $T \mathrel|\!\sim \phi$ iff $\mu(X) \models \phi$ ($\models$ classical validity), we see that $\mathrel|\!\sim$ is at least as strong as classical logic, $\vdash$, $T \vdash \phi$ implies $T \mathrel|\!\sim \phi$.

We thus have preferential structures (with various additional properties for the relation $\prec$), the structural semantics $\prec$, resulting algebraic semantics of the μ operator, and finally, resulting logical properties.

In the present chapter, we first discuss fundamental logical properties, relate logical, algebraic, and structural properties in Section 1.1, and summarize the correspondances in the tables of Section 1.2.6 and Section 1.2.7.

We then discuss general preferential structures, smooth, and ranked structures, see Section 1.3 and Section 1.4. Importance of domain closure properties for the smooth case is discussed in Section 1.5.

There might be no minimal models, only ever smaller ones, this is discussed in Section 1.6. It has somewhat surprising results, roughly, such structures are either trivial (equivalent to a minimal case), see Section 1.6.5.3, or too difficult (no fixed size characterization), see Section 1.7.3.

The importance of "definability preservation" of the μ operator, i.e., if X is a model set definable by a formula (or theory), then so will $\mu(X)$ be, is discussed in Section 1.7. Again, we have an impossibilty result if definability is not preserved, see again Section 1.7.3.

We conclude this chapter with various approaches and results in Section 1.8.

1.1 Introduction

Preferential structures are among the best examined semantics for nonmonotonic logics, and a main subject of research by the the author.

We develop the general theory in Section 1.2 to Section 1.7. Section 1.8 contains various results which complement the basic theory.

- Section 1.2 gives basic definitions and summaries of results,
- Section 1.3 discusses the general and smooth case,
- Section 1.4 discusses ranked structures,
- Section 1.5 discusses the smooth case without domain closure under $\cup$,
- Section 1.6 discusses the limit version of preferential structures, which, though intuitively appealing, basically either gives no new properties, or is provably too complicated.
- Section 1.7 discusses the importance of "definability preservation",
- Section 1.8 contains various results:
    - In Section 1.8.2, we discuss the role of copies (or non-injective labelling functions in KLM terminology) in preferential structures. ("KLM" stands for [KLM90] or its authors.)
    - In Section 1.8.3, we show that the KLM characterization cannot be extended to the infinite case.
    - In Section 1.8.4, we show how to obtain cumulativity by a topological construction, and not through smoothness of the structure.
    - In Section 1.8.5, we replace the partial orders between models of preferential structures by unions of total orders, and discuss the consequences.
    - In Section 1.8.6, we discuss a joint article with S. Berger and D. Lehmann on preferred update histories.
    - In Section 1.8.7, we reconstruct completeness proofs for preferential structures as done by KLM, to facilitate comparison with our own constructions.
    - In Section 1.8.8, we discuss preferential choice for branching time structures, and extend results by Katsuno and Mendelzon.

For more comments and motivation, see e.g. [Sch04].

1.2 Basic Definitions and Overview

1.2.1 Introduction

This section contains a multitude of definitions and results, most of which will be used over and again.

It also contains a large number of mostly small results, e.g. connections between different orders, which are put here, so they do not interrupt the flow of argumentation in the other chapters.

1.2.2 General Properties

1.2.2.1 Algebraic Properties

Notation 1.2.1

We use sometimes FOL as an abbreviation for first-order logic, and NML for non-monotonic logic. To avoid LaTeX complications in bigger expressions, we replace $\widetilde{xxxxx}$ by $\overline{xxxxx}$.

Definition 1.2.1

(1) We use $:=$ and $:\Leftrightarrow$ to define the left-hand side by the right-hand side, as in the following two examples:

$X := \{x\}$ defines X as the singleton with element x.

$X < Y :\Leftrightarrow \forall x \in X \forall y \in Y (x < y)$ extends the relation $<$ from elements to sets.

We sometimes write "wrt." for "with respect to", "s.th." for "such that", and "wlog." for "without loss of generality", when the full version seems too verbose.

(2) We use $\mathcal{P}$ to denote the power set operator.

$\Pi\{X_i : i \in I\} := \{g : g : I \to \bigcup\{X_i : i \in I\}, \forall i \in I.g(i) \in X_i\}$ is the general Cartesian product, $X \times X'$ is the binary Cartesian product.

$card(X)$ shall denote the cardinality of X, and V the set-theoretic universe we work in — the class of all sets.

Given a set of pairs $\mathcal{X}$, and a set X, we let $\mathcal{X} \upharpoonright X := \{\langle x, i \rangle \in \mathcal{X} : x \in X\}$. (When the context is clear, we will sometimes simply write X for $\mathcal{X} \upharpoonright X$.)

We will use the same notation $\upharpoonright$ to denote the restriction of functions and in particular of sequences to a subset of the domain.

If Σ is a set of sequences over an index set X, and $X' \subseteq X$, we will abuse notation and write $\Sigma \upharpoonright X'$ for $\{\sigma \upharpoonright X' : \sigma \in \Sigma\}$.

Concatenation of sequences, e.g., of σ and σ', will often be denoted by juxtaposition: $\sigma\sigma'$.

Compositions of functions will often be written $g \circ f$ etc., with $(g \circ f)(x) := g(f(x))$.

(3) $A \subseteq B$ will denote that A is a subset of B or equal to B, and $A \subset B$ that A is a proper subset of B; likewise for $A \supseteq B$ and $A \supset B$.

Given some fixed set U we work in, and $X \subseteq U$, $C(X) := U - X$.

(4) If $\mathcal{Y} \subseteq \mathcal{P}(X)$ for some X, we say that $\mathcal{Y}$ satisfies

($\cap$) iff it is closed under finite intersections,

($\bigcap$) iff it is closed under arbitrary intersections,

($\cup$) iff it is closed under finite unions,

($\bigcup$) iff it is closed under arbitrary unions,

(C) iff it is closed under complementation,

($-$) iff it is closed under set difference.

(5) We will sometimes write $A = B \parallel C$ for: $A = B$, or $A = C$, or $A = B \cup C$.

We make ample and tacit use of the Axiom of Choice.

Closure

Definition 1.2.2

Let $\mathcal{Y} \subseteq \mathcal{P}(Z)$ be given and closed under arbitrary intersections.

(1) For $A \subseteq Z$, let $\widehat{A} := \bigcap \{X \in \mathcal{Y} : A \subseteq X\}$.

(2) For $B \in \mathcal{Y}$, we call $A \subseteq B$ a small subset of B iff there is no $X \in \mathcal{Y}$ such that $B - A \subseteq X \subset B$.

(Context will disambiguate from other uses of "small".)

Intuitively, Z is the set of all models for $\mathcal{L}$, $\mathcal{Y}$ is $\mathbf{D}_\mathcal{L}$, and $\widehat{A} = M(Th(A))$, this is the intended application - $Th(A)$ is the set of formulas which hold in all $a \in A$, and $M(Th(A))$ is the set of models of $Th(A)$. Note that then $\widehat{\emptyset} = \emptyset$.

1.2 Basic Definitions and Overview

Fact 1.2.1

(1) If $\mathcal{Y} \subseteq \mathcal{P}(Z)$ is closed under arbitrary intersections and finite unions, $Z \in \mathcal{Y}$, $X, Y \subseteq Z$, then the following hold:

$(Cl\cup)\ \widehat{X \cup Y} = \widehat{X} \cup \widehat{Y}$

$(Cl\cap)\ \widehat{X \cap Y} \subseteq \widehat{X} \cap \widehat{Y}$, but usually not conversely,

$(Cl-)\ \widehat{A} - \widehat{B} \subseteq \widehat{A - B}$,

$(Cl =)\ X = Y \Rightarrow \widehat{X} = \widehat{Y}$, but not conversely,

$(Cl \subseteq 1)\ \widehat{X} \subseteq Y \Rightarrow X \subseteq Y$, but not conversely,

$(Cl \subseteq 2)\ X \subseteq \widehat{Y} \Rightarrow \widehat{X} \subseteq \widehat{Y}$.

(2) If, in addition, $X \in \mathcal{Y}$ and $CX := Z - X \in \mathcal{Y}$, then the following two properties hold, too:

$(Cl\cap +)\ \widehat{A} \cap X = \widehat{A \cap X}$,

$(Cl - +)\ \widehat{A} - X = \widehat{A - X}$.

(3) In the intended application, i.e., $\widehat{A} = M(Th(A))$, the following hold:

(3.1) $Th(X) = Th(\widehat{X})$,

(3.2) Even if $A = \widehat{A}$, $B = \widehat{B}$, it is not necessarily true that $\widehat{A - B} \subseteq \widehat{A} - \widehat{B}$.

Proof

$(Cl =), (Cl \subseteq 1), (Cl \subseteq 2), (3.1)$ are trivial.

$(CL\cup)$: Exercise, solution in the Appendix.

$(Cl\cap)$ Let $X', Y' \in \mathcal{Y}$, $X \subseteq X', Y \subseteq Y'$, then $X \cap Y \subseteq X' \cap Y'$, so $\widehat{X \cap Y} \subseteq \widehat{X} \cap \widehat{Y}$. For the converse, set $X := M_{\mathcal{L}} - \{m\}$, $Y := \{m\}$ in Example 1.2.1. ($M_{\mathcal{L}}$ is the set of all models of the language $\mathcal{L}$.)

(CL-): Exercise, solution in the Appendix.

$(Cl\cap +)\ \widehat{A} \cap X \supseteq \widehat{A \cap X}$ by $(Cl\cap)$. For "$\subseteq$": Let $A \cap X \subseteq A' \in \mathcal{Y}$, then by closure under $(\cup)$, $A \subseteq A' \cup CX \in \mathcal{Y}$, $(A' \cup CX) \cap X \subseteq A'$. So $\widehat{A} \cap X \subseteq \widehat{A \cap X}$.

$(Cl-+)$ $\widehat{A-X} = \widehat{A \cap CX} = \widehat{A} \cap CX = \widehat{A} - X$ by $(Cl \cap +)$.

(3.2) Set $A := M_{\mathcal{L}}$, $B := \{m\}$ for $m \in M_{\mathcal{L}}$ arbitrary, $\mathcal{L}$ infinite. So $A = \widehat{A}$, $B = \widehat{B}$, but $\widehat{A-B} = A \neq A - B$.

□

General Relations

Definition 1.2.3

As usual, $\prec^*$ will denote the transitive closure of the relation $\prec$.

If $<$, $\prec$, or a similar relation is given, $a \perp b$ will express that a and b are $<$-incomparable (or $\prec$-incomparable); context will tell.

Given any relation $<$, $\leq$ will stand for $<$ or $=$; conversely, given $\leq$, $<$ will stand for $\leq$, but not $=$; similarly for $\prec$, etc.

Definition 1.2.4

A child (or successor) of an element x in a tree t will be a direct child in t. A child of a child, etc. will be called an indirect child. Trees will be supposed to grow downwards, so the root is the top element.

Definition 1.2.5

A subsequence $\sigma_i : i \in I \subseteq \mu$ of a sequence $\sigma_i : i \in \mu$ is called cofinal, iff for all $i \in \mu$ there is $i' \in I$ $i \leq i'$.

Given two sequences σ_i and τ_i of the same length, then their Hamming distance is the quantity of i where they differ.

Definition 1.2.6

A binary relation $\leq$ on X is a preorder, iff $\leq$ is reflexive and transitive. If $\leq$ is in addition total, i.e. iff $\forall x, y \in X$ $x \leq y$ or $y \leq x$, then $\leq$ is a total preorder.

A binary relation $<$ on X is a total order, iff $<$ is transitive, irreflexive, i.e. $x \not< x$ for all $x \in X$, and for all $x, y \in X$ $x < y$ or $y < x$ or $x = y$.

Remark 1.2.2

If $\leq$ is a total preorder on X, $\approx$ the corresponding equivalence relation defined by $x \approx y$ iff $x \leq y$ and $y \leq x$, $[x]$ the $\approx$-equivalence class of x, and we define $[x] < [y]$ iff $x \leq y$, but not $y \leq x$, then $<$ is a total order on $\{[x] : x \in X\}$.

We give a generalized abstract nonsense result, taken from [LMS01], which must be part of the folklore:

1.2 Basic Definitions and Overview

Lemma 1.2.3

Given a set X and a binary relation R on X, there exists a total preorder S on X that extends R such that

$$\forall x, y \in X (xSy, ySx \Rightarrow xR^*y)$$

where R^* is the reflexive and transitive closure of R.

Proof

Define $x \equiv y$ iff xR^*y and yR^*x. The relation $\equiv$ is an equivalence relation. Let $[x]$ be the equivalence class of x under $\equiv$. Define $[x] \preceq [y]$ iff xR^*y. The definition of $\preceq$ does not depend on the representatives x and y chosen. The relation $\preceq$ on equivalence classes is a partial order. Let $\leq$ be any total order on these equivalence classes that extends $\preceq$. Define xSy iff $[x] \leq [y]$. The relation S is total (since $\leq$ is total) and transitive (since $\leq$ is transitive) and is therefore a total preorder. It extends R by the definition of $\preceq$ and the fact that $\leq$ extends $\preceq$. Suppose now xSy and ySx. We have $[x] \leq [y]$ and $[y] \leq [x]$ and therefore $[x] = [y]$ by antisymmetry. Therefore $x \equiv y$ and xR^*y.

□

In the following, the X_i are arbitrary (nonempty) sets, standing for the set of truth values, and I, J, L etc. are intuitively the set of propositional variables. So any element of $\Pi\{X_i : i \in J\}$ stands for a propositional model. Inessential variables in a set Σ are those which do not have any influence on the truth of the formula whose model set is Σ.

Definition 1.2.7

Let $\Sigma \subseteq \Pi := \Pi\{X_i : i \in J\}$. Define:

(1) For $\sigma, \sigma' \in \Sigma$, $J' \subseteq J$, define:

$\sigma \sim_{J'} \sigma' :\Leftrightarrow \forall x \in J' \sigma(x) = \sigma'(x)$.

(2) $I(\Sigma) := \{i \in J : \Sigma = \Sigma \upharpoonright (J - \{i\}) \times X_i\}$ (up to reordering) (they are the irrelevant or inessential i) and

$R(\Sigma) := J - I(\Sigma)$ (the relevant or essential i).

Fact 1.2.4

(1) $\Sigma = \Sigma \upharpoonright R(\Sigma) \times \Pi \upharpoonright I(\Sigma)$ (up to reordering)
(2) $\sigma \upharpoonright R(\Sigma) = \sigma' \upharpoonright R(\Sigma) \wedge \sigma \in \Sigma \Rightarrow \sigma' \in \Sigma$.

Proof

(1) Exercise, solution in the Appendix.
(2) Trivial.

□

Some General and Combinatorial Results

These facts are elementary, but useful, also for the intuition.

(Again, $I = J \cup J' \cup J''$, the J's disjoint, is, intuitively, the set of propositional variables of a fixed language $\mathcal{L}$, ΠI the set of models (i.e. the product of $\{0,1\}$ I times), etc.)

Fact 1.2.5

Let $I = J \cup J' \cup J''$, $A, B \subseteq \Pi I$. Then
$((A \upharpoonright J \cup J') \times B \upharpoonright J'') \upharpoonright J' \cup J'' = A \upharpoonright J' \times B \upharpoonright J''$.

Proof

Exercise, solution in the Appendix.

Fact 1.2.6

Let $I = J \cup J'$ - except in (8), where $I = J \cup J' \cup J''$ - with all J's pairwise disjoint.

(1) $A \subseteq B \to A \upharpoonright J \subseteq B \upharpoonright J$
(2) $A \subset B \not\to A \upharpoonright J \subset B \upharpoonright J$
(3) $A \upharpoonright J = B \upharpoonright J \not\to A \subseteq B$
(4) $A \upharpoonright J \subset B \upharpoonright J \not\to A \subseteq B$
(5) $A \cup B \upharpoonright J = A \upharpoonright J \cup B \upharpoonright J$
(6) $A \cap B \upharpoonright J \subseteq A \upharpoonright J \cap B \upharpoonright J$
(7) It is not always true that $A \upharpoonright J \cap B \upharpoonright J \subseteq A \cap B \upharpoonright J$
(8) Let $I = J \cup J' \cup J''$, and $A = \Pi J \times A \upharpoonright (J' \cup J'')$, $B = B \upharpoonright (J \cup J') \times \Pi J''$, then $A \upharpoonright J' \cap B \upharpoonright J' \subseteq A \cap B \upharpoonright J'$ holds, too.

Proof

J will be the first coordinate.

(1) Trivial.
(2) Consider $A = \{00\}$, $B = \{00, 01\}$.
(3) Consider $B = \{00\}$, $A = \{00, 01\}$.
(4) Consider $A = \{00, 01\}$, $B = \{00, 11\}$. Then $A \upharpoonright J = \{0\}$, $B \upharpoonright J = \{0, 1\}$.
(5) "$\subseteq$": Let $\sigma_J \in A \cup B \upharpoonright J$, so $\exists \tau \in A \cup B. \tau \upharpoonright J = \sigma_J$, so $\exists \tau \in A. \tau \upharpoonright J = \sigma_J$, or $\exists \tau \in B. \tau \upharpoonright J = \sigma_J$, so $\sigma_J \in A \upharpoonright J \cup B \upharpoonright J$.

"$\supseteq$" follows from (1).

1.2 Basic Definitions and Overview

(6) By (1).

(7) $A = \{01\}$, $B = \{00\}$, then $A \upharpoonright J = B \upharpoonright J = \{0\}$, but $A \cap B = \emptyset$, so $A \cap B \upharpoonright J = \emptyset$.

(8) Let $\sigma_{J'} \in A \upharpoonright J' \cap B \upharpoonright J'$, then there are $\sigma_J^A \sigma_{J'} \sigma_{J''}^A \in A$, $\sigma_J^B \sigma_{J'} \sigma_{J''}^B \in B$, so $\sigma_J^B \sigma_{J'} \sigma_{J''}^A \in A \cap B$, and $\sigma_{J'} \in (A \cap B) \upharpoonright J'$.

□

The following fact and the argument are rather trivial, it mainly serves to illustrate a central idea, which is used several times.

Fact 1.2.7

Let $I = J \cup J' \cup J''$.

Let $A = (A \upharpoonright J \cup J') \times (A \upharpoonright J'')$, $B = (B \upharpoonright J) \times (B \upharpoonright J' \cup J'')$, $A \subseteq B$.

Let $\sigma = \sigma_J \sigma_{J'} \sigma_{J''}$ such that $\sigma_J \in B \upharpoonright J$, $\sigma_{J'} \in A \upharpoonright J'$, $\sigma_{J''} \in A \upharpoonright J''$.

Then $\sigma \in B$.

Proof

Exercise, solution in the Appendix.

1.2.2.2 Logical Properties

Definition 1.2.8

(1) V will be the set of truth values when there are more than the classical ones, TRUE and FALSE.

We work here mostly in a classical propositional language $\mathcal{L}$. A theory T will be an arbitrary set of formulas. Formulas will often be named ϕ, ψ, etc., theories T, S, etc.

$T \vee T'$ will be the set $\{\phi \vee \phi' : \phi \in T, \phi' \in T'\}$.

$v(\mathcal{L})$ or simply L will be the set of propositional variables of $\mathcal{L}$.

$F(\mathcal{L})$ will be the set of formulas of $\mathcal{L}$.

A propositional model m will be a function from the set of propositional variables to the set of truth values V when we have more than two truth values.

$M_{\mathcal{L}}$, or simply M when the context is clear, will be the set of (classical) models for $\mathcal{L}$; $M(T)$ or M_T is the set of models of T; likewise for $M(\phi)$ for a formula ϕ.

(2) $\boldsymbol{D_{\mathcal{L}}} := \{M(T) : T \text{ a theory in } \mathcal{L}\}$, the set of definable model sets.

Note that, in classical propositional logic, $\emptyset, M_{\mathcal{L}} \in \boldsymbol{D}_{\mathcal{L}}$, $\boldsymbol{D}_{\mathcal{L}}$ contains singletons, and is closed under arbitrary intersections and finite unions.

An operation $f : \mathcal{Y} \to \mathcal{P}(M_{\mathcal{L}})$ for $\mathcal{Y} \subseteq \mathcal{P}(M_{\mathcal{L}})$ is called definability preserving, (dp) or (μdp) in short, iff for all $X \in \boldsymbol{D}_{\mathcal{L}} \cap \mathcal{Y}$, $f(X) \in \boldsymbol{D}_{\mathcal{L}}$.

We will also use (μdp) for binary functions $f : \mathcal{Y} \times \mathcal{Y} \to \mathcal{P}(M_{\mathcal{L}})$ — as needed for theory revision — with the obvious meaning.

(3) $\vdash$ will be classical derivability, and

$\overline{T} := \{\phi : T \vdash \phi\}$, the closure of T under $\vdash$.

(4) $Con(.)$ will stand for classical consistency, so $Con(\phi)$ will mean that ϕ is classically consistent; likewise for $Con(T)$. $Con(T, T')$ will stand for $Con(T \cup T')$, etc.

(5) Given a fixed language, $K_{\bot}$ will denote an inconsistent theory in that language.

(6) Given a consequence relation $\mathrel{|\!\sim}$, we define

$\overline{\overline{T}} := \{\phi : T \mathrel{|\!\sim} \phi\}$.

(There is no fear of confusion with $\overline{T}$, as it just is not useful to close twice under classical logic.)

(7) If $X \subseteq M_{\mathcal{L}}$, then $Th(X) := \{\phi : X \models \phi\}$; likewise for $Th(m)$, $m \in M_{\mathcal{L}}$. ($\models$ will usually be classical validity.)

Fact 1.2.8 Let $\mathcal{L}$ be a fixed propositional language, $\boldsymbol{D}_{\mathcal{L}} \subseteq X$, $\mu : X \to \mathcal{P}(M_{\mathcal{L}})$, for a $\mathcal{L}$-theory T. Suppose $\overline{\overline{T}} = Th(\mu(M_T))$, let T, T' be arbitrary theories, then:

(1) $\mu(M_T) \subseteq M_{\overline{\overline{T}}}$,

(2) $M_T \cup M_{T'} = M_{T \vee T'}$ and $M_{T \cup T'} = M_T \cap M_{T'}$, moreover, if T and T' are deductively closed, then also $M_T \cup M_{T'} = M_{T \cap T'}$,

(3) $\mu(M_T) = \emptyset \Leftrightarrow \bot \in \overline{\overline{T}}$.

If μ is definability preserving or $\mu(M_T)$ is finite, then the following also hold:

(4) $\mu(M_T) = M_{\overline{\overline{T}}}$,

(5) $T' \vdash \overline{\overline{T}} \Leftrightarrow M_{T'} \subseteq \mu(M_T)$,

(6) $\mu(M_T) = M_{T'} \Leftrightarrow \overline{\overline{T'}} = \overline{\overline{T}}$.

$\square$

Fact 1.2.9

Let $A, B \subseteq M_{\mathcal{L}}$.
Then $Th(A \cup B) = Th(A) \cap Th(B)$.

1.2 Basic Definitions and Overview

Proof

Exercise, solution in the Appendix.

Fact 1.2.10

Let $X \subseteq M_{\mathcal{L}}, \phi, \psi$ formulas.

(1) $X \cap M(\phi) \models \psi$ iff $X \models \phi \to \psi$.
(2) $X \cap M(\phi) \models \psi$ iff $M(Th(X)) \cap M(\phi) \models \psi$.
(3) $Th(X \cap M(\phi)) = \overline{Th(X) \cup \{\phi\}}$
(4) $X \cap M(\phi) = \emptyset \Leftrightarrow M(Th(X)) \cap M(\phi) = \emptyset$
(5) $Th(M(T) \cap M(T')) = \overline{T \cup T'}$.

Proof

(1) "$\Rightarrow$": $X = (X \cap M(\phi)) \cup (X \cap M(\neg\phi))$. In both parts holds $\neg\phi \vee \psi$, so $X \models \phi \to \psi$. "$\Leftarrow$": Trivial.
(2) $X \cap M(\phi) \models \psi$ (by (1)) iff $X \models \phi \to \psi$ iff $M(Th(X)) \models \phi \to \psi$ iff (again by (1)) $M(Th(X)) \cap M(\phi) \models \psi$.
(3) $\psi \in Th(X \cap M(\phi)) \Leftrightarrow X \cap M(\phi) \models \psi \Leftrightarrow_{(2)} M(Th(X) \cup \{\phi\}) = M(Th(X)) \cap M(\phi) \models \psi \Leftrightarrow Th(X) \cup \{\phi\} \vdash \psi$.
(4) $X \cap M(\phi) = \emptyset \Leftrightarrow X \models \neg\phi \Leftrightarrow M(Th(X)) \models \neg\phi \Leftrightarrow M(Th(X)) \cap M(\phi) = \emptyset$.
(5) $M(T) \cap M(T') = M(T \cup T')$.

□

Fact 1.2.11

If $X = M(T)$, then $M(Th(X)) = X$.

Proof

$X \subseteq M(Th(X))$ is trivial. $Th(M(T)) = \overline{T}$ is trivial by classical soundness and completeness. So $M(Th(M(T))) = M(\overline{T}) = M(T) = X$.

□

Example 1.2.1

If $v(\mathcal{L})$ is infinite, and m any model for $\mathcal{L}$, then $M := M_{\mathcal{L}} - \{m\}$ is not definable by any theory T. (Proof: Suppose it were, and let ϕ hold in M, but not in m, so in m $\neg\phi$ holds, but as ϕ is finite, there is a model m' in M which coincides on all propositional variables of ϕ with m, so in m' $\neg\phi$ holds, too, a contradiction.) Thus, in the infinite case, $\mathcal{P}(M_{\mathcal{L}}) \neq \boldsymbol{D}_{\mathcal{L}}$.

(There is also a simple cardinality argument, which shows that almost no model sets are definable, but it is not constructive and thus less instructive than above argument. We give it nonetheless: Let $\kappa := card(v(\mathcal{L}))$. Then there are κ many formulas, so 2^κ many theories, and thus 2^κ many definable model sets. But there are 2^κ many models, so $(2^\kappa)^\kappa$ many model sets.)

□

Fact 1.2.12

Let $\mu : \boldsymbol{D_\mathcal{L}} \to \boldsymbol{D_\mathcal{L}}$, and $\overline{\overline{T}} = Th(\mu(M(T)))$ for all theories T. Then:

(1) $\mu(M(T)) = M(\overline{\overline{T}})$,
(2) $M(T \vee T') = M(T) \cup M(T')$,
(3) $Th(X \cup Y) = \overline{Th(X) \vee Th(Y)}$,
(4) If $\widehat{U} := M(Th(U))$, then $\widehat{X \cup Y} = \widehat{X} \cup \widehat{Y}$ for $X, Y \in \boldsymbol{D_\mathcal{L}}$,
(5) $\mu(M(T) \cup M(T')) \subseteq M(T) \Leftrightarrow T \subseteq \overline{\overline{T \vee T'}}$,
(6) $M(\overline{\overline{T \vee T'}}) \subseteq M(\overline{\overline{T}} \vee \overline{\overline{T'}}) \Leftrightarrow \overline{\overline{T}} \vee \overline{\overline{T'}} \subseteq \overline{\overline{T \vee T'}}$.

Proof

Exercise, solution see [Sch04], Fact 5.2.10.

This might be the right place to add the following short remark on formula sets.

We show here that, independent of the cardinality of the language, one can define only countably many inconsistent formulas.

The problem is due to D. Makinson (personal communication).

Example 1.2.2

There is a countably infinite set of formulas s.t. the defined model sets are pairwise disjoint.

Let $p_i : i \in \omega$ be propositional variables.

Consider $\phi_i := \bigwedge\{\neg p_j : j < i\} \wedge p_i$ for $i \in \omega$.

Obviously, $M(\phi_i) \neq \emptyset$ for all i.

Let $i < i'$; we show $M(\phi_i) \cap M(\phi_{i'}) = \emptyset$. $M(\phi_{i'}) \models \neg p_i$, $M(\phi_i) \models p_i$.

□

Fact 1.2.13

Any set X of consistent formulas with pairwise disjoint model sets is at most countable

1.2 Basic Definitions and Overview

Proof

Let such X be given.

(1) We may assume that X consists of conjunctions of propositional variables or their negations.

Proof: Rewrite all $\phi \in X$ as disjunctions of conjunctions ϕ_j. At least one of the conjunctions ϕ_j is consistent. Replace ϕ by one such ϕ_j. Consistency is preserved, as is pairwise disjointness.

(2) Let X be such a set of formulas. Let $X_i \subseteq X$ be the set of formulas in X with length i, i.e., a consistent conjunction of i many propositional variables or their negations, $i > 0$.

As the model sets for X are pairwise disjoint, the model sets for all $\phi \in X_i$ have to be disjoint.

(3) It suffices now to show that each X_i is at most countable; we even show that each X_i is finite.

Proof by induction:

Consider $i = 1$. Let $\phi, \phi' \in X_1$. Let ϕ be p or $\neg p$. If ϕ' is not $\neg\phi$, then ϕ and ϕ' have a common model. So one must be p, the other $\neg p$. But these are all possibilities, so $card(X_1)$ is finite.

Let the result be shown for $k < i$.

Consider now X_i. Take arbitrary $\phi \in X_i$. Without loss of generality, let $\phi = p_1 \wedge \ldots \wedge p_i$. Take arbitrary $\phi' \neq \phi$. As $M(\phi) \cap M(\phi') = \emptyset$, ϕ' must be a conjunction containing one of $\neg p_k$, $1 \leq k \leq i$. Consider now $X_{i,k} := \{\phi' \in X_i : \phi' \text{ contains } \neg p_k\}$. Thus $X_i = \{\phi\} \cup \bigcup\{X_{i,k} : 1 \leq k \leq i\}$. Note that all $\psi, \psi' \in X_{i,k}$ agree on $\neg p_k$, so the situation in $X_{i,k}$ is isomorphic to X_{i-1}. So, by induction hypothesis, $card(X_{i,k})$ is finite, as all $\phi' \in X_{i,k}$ have to be mutually inconsistent. Thus, $card(X_i)$ is finite. (Note that we did not use the fact that elements from different $X_{i,k}$, $X_{i,k'}$ also have to be mutually inconsistent; our rough proof suffices.)

$\square$

Note that the proof depends very little on logic. We needed normal forms, and used two truth values. Obviously, we can easily generalize to finitely many truth values.

Many-Valued Logic

In two-valued logic, we have a correspondence between sets and logic, e.g., we can speak about the set of models of a formula. This has now to be replaced by a many-valued function (or, alternatively, but generally not pursued here, by many-valued sets).

We will later also modify the definition of a preferential structure (see Definition 1.2.14); attacking elements need to be at least as "good" (with respect to an order on the truth values) as attacked elements. We will later be able to show that our results on interpolation carry over to the many-valued case without any problems; see Chapter 6.

Motivation

Preferential logics offer, indirectly, four truth values: classically TRUE and FALSE and defeasibly TRUE and FALSE. Inheritance systems offer, through specificity, arbitrarily many truth values, with a partial order. Thus, it is a bit surprising that many-valued logics seem to have received little attention so far in the context of non-monotonic logics. Finally, finite Goedel logics offer arbitrarily many truth values, with a total order. This motivates us in our context to consider the following:

Definition 1.2.9

(1) We assume here a finite set of truth values, V, to be given, with a partial order $\leq$. We assume that a minimal and a maximal element exist, which will be denoted 0 and 1, or TRUE and FALSE, or min and max, depending on context. In many cases we will assume that sup and inf (equivalently, max and min, as V is supposed to be finite) exist for any subset of V. This will not always be necessary, but often it will be convenient.

Inf will correspond to classical $\wedge$, sup to classical $\vee$. Given $x \in V$, Cx will be $inf\{y \in V : sup\{x,y\} = 1\}$; it corresponds to classical $\neg$. We will not assume that classical $\neg$ is always part of the language.

(2) A model is a function $m : L \to V$.

In classical logic, a formula ϕ defines a model set $M(\phi) \subseteq M$; equivalently a function $f_\phi : M \to \{0,1\}$ with $f_\phi(m) = 1 :\Leftrightarrow m \models \phi$. A straightforward generalization is to define in the many-valued case $f_\phi : M \to V$.

This definition should respect the following postulates:

(2.1) $f_p(m) = m(p)$ for $p \in L$.

This postulate is the basis for a seemingly trivial property, which has far-reaching consequences: If ϕ contains only variables in $L' \subseteq L$, then its truth value is the same in L-models and L'-models, whenever they agree on L'. Of course, validity of the operators has to be truth functional, and again not to depend on other variables.

(2.2) $f_{\phi \wedge \psi}(m) = inf\{f_\phi(m), f_\psi(m)\}$ and $f_{\phi \vee \psi}(m) = sup\{f_\phi(m), f_\psi(m)\}$.

For a set T of formulas, we define $f_T(m) := inf\{f_\phi(m) : \phi \in T\}$.

1.2 Basic Definitions and Overview

(3) In general, a model set corresponds now to an arbitrary function $f : M \to V$. Such an f is called (formula) definable iff there is ϕ such that $f = f_\phi$. The definition of theory definable is analogous.

$\mathcal{P}(M)$ is replaced by V^M, the set of all functions from M to V; $\mathcal{D}$ will denote the set of all definable functions from M to V.

(4) Semantical consequence should respect $\leq$:

(4.1) For $f, g : M \to V$, we write $f \models g$ or $f \leq g$ iff $\forall m \in M. f(m) \leq g(m)$,

(4.2) we write $\phi \models \psi$ iff $f_\phi \models f_\psi$, and

(4.3) we assume that $\to$ is compatible with $\models$:

$f_{\phi \to \psi}(m) = TRUE$ iff $f_\phi(m) \leq f_\psi(m)$, so

$f_{\phi \to \psi} =$ (constant) TRUE iff $\phi \models \psi$.

In classical logic, the set of definable model sets satisfies certain closure conditions. We will examine them now, and generalize them.

(1) We will assume that there are formulas TRUE and FALSE (by abuse of language) such that $f_{TRUE}(m) = TRUE$ and $f_{FALSE}(m) = FALSE$ for all m; thus we have at least the two definable constant functions TRUE and FALSE, again by abuse of language.

Not necessarily all constant functions are definable.

(2) For each $x \in L$ there is f_x defined by $f_x(m) := m(x)$.

(3) $\mathcal{D}$ is closed under finite sup and finite inf.

(4) $\mathcal{D}$ will not necessarily be closed under complementation:

Given $f \in \mathcal{D}$, the complement $\mathcal{C}(f)$ is defined as above by

$\mathcal{C}(f)(m) := inf\{v \in V : sup\{v, f(m)\} = TRUE\} = \mathcal{C}(f(m))$.

(5) In classical logic, $\mathcal{D}$ is closed under simplification: If $X \subseteq M$ is a definable model set, $L' \subseteq L$, then $X' := \{m \in M : \exists m' \in X. m' \restriction L' = m \restriction L'\}$ is definable. This is a consequence of the existence of the standard normal forms; consider, e.g., the formula $p \wedge q$, with $L = \{p, q\}$, $L' = \{p\}$; then $X' = \{m, m'\}$, where $m(p) = m(q) = 1$, $m'(p) = 1$, $m'(q) = 0$, and the new formula is p. We "neglect" or "forget" q, and take the projection. It is also a sufficient condition for syntactic interpolation; see Chapter 6. (The following Example 1.2.3 shows that two different formulas might have the same model function, but should have different projections.)

We have to define the analogon to X' in many-valued logic.

Note that

- if $m \restriction L' = m' \restriction L'$, then $m \in X' \Leftrightarrow m' \in X'$,
- $m \in X'$ iff there is $m' \in X.m \restriction L' = m' \restriction L'$; thus $f_{X'}(m) = sup\{f_X(m') : m \restriction L' = m' \restriction L'\}$.

So we impose the same conditions: Let f and $L' \subseteq L$ be given, we look for suitable f'.

(5.1) f' has to be indifferent to $L - L'$: if $m \restriction L' = m' \restriction L'$, then we should have $f'(m) = f'(m')$.

(5.2) $f'(m) = sup\{f(m') : m \restriction L' = m' \restriction L'\}$.

(6) When the model set has additional structure, we can ask whether the resulting model choice functions preserve definability:

(6.1) the case of preferential structures is treated below in Definition 1.2.14,

(6.2) we can ask the same question, e.g., for modal structures: is the set of all models reachable from some model definable, etc.

Example 1.2.3

This example shows that two different formulas ϕ and ϕ' may define the same $f_\phi = f_{\phi'}$, but neglecting a certain variable should give different results. We give two variants.

(1) Set $\phi := p \vee (q \vee \neg q)$, $\psi := \neg p \vee (q \vee \neg q)$. So $f_\phi = f_\psi$, but neglecting q should result in p in the first case, in $\neg p$ in the second case.

(2) We work with three truth values, 0 for FALSE, 2 for TRUE, $\wedge$ is as usual interpreted by inf. Define two new unary operators $K(x) := 1$ (constant), $M(x) := min\{1, x\}$.

a	b	$\phi = K(a) \wedge b$	$\phi' = K(a) \wedge M(b)$
0	0	0	0
0	1	1	1
0	2	1	1
1	0	0	0
1	1	1	1
1	2	1	1
2	0	0	0
2	1	1	1
2	2	1	1

1.2 Basic Definitions and Overview

So they define the same model function $f : M \to V$. But when we forget about a, the first should just be b, but the second should be $M(b)$.

$\square$

We may consider more systematically other operators, under which the definable model sets should be closed:

(1) Constants for each truth value, like $FALSE = p \wedge \neg p$, $TRUE = p \vee \neg p$ in classical logic.

(2) Complementation, if the complement is defined on the truth value set. (In classical logic, this is, of course, negation.)

   This might for instance be interesting for argumentation, where arguments, or their sources, are the truth values.

(3) Functions similar to the basic operations SHL and SHR of computer science: suppose a linear order is given on the truth values $0, \ldots, n$; then $SHR(p) := p + 1$ if $p < n$, and 0 if $p = n$, etc.

   The function J of finite Goedel logics (see Definition 6.4.3) has some similarity to such shift operations.

(Table 1.1 also contains further material which will become clearer only later.)

Table 1.1 summarizes the situation for the two-valued and the many-valued case.

For the definition of a (semantical) interpolant for many-valued logics, the reader is referred directly to Definition 6.2.1 in Section 6.2.

1.2.3 General Nonmonotonic Logic

The tables contain definitions and statements. Thus, we may refer to them as definitions, and also prove the statements they contain.

We discuss algebraic and logical properties.

Definitions and correspondences are presented in Table 1.2 and Table 1.3, the correspondences repeated for easier reading in Table 1.5, proofs of the correspondences are found in Proposition 1.2.18.

The correspondences to the Size Rules, and their definitions, are discussed in Section 3.2.

Table 1.1 Notation and Definitions

	two-valued $\{0,1\}$	many-valued $(V, \leq)$
Propositional language L, $L' \subseteq L$, propositional variables $s, \ldots$		
definability of f	$\exists \phi : f_\phi = f$	
model m	$m : L \to \{0,1\}$	$m : L \to V$
M set of all L-models		
(for $\Gamma \subseteq M$) $\Gamma \upharpoonright L'$	$\Gamma \upharpoonright L' := \{m \upharpoonright L' : m \in \Gamma\}$	
$m \upharpoonright L'$	like m, but restricted to L'	
$m \sim_{L'} m'$	$m \sim_{L'} m'$ iff $\forall s \in L'.m(s) = m'(s)$	
model "set" of formula ϕ	$M(\phi) \subseteq M$, $f_\phi : M \to \{0,1\}$	$f_\phi : M \to V$
semantical equivalence of ϕ, ψ	$f_\phi = f_\psi$	
general model set	$M' \subseteq M$, $f : M \to \{0,1\}$	$f : M \to V$
f insensitive to L'	$\forall m, m' \in M.(m \sim_{L-L'} m' \Rightarrow f(m) = f(m'))$	
(ir)relevant	$s \in L$ is irrelevant for f iff f is insensitive to s,	
	$I(f) := \{s \in L : s \text{ is irrelevant for } f\}$, $R(f) := L - I(f)$	
	$I(\phi) := I(f_\phi)$, $R(\phi) := R(f_\phi)$	
$f^+(m, L')$, $f^-(m, L')$	$f^+(m, L') := max\{f(m') : m' \in M, m \sim_{L'} m'\}$	
	$f^-(m, L') := min\{f(m') : m' \in M, m \sim_{L'} m'\}$	
$f \leq g$	$\forall m \in M.f(m) \leq g(m)$	

1.2.3.1 Algebraic Properties

Fact 1.2.14

Table 1.4, "Interdependencies of algebraic rules", is to be read as follows: If the left hand side holds for some function $f : \mathcal{Y} \to \mathcal{P}(U)$, and the auxiliary properties noted in the middle also hold for f or $\mathcal{Y}$, then the right hand side will hold, too - and conversely.

"sing." will stand for: "$\mathcal{Y}$ contains singletons"

The numbers correspond to the numbers in Table 1.4.

Proof

All sets are to be in $\mathcal{Y}$.

(1.1) $(\mu PR) + (\cap) + (\mu \subseteq) \Rightarrow (\mu PR')$:

1.2 Basic Definitions and Overview

By $X \cap Y \subseteq X$ and (μPR), $f(X) \cap X \cap Y \subseteq f(X \cap Y)$. By $(\mu \subseteq)$ $f(X) \cap Y = f(X) \cap X \cap Y$.

(1.2) $(\mu PR') \Rightarrow (\mu PR)$:

Let $X \subseteq Y$, so $X = X \cap Y$, so by $(\mu PR')$ $f(Y) \cap X \subseteq f(X \cap Y) = f(X)$.

(2.1) $(\mu PR) + (\mu \subseteq) \Rightarrow (\mu OR)$:

$f(X \cup Y) \subseteq X \cup Y$ by $(\mu \subseteq)$, so $f(X \cup Y) = (f(X \cup Y) \cap X) \cup (f(X \cup Y) \cap Y) \subseteq f(X) \cup f(Y)$.

(2.2) $(\mu OR) + (\mu \subseteq) + (-) \Rightarrow (\mu PR)$:

Let $X \subseteq Y$, $X' := Y - X$. $f(Y) \subseteq f(X) \cup f(X')$ by (μOR), so $f(Y) \cap X \subseteq (f(X) \cap X) \cup (f(X') \cap X) =_{(\mu \subseteq)} f(X) \cup \emptyset = f(X)$.

(2.3) $(\mu PR) + (\mu \subseteq) \Rightarrow (\mu wOR)$:

Trivial by (2.1).

(2.4) $(\mu wOR) + (\mu \subseteq) + (-) \Rightarrow (\mu PR)$:

Let $X \subseteq Y$, $X' := Y - X$. $f(Y) \subseteq f(X) \cup X'$ by (μwOR), so $f(Y) \cap X \subseteq (f(X) \cap X) \cup (X' \cap X) =_{(\mu \subseteq)} f(X) \cup \emptyset = f(X)$.

(3) $(\mu PR) \Rightarrow (\mu CUT)$:

$f(X) \subseteq Y \subseteq X \Rightarrow f(X) \subseteq f(X) \cap Y \subseteq f(Y)$ by (μPR).

(4) $(\mu \subseteq) + (\mu \subseteq \supseteq) + (\mu CUM) + (\mu RatM) + (\cap) \not\Rightarrow (\mu PR)$:

This is shown in Example 1.2.5.

(5.1) $(\mu CM) + (\cap) + (\mu \subseteq) \Rightarrow (\mu ResM)$:

Let $f(X) \subseteq A \cap B$, so $f(X) \subseteq A$, so by $(\mu \subseteq)$ $f(X) \subseteq A \cap X \subseteq X$, so by (μCM) $f(A \cap X) \subseteq f(X) \subseteq B$.

(5.2) $(\mu ResM) \Rightarrow (\mu CM)$:

We consider here the infinitary version, where all sets can be model sets of infinite theories. Let $f(X) \subseteq Y \subseteq X$, so $f(X) \subseteq Y \cap f(X)$, so by $(\mu ResM)$ $f(Y) = f(X \cap Y) \subseteq f(X)$.

(6) $(\mu CM) + (\mu CUT) \Leftrightarrow (\mu CUM)$:

Trivial.

(7) $(\mu \subseteq) + (\mu \subseteq \supseteq) \Rightarrow (\mu CUM)$:

Suppose $f(D) \subseteq E \subseteq D$. So by $(\mu \subseteq)$ $f(E) \subseteq E \subseteq D$, so by $(\mu \subseteq \supseteq)$ $f(D) = f(E)$.

(8) $(\mu \subseteq) + (\mu CUM) + (\cap) \Rightarrow (\mu \subseteq \supseteq)$:

Let $f(D) \subseteq E$, $f(E) \subseteq D$, so by $(\mu \subseteq)$ $f(D) \subseteq D \cap E \subseteq D$, $f(E) \subseteq D \cap E \subseteq E$. As $f(D \cap E)$ is defined, so $f(D) = f(D \cap E) = f(E)$ by (μCUM).

(9) $(\mu \subseteq) + (\mu CUM) \not\Rightarrow (\mu \subseteq \supseteq)$:

This is shown in Example 1.2.4.

(10) $(\mu RatM) + (\mu PR) \Rightarrow (\mu =)$:

Trivial.

The following statements (up to (19)) are left as exercises, solutions can be found in [Sch04], Fact 3.10.9 for (11) to (19).

(20) $(\mu \subseteq) + (\mu PR) + (\mu =) \not\Rightarrow (\mu \parallel)$:

See Example 1.2.6.

(21) $(\mu \subseteq) + (\mu PR) + (\mu \parallel) \not\Rightarrow (\mu =)$:

See Example 1.2.7.

(22) $(\mu \subseteq) + (\mu PR) + (\mu \parallel) + (\mu =) + (\mu \cup) \not\Rightarrow (\mu \in)$:

See Example 1.2.8.

Thus, by Fact 1.4.7, the conditions do not assure representability by ranked structures.

□

Remark 1.2.15

Note that $(\mu =')$ is very close to $(RatM)$: $(RatM)$ says: $\alpha \hspace{2pt}\vert\!\sim\hspace{2pt} \beta, \alpha \hspace{2pt}\not\vert\!\sim\hspace{2pt} \neg\gamma \Rightarrow \alpha \wedge \gamma \hspace{2pt}\vert\!\sim\hspace{2pt} \beta$. Or, $f(A) \subseteq B, f(A) \cap C \neq \emptyset \Rightarrow f(A \cap C) \subseteq B$ for all A, B, C. This is not quite, but almost: $f(A \cap C) \subseteq f(A) \cap C$ (it depends how many B there are, if $f(A)$ is some such B, the fit is perfect).

Example 1.2.4

We show here $(\mu \subseteq) + (\mu CUM) \not\Rightarrow (\mu \subseteq \supseteq)$.

Consider $X := \{a, b, c\}$, $Y := \{a, b, d\}$, $f(X) := \{a\}$, $f(Y) := \{a, b\}$, $\mathcal{Y} := \{X, Y\}$. (If $f(\{a, b\})$ were defined, we would have $f(X) = f(\{a, b\}) = f(Y)$, *contradiction*.)

Obviously, $(\mu \subseteq)$ and (μCUM) hold, but not $(\mu \subseteq \supseteq)$.

□

Example 1.2.5

We show here $(\mu \subseteq) + (\mu \subseteq \supseteq) + (\mu CUM) + (\mu RatM) + (\cap) \not\Rightarrow (\mu PR)$.

Let $U := \{a, b, c\}$. Let $\mathcal{Y} = \mathcal{P}(U)$. So $(\cap)$ is trivially satisfied. Set $f(X) := X$ for all $X \subseteq U$ except for $f(\{a, b\}) = \{b\}$. Obviously, this cannot be represented by a preferential structure and (μPR) is false for U and $\{a, b\}$. But it satisfies $(\mu \subseteq), (\mu CUM), (\mu RatM)$. $(\mu \subseteq)$ is trivial. (μCUM) : Let $f(X) \subseteq Y \subseteq X$. If $f(X) = X$, we are done. Consider $f(\{a, b\}) = \{b\}$. If $\{b\} \subseteq Y \subseteq \{a, b\}$, then $f(Y) = \{b\}$, so we are done again. It is shown in Fact 1.2.14, (8) that $(\mu \subseteq \supseteq)$

1.2 Basic Definitions and Overview

follows. $(\mu RatM)$: Suppose $X \subseteq Y$, $X \cap f(Y) \neq \emptyset$, we have to show $f(X) \subseteq f(Y) \cap X$. If $f(Y) = Y$, the result holds by $X \subseteq Y$, so it does if $X = Y$. The only remaining case is $Y = \{a, b\}$, $X = \{b\}$, and the result holds again.

□

Example 1.2.6

The example shows that $(\mu \subseteq) + (\mu PR) + (\mu =) \not\Rightarrow (\mu \parallel)$.

Consider the following structure without transitivity: $U := \{a, b, c, d\}$, c and d have ω many copies in descending order $c_1 \succeq c_2 \ldots$, etc. a, b have one single copy each. $a \succeq b$, $a \succeq d_1$, $b \succeq a$, $b \succeq c_1$.

$(\mu \parallel)$ does not hold: $f(U) = \emptyset$, but $f(\{a, c\}) = \{a\}$, $f(\{b, d\}) = \{b\}$.

(μPR) holds as in all preferential structures.

$(\mu =)$ holds: If it were to fail, then for some $A \subseteq B$, $f(B) \cap A \neq \emptyset$, so $f(B) \neq \emptyset$. But the only possible cases for B are now: ($a \in B$, $b, d \notin B$) or ($b \in B$, $a, c \notin B$). Thus, B can be $\{a\}$, $\{a, c\}$, $\{b\}$, $\{b, d\}$ with $f(B) = \{a\}, \{a\}, \{b\}, \{b\}$. If $A = B$, then the result will hold trivially. Moreover, A has to be $\neq \emptyset$. So the remaining cases of B where it might fail are $B = \{a, c\}$ and $\{b, d\}$, and by $f(B) \cap A \neq \emptyset$, the only cases of A where it might fail, are $A = \{a\}$ or $\{b\}$ respectively. So the only cases remaining are: $B = \{a, c\}$, $A = \{a\}$ and $B = \{b, d\}$, $A = \{b\}$. In the first case, $f(A) = f(B) = \{a\}$, in the second $f(A) = f(B) = \{b\}$, but $(\mu =)$ holds in both.

□

Example 1.2.7

$(\mu \subseteq) + (\mu PR) + (\mu \parallel) \not\Rightarrow (\mu =)$.

For details, see [Sch04], Fact 3.10.10 (2).

Example 1.2.8

$(\mu \subseteq) + (\mu PR) + (\mu \parallel) + (\mu =) + (\mu \cup) \not\Rightarrow (\mu \in)$.

For details, see [Sch04], Fact 3.10.10 (3).

We turn to interdependencies of the different μ-conditions. Again, we will sometimes use preferential structures in our arguments.

Fact 1.2.16

If (μwOR) and $(\mu \subseteq)$ hold for f, then $f(X \cup Y) \subseteq f(X) \cup f(Y) \cup (X \cap Y)$

Proof

$f(X \cup Y) \subseteq f(X) \cup Y$, $f(X \cup Y) \subseteq X \cup f(Y)$, so $f(X \cup Y) \subseteq (f(X) \cup Y) \cap (X \cup f(Y)) = f(X) \cup f(Y) \cup (X \cap Y)$

□

1.2.3.2 Logical Properties

Definition 1.2.10

The definitions are given in Table 1.2, "Logical rules, definitions and connections Part I" and Table 1.3, "Logical rules, definitions and connections Part II", which also show connections between different versions of rules, the semantics, and rules about size. (The tables are split in two, as they would not fit onto one page otherwise.)

Explanation of the tables:

(1) The first table gives the basic properties, the second table those for Cumulativity and Rational Monotony.
(2) The difference between the first two columns is that the first column treats the formula version of the rule, the second the more general theory (i.e., set of formulas) version.
(3) "Corr." stands for "Correspondence".
(4) The numbers in the first column "Corr.", refer to Proposition 1.2.18, published as Proposition 21 in [GS08c], those in the second column "Corr." to Proposition 3.2.13.
(5) The third column, "Corr.", is to be understood as follows:

Let a logic $\mid\!\sim$ satisfy (LLE) and (CCL), and define a function $f : \boldsymbol{D_\mathcal{L}} \to \boldsymbol{D_\mathcal{L}}$ by $f(M(T)) := M(\overline{\overline{T}})$. Then f is well defined and satisfies (μdp), and $\overline{\overline{T}} = Th(f(M(T)))$.

If $\mid\!\sim$ satisfies a rule on the left-hand side, then — provided the additional properties noted in the middle for $\Rightarrow$ hold, too — f will satisfy the property on the right-hand side.

Conversely:

If $f : \mathcal{Y} \to \mathcal{P}(M_\mathcal{L})$ is a function, with $\boldsymbol{D_\mathcal{L}} \subseteq \mathcal{Y}$, and we define a logic $\mid\!\sim$ by $\overline{\overline{T}} := Th(f(M(T)))$, then $\mid\!\sim$ satisfies (LLE) and (CCL). If f satisfies (μdp), then $f(M(T)) = M(\overline{\overline{T}})$.

If f satisfies a property on the right-hand side, then — provided the additional properties noted in the middle for $\Leftarrow$ hold, too — $\mid\!\sim$ will satisfy the property on the left-hand side.

1.2 Basic Definitions and Overview

(6) We use the following abbreviations for those supplementary conditions in the "Correspondence" columns:

"$T = \phi$" means that if one of the theories (the one named the same way in Definition 1.2.10) is equivalent to a formula, we do not need (μdp).

$-(\mu dp)$ stands for "without (μdp)".

(7) $A = B \parallel C$ will abbreviate $A = B$, or $A = C$, or $A = B \cup C$.

Further comments:

(1) (PR) is also called infinite conditionalization We choose this name for its central role for preferential structures (PR) or (μPR).

(2) The system of rules (AND) (OR) (LLE) (RW) (SC) (CP) (CM) (CUM) is also called system P (for preferential). Adding $(RatM)$ gives the system R (for rationality or rankedness).

Roughly, smooth preferential structures generate logics satisfying system P, while ranked structures generate logics satisfying system R.

(3) A logic satisfying (REF), $(ResM)$, and (CUT) is called a consequence relation.

(4) (LLE) and (CCL) will hold automatically, whenever we work with model sets.

(5) (AND) is obviously closely related to filters, and corresponds to closure under finite intersections. (RW) corresponds to upward closure of filters.

More precisely, validity of both depend on the definition, and the direction we consider.

Given f and $(\mu \subseteq)$, $f(X) \subseteq X$ generates a principal filter, $\{X' \subseteq X : f(X) \subseteq X'\}$, with the definition, if $X = M(T)$, then $T \mathrel|\!\sim \phi$ iff $f(X) \subseteq M(\phi)$. Validity of (AND) and (RW) are then trivial.

Conversely, we can define for $X = M(T)$

$$\mathcal{X} := \{X' \subseteq X : \exists \phi (X' = X \cap M(\phi) \text{ and } T \mathrel|\!\sim \phi)\}.$$

(AND) then makes $\mathcal{X}$ closed under finite intersections, and (RW) makes $\mathcal{X}$ upward closed. This is in the infinite case usually not yet a filter, as not all subsets of X need to be definable this way. In this case, we complete $\mathcal{X}$ by adding all X'' such that there is $X' \subseteq X'' \subseteq X$, $X' \in \mathcal{X}$.

Alternatively, we can define

$$\mathcal{X} := \{X' \subseteq X : \bigcap \{X \cap M(\phi) : T \mathrel|\!\sim \phi\} \subseteq X'\}.$$

(6) (SC) corresponds to the choice of a subset.

(7) (CP) is somewhat delicate, as it presupposes that the chosen model set is nonempty. This might fail in the presence of ever better choices, without ideal ones; the problem is addressed by the limit versions.

(8) (PR) is an infinitary version of one half of the deduction theorem: Let T stand for ϕ, T' for ψ, and $\phi \wedge \psi \hspace{0.1cm}\vert\!\sim \sigma$, so $\phi \hspace{0.1cm}\vert\!\sim \psi \to \sigma$, but $(\psi \to \sigma) \wedge \psi \vdash \sigma$.

(9) (CUM) (whose more interesting half in our context is (CM)) may best be seen as a normal use of lemmas: We have worked hard and found some lemmas. Now we can take a rest, and come back again with our new lemmas. Adding them to the axioms will neither add new theorems, nor prevent old ones from holding. (This is, of course, a meta-level argument concerning an object-level rule. But also object level rules should — at least generally — have an intuitive justification, which will then come from a meta-level argument.)

The proof of the statements of the tables - together with the subsequent examples - requires some knowledge of preferential structures, which will be introduced "officially" only in Section 1.3. We chose to give those results here, so the reader will have immediately a global picture, and can come back later, if desired, and read the proofs and (counter)examples.

Fact 1.2.17

$(CUT) \not\Rightarrow (PR)$

Proof

We give two proofs:

(1) If $(CUT) \Rightarrow (PR)$, then by $(\mu PR) \Rightarrow$ (by Fact 1.2.14 (3)) $(\mu CUT) \Rightarrow$ (by Proposition 1.2.18 (7.2) $(CUT) \Rightarrow (PR)$ we would have a proof of $(\mu PR) \Rightarrow (PR)$ without (μdp), which is impossible, as shown by Example 1.7.1.

(2) Reconsider Example 1.2.5, and say $a \models p \wedge q$, $b \models p \wedge \neg q$, $c \models \neg p \wedge q$. It is shown there that (μCUM) holds, so (μCUT) holds, so by Proposition 1.2.18 (7.2) (CUT) holds, if we define $\overline{\overline{T}} := Th(f(M(T)))$. Set $T := \{p \vee (\neg p \wedge q)\}$, $T' := \{p\}$, then $\overline{\overline{T \cup T'}} = \overline{\overline{T'}} = \overline{\{p \wedge \neg q\}}$, $\overline{\overline{T}} = \overline{T}$, $\overline{T \cup T'} = \overline{T'} = \overline{\{p\}}$, so (PR) $fails$.

$\square$

1.2.3.3 Connections Between Algebraic and Logical Properties

Proposition 1.2.18

Table 1.5 "Logical and algebraic rules" is to be read as follows:

The definitions are given in Definition 1.2.10.

Let a logic $\hspace{0.1cm}\vert\!\sim$ satisfy (LLE) and (CCL), and define a function $f : D_\mathcal{L} \to D_\mathcal{L}$ by $f(M(T)) := M(\overline{\overline{T}})$. Then f is well defined, satisfies (μdp), and $\overline{\overline{T}} = Th(f(M(T)))$.

1.2 Basic Definitions and Overview

If $\mathrel{|\!\sim}$ satisfies a rule in the left hand side, then - provided the additional properties noted in the middle for $\Rightarrow$ hold, too - f will satisfy the property in the right hand side.

Conversely, if $f : \mathcal{Y} \to \mathcal{P}(M_\mathcal{L})$ is a function, with $\boldsymbol{D}_\mathcal{L} \subseteq \mathcal{Y}$, and we define a logic $\mathrel{|\!\sim}$ by $\overline{\overline{T}} := Th(f(M(T)))$, then $\mathrel{|\!\sim}$ satisfies (LLE) and (CCL). If f satisfies (μdp), then $f(M(T)) = M(\overline{\overline{T}})$.

If f satisfies a property in the right hand side, then - provided the additional properties noted in the middle for $\Leftarrow$ hold, too - $\mathrel{|\!\sim}$ will satisfy the property in the left hand side.

If "$T = \phi$" is noted in the table, this means that, if one of the theories (the one named the same way in Definition 1.2.10) is equivalent to a formula, we do not need (μdp).

Proof

Some of these statements were already shown in [Sch04], see e.g. Proposition 3.4.2, but often in a slightly different form and notation, so we give them here.

Set $f(T) := f(M(T))$, note that $f(T \cup T') := f(M(T \cup T')) = f(M(T) \cap M(T'))$.

We show first the general framework.

Let $\mathrel{|\!\sim}$ satisfy (LLE) and (CCL). Let $f : \boldsymbol{D}_\mathcal{L} \to \boldsymbol{D}_\mathcal{L}$ be defined by $f(M(T)) := M(\overline{\overline{T}})$. If $M(T) = M(T')$, then $\overline{T} = \overline{T'}$, so by (LLE) $\overline{\overline{T}} = \overline{\overline{T'}}$, so $f(M(T)) = f(M(T'))$, so f is well defined and satisfies (μdp). By (CCL) $Th(M(\overline{\overline{T}})) = \overline{\overline{T}}$.

Let f be given, and $\mathrel{|\!\sim}$ be defined by $\overline{\overline{T}} := Th(f(M(T)))$. Obviously, $\mathrel{|\!\sim}$ satisfies (LLE) and (CCL) (and thus (RW)). If f satisfies (μdp), then $f(M(T)) = M(T')$ for some T', and $f(M(T)) = M(Th(f(M(T)))) = M(\overline{\overline{T}})$ by Fact 1.2.11. (We will use Fact 1.2.11 now without further mentioning.)

Next we show the following fact:

(a)

If f satisfies (μdp), or T' is equivalent to a formula, then $Th(f(T) \cap M(T')) = \overline{\overline{T}} \cup T'$.

Case 1, f satisfies (μdp). $Th(f(M(T)) \cap M(T')) = Th(M(\overline{\overline{T}}) \cap M(T') = \overline{\overline{T}} \cup T'$ by Fact 1.2.10 (5).

Case 2, T' is equivalent to ϕ'. $Th(f(M(T)) \cap M(\phi')) = \overline{Th(f(M(T))) \cup \{\phi'\}} = \overline{\overline{T}} \cup \{\phi'\}$ by Fact 1.2.10 (3).

We now prove the individual properties.

(1.1) $(OR) \Rightarrow (\mu OR)$

Let $X = M(T), Y = M(T')$. $f(X \cup Y) = f(M(T) \cup M(T')) = f(M(T \vee T'))$
$:= M(\overline{\overline{T \vee T'}}) \subseteq_{(OR)} M(\overline{\overline{T}} \cap \overline{\overline{T'}}) =_{(CCL)} M(\overline{\overline{T}}) \cup M(\overline{\overline{T'}}) =: f(X) \cup f(Y)$.

(1.2) $(\mu OR) \Rightarrow (OR)$

$\overline{\overline{T \vee T'}} := Th(f(M(T \vee T'))) = Th(f(M(T) \cup M(T'))) \supseteq_{(\mu OR)}$
$Th(f(M(T)) \cup f(M(T'))) = $ (by Fact 1.2.9) $Th(f(M(T))) \cap Th(f(M(T')))$
$=: \overline{\overline{T}} \cap \overline{\overline{T'}}$.

(2) By $\neg Con(T, T') \Leftrightarrow M(T) \cap M(T') = \emptyset$, we can use directly the proofs for 1.

(3.1) $(wOR) \Rightarrow (\mu wOR)$

Let $X = M(T), Y = M(T')$. $f(X \cup Y) = f(M(T) \cup M(T')) = f(M(T \vee T'))$
$:= M(\overline{\overline{T \vee T'}}) \subseteq_{(wOR)} M(\overline{\overline{T}} \cap \overline{\overline{T'}}) =_{(CCL)} M(\overline{\overline{T}}) \cup M(\overline{\overline{T'}}) =: f(X) \cup Y$.

(3.2) $(\mu wOR) \Rightarrow (wOR)$

$\overline{\overline{T \vee T'}} := Th(f(M(T \vee T'))) = Th(f(M(T) \cup M(T'))) \supseteq_{(\mu wOR)}$
$Th(f(M(T)) \cup M(T')) = $ (by Fact 1.2.9) $Th(f(M(T))) \cap Th(M(T')) =:$
$\overline{\overline{T}} \cap \overline{T'}$.

(4.1) $(SC) \Rightarrow (\mu \subseteq)$

Trivial.

(4.2) $(\mu \subseteq) \Rightarrow (SC)$

Trivial.

(5.1) $(CP) \Rightarrow (\mu \emptyset)$

Trivial.

(5.2) $(\mu \emptyset) \Rightarrow (CP)$

Trivial.

(6.1) $(PR) \Rightarrow (\mu PR)$:

Exercise, solution in the Appendix.

(6.2) $(\mu PR) + (\mu dp) + (\mu \subseteq) \Rightarrow (PR)$:

$f(T) \cap M(T') =_{(\mu \subseteq)} f(T) \cap M(T) \cap M(T') = f(T) \cap M(T \cup T') \subseteq_{(\mu PR)}$
$f(T \cup T')$, so $\overline{\overline{T \cup T'}} = Th(f(T \cup T')) \subseteq Th(f(T) \cap M(T')) = \overline{\overline{T}} \cup T'$ by
(a) above and (μdp).

(6.3) $(\mu PR) \not\Rightarrow (PR)$ without (μdp):

(μPR) holds in all preferential structures (see Definition 1.2.11) by Fact 1.3.2. Example 1.7.1 shows that (DP) may fail in the resulting logic.

1.2 Basic Definitions and Overview

(6.4) $(\mu PR) + (\mu \subseteq) \Rightarrow (PR)$ if T' is classically equivalent to a formula:

It was shown in the proof of (6.2) that $f(T) \cap M(\phi') \subseteq f(T \cup \{\phi'\})$, so $\overline{\overline{T \cup \{\phi'\}}} = Th(f(T \cup \{\phi'\})) \subseteq Th(f(T) \cap M(\phi')) = \overline{\overline{T}} \cup \{\phi'\}$ by (a) above.

(6.5) $(\mu PR') \Rightarrow (PR)$, if T' is classically equivalent to a formula:

$f(M(T)) \cap M(\phi') \subseteq_{(\mu PR')} f(M(T) \cap M(\phi')) = f(M(T \cup \{\phi'\}))$. So again $\overline{\overline{T \cup \{\phi'\}}} = Th(f(T \cup \{\phi'\})) \subseteq Th(f(T) \cap M(\phi')) = \overline{\overline{T}} \cup \{\phi'\}$ by (a) above.

(7.1) $(CUT) \Rightarrow (\mu CUT)$

So let $X = M(T), Y = M(T')$, and $f(T) := M(\overline{\overline{T}}) \subseteq M(T') \subseteq M(T)$ $\Rightarrow \overline{T} \subseteq \overline{T'} \subseteq \overline{\overline{T}} =_{(LLE)} \overline{(\overline{T})} \Rightarrow$ (by (CUT)) $\overline{\overline{T}} = \overline{(\overline{T})} \supseteq \overline{(\overline{T'})} = \overline{T'} \Rightarrow$ $f(T) = M(\overline{\overline{T}}) \subseteq M(\overline{\overline{T'}}) = f(T')$, thus $f(X) \subseteq f(Y)$.

(7.2) $(\mu CUT) \Rightarrow (CUT)$

Exercise, solution in the Appendix.

(8.1) $(CM) \Rightarrow (\mu CM)$

So let $X = M(T), Y = M(T')$, and $f(T) := M(\overline{\overline{T}}) \subseteq M(T') \subseteq M(T) \Rightarrow$ $\overline{T} \subseteq \overline{T'} \subseteq \overline{\overline{T}} =_{(LLE)} \overline{(\overline{T})} \Rightarrow$ (by $(LLE), (CM)$) $\overline{\overline{T}} = \overline{(\overline{T})} \subseteq \overline{(\overline{T'})} = \overline{T'} \Rightarrow$ $f(T) = M(\overline{\overline{T}}) \supseteq M(\overline{\overline{T'}}) = f(T')$, thus $f(X) \supseteq f(Y)$.

(8.2) $(\mu CM) \Rightarrow (CM)$

Let $T \subseteq \overline{T'} \subseteq \overline{\overline{T}}$. Thus by (μCM) and $f(T) \subseteq M(\overline{\overline{T}}) \subseteq M(T') \subseteq M(T)$, so $f(T) \supseteq f(T')$ by (μCM), so $\overline{\overline{T}} = Th(f(T)) \subseteq Th(f(T')) = \overline{\overline{T'}}$.

(9.1) $(ResM) \Rightarrow (\mu ResM)$

Let $f(X) := M(\overline{\overline{\Delta}})$, $A := M(\alpha)$, $B := M(\beta)$. So $f(X) \subseteq A \cap B \Rightarrow$ $\Delta \hspace{0.1em}\vert\hspace{-0.3em}\sim \alpha, \beta \Rightarrow_{(ResM)} \Delta, \alpha \hspace{0.1em}\vert\hspace{-0.3em}\sim \beta \Rightarrow M(\overline{\overline{\Delta, \alpha}}) \subseteq M(\beta) \Rightarrow f(X \cap A) \subseteq B$.

(9.2) $(\mu ResM) \Rightarrow (ResM)$

Let $f(X) := M(\overline{\overline{\Delta}}), A := M(\alpha), B := M(\beta)$. So $\Delta \hspace{0.1em}\vert\hspace{-0.3em}\sim \alpha, \beta \Rightarrow f(X) \subseteq A \cap B \Rightarrow_{(\mu ResM)} f(X \cap A) \subseteq B \Rightarrow \Delta, \alpha \hspace{0.1em}\vert\hspace{-0.3em}\sim \beta$.

(10.1) $(\subseteq \supseteq) \Rightarrow (\mu \subseteq \supseteq)$

Let $f(T) \subseteq M(T'), f(T') \subseteq M(T)$. So $Th(M(T')) \subseteq Th(f(T))$, $Th(M(T)) \subseteq Th(f(T'))$, so $T' \subseteq \overline{T'} \subseteq \overline{\overline{T}}, T \subseteq \overline{T} \subseteq \overline{\overline{T'}}$, so by $(\subseteq \supseteq)$ $\overline{\overline{T}} = \overline{\overline{T'}}$, so $f(T) := M(\overline{\overline{T}}) = M(\overline{\overline{T'}}) =: f(T')$.

(10.2) $(\mu \subseteq \supseteq) \Rightarrow (\subseteq \supseteq)$

Let $T \subseteq \overline{\overline{T'}}$ and $T' \subseteq \overline{\overline{T}}$. So by (CCL) $Th(M(T)) = \overline{T} \subseteq \overline{\overline{T'}} = Th(f(T'))$. But $Th(M(T)) \subseteq Th(X) \Rightarrow X \subseteq M(T) : X \subseteq M(Th(X)) \subseteq M(Th(M(T))) = M(T)$. So $f(T') \subseteq M(T)$, likewise $f(T) \subseteq M(T')$, so by $(\mu \subseteq \supseteq)$ $f(T) = f(T')$, so $\overline{\overline{T}} = \overline{\overline{T'}}$.

(11.1) $(CUM) \Rightarrow (\mu CUM)$:

So let $X = M(T)$, $Y = M(T')$, and $f(T) := M(\overline{\overline{T}}) \subseteq M(T') \subseteq M(T) \Rightarrow T \subseteq T' \subseteq \overline{\overline{T}} =_{(LLE)} \overline{\overline{(T)}} \Rightarrow \overline{\overline{T}} = \overline{\overline{(T)}} = \overline{\overline{(T')}} = \overline{\overline{T'}} \Rightarrow f(T) = M(\overline{\overline{T}}) = M(\overline{\overline{T'}}) = f(T')$, thus $f(X) = f(Y)$.

(11.2) $(\mu CUM) \Rightarrow (CUM)$:

Let $T \subseteq T' \subseteq \overline{\overline{T}}$. Thus by (μCUM) and $f(T) \subseteq M(\overline{\overline{T}}) \subseteq M(T') \subseteq M(T)$, so $f(T) = f(T')$, so $\overline{\overline{T}} = Th(f(T)) = Th(f(T')) = \overline{\overline{T'}}$.

(12.1) $(RatM) \Rightarrow (\mu RatM)$

Let $X = M(T)$, $Y = M(T')$, and $X \subseteq Y$, $X \cap f(Y) \neq \emptyset$, so $T \vdash T'$ and $M(T) \cap f(M(T')) \neq \emptyset$, so $Con(T, \overline{\overline{T'}})$, so $\overline{\overline{T'}} \cup T \subseteq \overline{\overline{T}}$ by $(RatM)$, so $f(X) = f(M(T)) = M(\overline{\overline{T}}) \subseteq M(\overline{\overline{T'}} \cup T) = M(\overline{\overline{T'}}) \cap M(T) = f(Y) \cap X$.

(12.2) $(\mu RatM) + (\mu dp) \Rightarrow (RatM)$:

Let $X = M(T)$, $Y = M(T')$, $T \vdash T'$, $Con(T, \overline{\overline{T'}})$, so $X \subseteq Y$ and by (μdp) $X \cap f(Y) \neq \emptyset$, so by $(\mu RatM)$ $f(X) \subseteq f(Y) \cap X$, so $\overline{\overline{T}} = \overline{\overline{T \cup T'}} = Th(f(T \cup T')) \supseteq Th(f(T') \cap M(T)) = \overline{\overline{T'}} \cup T$ by (a) above and (μdp).

(12.3) $(\mu RatM) \not\Rightarrow (RatM)$ without (μdp):

$(\mu RatM)$ holds in all ranked preferential structures (see Definition 1.2.13) by Fact 1.4.7. Example 1.2.9 (page 32) (2) shows that $(RatM)$ may fail in the resulting logic.

(12.4) $(\mu RatM) \Rightarrow (RatM)$ if T is classically equivalent to a formula:

Exercise, solution in the Appendix.

(13.1) $(RatM =) \Rightarrow (\mu =)$

Let $X = M(T)$, $Y = M(T')$, and $X \subseteq Y$, $X \cap f(Y) \neq \emptyset$, so $T \vdash T'$ and $M(T) \cap f(M(T')) \neq \emptyset$, so $Con(T, \overline{\overline{T'}})$, so $\overline{\overline{T'}} \cup T = \overline{\overline{T}}$ by $(RatM =)$, so $f(X) = f(M(T)) = M(\overline{\overline{T}}) = M(\overline{\overline{T'}} \cup T) = M(\overline{\overline{T'}}) \cap M(T) = f(Y) \cap X$.

(13.2) $(\mu =) + (\mu dp) \Rightarrow (RatM =)$

Let $X = M(T)$, $Y = M(T')$, $T \vdash T'$, $Con(T, \overline{\overline{T'}})$, so $X \subseteq Y$ and by (μdp) $X \cap f(Y) \neq \emptyset$, so by $(\mu =)$ $f(X) = f(Y) \cap X$. So $\overline{\overline{T'}} \cup T = \overline{\overline{T}}$ (a) above and (μdp).

(13.3) $(\mu =) \not\Rightarrow (RatM =)$ without (μdp):

$(\mu =)$ holds in all ranked preferential structures (see Definition 1.2.13) by Fact 1.4.7. Example 1.2.9 (page 32) (1) shows that $(RatM =)$ may fail in the resulting logic.

1.2 Basic Definitions and Overview

(13.4) $(\mu =) \Rightarrow (RatM =)$ if T is classically equivalent to a formula:

The proof is almost identical to the one for (12.4). Again, the prerequisites of $(\mu =)$ are satisfied, so $f(M(\phi)) = f(M(T')) \cap M(\phi)$. Thus, $\overline{\overline{T' \cup \{\phi\}}} = \overline{\overline{\phi}}$ by (a) above.

Of the last four, we show (14), (15), (17), the proof for (16) is similar to the one for (17).

(14.1) $(Log =') \Rightarrow (\mu =')$:

$f(M(T')) \cap M(T) \neq \emptyset \Rightarrow Con(\overline{\overline{T'}} \cup T) \Rightarrow_{(Log=')} \overline{\overline{T \cup T'}} = \overline{\overline{T' \cup T}} \Rightarrow f(M(T \cup T')) = f(M(T')) \cap M(T)$.

(14.2) $(\mu =') + (\mu dp) \Rightarrow (Log =')$:

$Con(\overline{\overline{T'}} \cup T) \Rightarrow_{(\mu dp)} f(M(T')) \cap M(T) \neq \emptyset \Rightarrow f(M(T' \cup T)) = f(M(T') \cap M(T)) =_{(\mu=')} f(M(T')) \cap M(T)$, so $\overline{\overline{T' \cup T}} = \overline{\overline{T' \cup T}}$ by (a) above and (μdp).

(14.3) $(\mu =') \not\Rightarrow (Log =')$ without (μdp) :

By Fact 1.4.7 $(\mu =')$ holds in ranked structures. Consider Example 1.2.9 (2). There, $Con(T, \overline{\overline{T'}})$, $T = T \cup T'$, and it was shown that $\overline{\overline{T'}} \cup T \not\subseteq \overline{\overline{T}} = \overline{\overline{T \cup T'}}$

(14.4) $(\mu =') \Rightarrow (Log =')$ if T is classically equivalent to a formula:

$Con(\overline{\overline{T'}} \cup \{\phi\}) \Rightarrow \emptyset \neq M(\overline{\overline{T'}}) \cap M(\phi) \Rightarrow f(T') \cap M(\phi) \neq \emptyset$ by Fact 1.2.10 (4). So $f(M(T' \cup \{\phi\})) = f(M(T') \cap M(\phi)) = f(M(T')) \cap M(\phi)$ by $(\mu =')$, so $\overline{\overline{T' \cup \{\phi\}}} = \overline{\overline{T'}} \cup \{\phi\}$ by (a) above.

(15.1) $(Log \|) \Rightarrow (\mu \|)$:

Trivial.

(15.2) $(\mu \|) \Rightarrow (Log \|)$:

Trivial.

(16) $(Log \cup) \Leftrightarrow (\mu \cup)$: Analogous to the proof of (17).

(17.1) $(Log \cup') + (\mu \subseteq) + (\mu =) \Rightarrow (\mu \cup')$:

Exercise, solution in the Appendix.

(17.2) $(\mu \cup') + (\mu dp) \Rightarrow (Log \cup')$:

Exercise, solution in the Appendix.

(17.3) and (16.3) are solved by Example 1.2.9 (3).

□

Example 1.2.9

(1) $(\mu =)$ without (μdp) does not imply $(RatM =)$:

Take $\{p_i : i \in \omega\}$ and put $m := m_{\bigwedge p_i}$, the model which makes all p_i true, in the top layer, all the other in the bottom layer. Let $m' \neq m$, $T' := \emptyset$, $T := Th(m, m')$. Then Then $\overline{\overline{T'}} = T'$, so $Con(\overline{\overline{T'}}, T)$, $\overline{\overline{T}} = Th(m')$, $\overline{\overline{T'}} \cup T = T$.

So $(RatM =)$ fails, but $(\mu =)$ holds in all ranked structures.

(2) $(\mu RatM)$ without (μdp) does not imply $(RatM)$:

Take $\{p_i : i \in \omega\}$ and let $m := m_{\bigwedge p_i}$, the model which makes all p_i true. Let $X := M(\neg p_0) \cup \{m\}$ be the top layer, put the rest of $M_{\mathcal{L}}$ in the bottom layer. Let $Y := M_{\mathcal{L}}$. The structure is ranked, as shown in Fact 1.4.7, $(\mu RatM)$ holds.

Let $T' := \emptyset$, $T := Th(X)$. We have to show that $Con(T, \overline{\overline{T'}})$, $T \vdash T'$, but $\overline{\overline{T'}} \cup T \not\subseteq \overline{\overline{T}}$. $\overline{\overline{T'}} = Th(M(p_0) - \{m\}) = \overline{p_0}$. $T = \overline{\{\neg p_0\} \vee Th(m)}$, $\overline{\overline{T}} = T$. So $Con(T, \overline{\overline{T'}})$. $M(\overline{\overline{T'}}) = M(p_0)$, $M(T) = X$, $M(\overline{\overline{T'}} \cup T) = M(\overline{\overline{T'}}) \cap M(T) = \{m\}$, $m \models p_1$, so $p_1 \in \overline{\overline{T'}} \cup T$, but $X \not\models p_1$.

(3) This example shows that we need (μdp) to go from $(\mu\cup)$ to $(Log\cup)$ and from $(\mu\cup')$ to $(Log\cup')$.

Let $v(\mathcal{L}) := \{p, q\} \cup \{p_i : i < \omega\}$. Let m make all variables true.

Put all models of $\neg p$, and m, in the upper layer, all other models in the lower layer. This is ranked, so by Fact 1.4.7 $(\mu\cup)$ and $(\mu\cup')$ hold. Set $X := M(\neg q) \cup \{m\}$, $X' := M(q) - \{m\}$, $T := Th(X) = \neg q \vee Th(m)$, $T' := Th(X') = \overline{q}$. Then $\overline{\overline{T}} = \overline{p \wedge \neg q}$, $\overline{\overline{T'}} = \overline{p \wedge q}$. We have $Con(\overline{\overline{T'}}, T)$, $\neg Con(\overline{\overline{T'}}, \overline{\overline{T}})$. But $\overline{T \vee T'} = \overline{p} \neq \overline{\overline{T}} = \overline{p \wedge \neg q}$ and $Con(\overline{T \vee T'}, T')$, so $(Log\cup)$ and $(Log\cup')$ fail.

□

1.2.4 Preferential Structures

Table 1.6, "Preferential representation", summarizes the more difficult half of a full representation result for preferential structures. It shows equivalence between certain abstract conditions for model choice functions and certain preferential structures. They are shown in the respective representation theorems.

The other half - equivalence between certain logical rules and certain abstract conditions for model choice functions - are summarized in Definition 1.2.10 and shown in Proposition 1.2.18 (going via the μ-functions).

1.2 Basic Definitions and Overview

"Singletons" means that the domain must contain all singletons, "1 copy" or "≥ 1 copy" means that the structure may contain only one copy for each point, or several, and "$(\mu\emptyset)$" for the preferential structure means that the μ-function of the structure has to satisfy this property.

We call a characterization "normal" iff it is a universally quantified Boolean combination (of any fixed, but perhaps infinite, length) of rules of the usual form. We do not go into details here.

In the second column from the left, "⇒" means, for instance for the smooth case, that for any $\mathcal{Y}$ closed under finite unions, and any choice function f which satisfies the conditions in the left-hand column, there is a (here $\mathcal{Y}$-smooth) preferential structure $\mathcal{X}$ which represents it, i.e., for all $Y \in \mathcal{Y}$, $f(Y) = \mu_\mathcal{X}(Y)$, where $\mu_\mathcal{X}$ is the model choice function of the structure $\mathcal{X}$. The inverse arrow ⇐ means that the model choice function for any smooth $\mathcal{X}$ defined on such a $\mathcal{Y}$ will satisfy the conditions on the left.

1.2.4.1 Algebraic Properties

Recall Definition 1.2.4.

The Minimal Version

The following two definitions make preferential structures precise. We first give the algebraic definition, and then the definition of the consequence relation generated by an preferential structure. In the algebraic definition, the set U is an arbitrary set, in the application to logic, this will be the set of classical models of the underlying propositional language.

In both cases, we first present the simpler variant without copies, and then the one with copies. (Note that e.g., [KLM90], [LM92] use labelling functions instead, the version without copies corresponds to injective labelling functions, the one with copies to the general case. These are just different ways of speaking.) We will discuss the difference between the version without and the version with copies below, where we show that the version with copies is strictly more expressive than the version without copies, and that transitivity of the relation adds new properties in the case without copies. In the general case with copies, transitivity can be added without changing properties.

We give here the "minimal version", the much more complicated "limit version" is presented and discussed in Section 1.6. Recall the intuition that the relation $\prec$ expresses "normality" or "importance" - the $\prec$-smaller, the more normal or important. The smallest elements are those which count.

Definition 1.2.11

Fix $U \neq \emptyset$, and consider arbitrary X. Note that this X has not necessarily anything to do with U, or the $\mathcal{U}$ below. Thus, the functions $\mu_{\mathcal{M}}$ below are in principle functions from V to V, where V is the set-theoretical universe we work in.

Note that we work here often with copies of elements (or models). In other areas of logic, most authors work with valuation functions. Both definitions — copies and valuation functions — are equivalent; a copy $\langle x, i \rangle$ can be seen as a state $\langle x, i \rangle$ with valuation x. In the beginning of research on preferential structures, the notion of copies was widely used, whereas, e.g., [KLM90] used that of valuation functions. There is perhaps a weak justification for the former terminology. In modal logic, even if two states have the same valid classical formulas, they might still be distinguishable by their valid modal formulas. But this depends on the modality being in the object language. In most work on preferential stuctures, the consequence relation is outside the object language, so different states with same valuation are in a stronger sense copies of each other.

(1) Preferential models or structures.

   (1.1) The version without copies:

   A pair $\mathcal{M} := \langle U, \prec \rangle$ with U an arbitrary set and $\prec$ an arbitrary binary relation on U is called a preferential model or structure.

   (1.2) The version with copies:

   A pair $\mathcal{M} := \langle \mathcal{U}, \prec \rangle$ with $\mathcal{U}$ an arbitrary set of pairs and $\prec$ an arbitrary binary relation on $\mathcal{U}$ is called a preferential model or structure.

   If $\langle x, i \rangle \in \mathcal{U}$, then x is intended to be an element of U, and i the index of the copy.

   We sometimes also need copies of the relation $\prec$. We will then replace $\prec$ by one or several arrows α attacking non-minimal elements, e.g., $x \prec y$ will be written $\alpha : x \to y$, $\langle x, i \rangle \prec \langle y, i \rangle$ will be written $\alpha : \langle x, i \rangle \to \langle y, i \rangle$, and finally we might have $\langle \alpha, k \rangle : x \to y$ and $\langle \alpha, k \rangle : \langle x, i \rangle \to \langle y, i \rangle$.

(2) Minimal elements, the functions $\mu_{\mathcal{M}}$

   (2.1) The version without copies:

   Let $\mathcal{M} := \langle U, \prec \rangle$, and define

   $\mu_{\mathcal{M}}(X) := \{ x \in X : x \in U \wedge \neg \exists x' \in X \cap U. x' \prec x \}$.

   $\mu_{\mathcal{M}}(X)$ is called the set of minimal elements of X (in $\mathcal{M}$).

   Thus, $\mu_{\mathcal{M}}(X)$ is the set of elements such that there is no smaller one in X.

   (2.2) The version with copies:

   Let $\mathcal{M} := \langle \mathcal{U}, \prec \rangle$ be as above. Define

1.2 Basic Definitions and Overview

$$\mu_{\mathcal{M}}(X) := \{x \in X : \exists \langle x, i \rangle \in \mathcal{U}. \neg \exists \langle x', i' \rangle \in \mathcal{U}(x' \in X \wedge \langle x', i' \rangle' \prec \langle x, i \rangle)\}.$$

Thus, $\mu_{\mathcal{M}}(X)$ is the projection on the first coordinate of the set of elements such that there is no smaller one in X.

Again, by abuse of language, we say that $\mu_{\mathcal{M}}(X)$ is the set of minimal elements of X in the structure. If the context is clear, we will write just μ.

We sometimes say that $\langle x, i \rangle$ "kills" or "minimizes" $\langle y, j \rangle$ if $\langle x, i \rangle \prec \langle y, j \rangle$. By abuse of language we also say a set X kills or minimizes a set Y if for all $\langle y, j \rangle \in \mathcal{U}$, $y \in Y$ there is $\langle x, i \rangle \in \mathcal{U}$, $x \in X$ s.t. $\langle x, i \rangle \prec \langle y, j \rangle$.

$\mathcal{M}$ is also called injective or 1-copy iff there is always at most one copy $\langle x, i \rangle$ for each x. Note that the existence of copies corresponds to a non-injective labelling function — as is often used in nonclassical logic, e.g., modal logic.

We say that $\mathcal{M}$ is transitive, irreflexive, etc., iff $\prec$ is.

Note that $\mu(X)$ might well be empty, even if X is not.

We now show that every preferential structure has an equivalent irreflexive one – perhaps by adding copies.

Lemma 1.2.19

For any preferential structure $\mathcal{Z} = \langle \mathcal{X}, \prec \rangle$, there is a preferential structure $\mathcal{Z}' := \langle \mathcal{X}', \prec' \rangle$ s.t.

(1) $\mu_{\mathcal{Z}} = \mu_{\mathcal{Z}'}$,

(2) $\mathcal{Z}'$ is irreflexive,

(3) if $\mathcal{Z}$ is transitive, then so is $\mathcal{Z}'$.

Proof

Let $\mathcal{X}' := \{\langle x, \langle i, n \rangle \rangle : \langle x, i \rangle \in \mathcal{X}, n \in \omega\}$ and $\langle x', \langle i', n' \rangle \rangle \prec' \langle x, \langle i, n \rangle \rangle$ iff

(i) $n' > n$ and

(ii) $\langle x', i' \rangle \prec \langle x, i \rangle$.

(1) Let Y be any set, we have to show $\mu_{\mathcal{Z}}(Y) = \mu_{\mathcal{Z}'}(Y)$.

"$\subseteq$": Suppose $y \in \mu_{\mathcal{Z}}(Y)$, but $y \notin \mu_{\mathcal{Z}'}(Y)$. Take $\langle y, i \rangle \in \mathcal{X}$ s.t. there is no $\langle y', i' \rangle \in \mathcal{X} \upharpoonright Y$, $\langle y', i' \rangle \prec \langle y, i \rangle$. Consider $u := \langle y, \langle i, 0 \rangle \rangle \in \mathcal{X}' \upharpoonright Y$. By $y \notin \mu_{\mathcal{Z}'}$, there is $u' := \langle y', \langle i', n' \rangle \rangle \in \mathcal{X}' \upharpoonright Y$, $u' \prec' u$, but then $< y', i' > \prec \langle y, i \rangle$, contradiction. "$\supseteq$": Suppose $y \in \mu_{\mathcal{Z}'}(Y)$, but $y \notin \mu_{\mathcal{Z}}(Y)$. Take $u := \langle y, \langle i, n \rangle \rangle \in \mathcal{X}' \upharpoonright Y$ s.t. there is no $u' := \langle y', \langle i', n' \rangle \rangle \in \mathcal{X}' \upharpoonright Y$, $u' \prec' u$. Then $\langle y, i \rangle \in \mathcal{X} \upharpoonright Y$, so there is $\langle y', i' \rangle \in \mathcal{X} \upharpoonright Y$ s.t. $\langle y', i' \rangle \prec \langle y, i \rangle$. But then $< y', \langle i', n+1 \rangle > \prec' \langle y, \langle i, n \rangle \rangle$, contradiction.

(2) is trivial by the condition $n' > n$.

(3) Let $\langle x'', \langle i'', n'' \rangle \rangle \prec' \langle x', \langle i', n' \rangle \rangle \prec' \langle x, \langle i, n \rangle \rangle$. Then $\langle x'', i'' \rangle \prec \langle x', i' \rangle \prec \langle x, i \rangle$, so by transitivity of $\prec$, $\langle x'', i'' \rangle \prec \langle x, i \rangle$. Moreover, $n'' > n' > n$, so $n'' > n$, and thus $\langle x'', \langle i'', n'' \rangle \rangle \prec' \langle x, \langle i, n \rangle \rangle$.

□

We define now two additional properties of the relation, smoothness and rankedness.

Definition 1.2.12

Let $\mathcal{Y} \subseteq \mathcal{P}(U)$. (In applications to logic, $\mathcal{Y}$ will be $\boldsymbol{D_\mathcal{L}}$.)

A preferential structure $\mathcal{M}$ is called $\mathcal{Y}$-smooth iff for every $X \in \mathcal{Y}$ every element $x \in X$ is either minimal in X or above an element which is minimal in X. More precisely:

(1) The version without copies:

If $x \in X \in \mathcal{Y}$, then either $x \in \mu(X)$ or there is $x' \in \mu(X).x' \prec x$.

(2) The version with copies:

If $x \in X \in \mathcal{Y}$, and $\langle x, i \rangle \in \mathcal{U}$, then either there is no $\langle x', i' \rangle \in \mathcal{U}$, $x' \in X$, $\langle x', i' \rangle \prec \langle x, i \rangle$ or there is a $\langle x', i' \rangle \in \mathcal{U}$, $\langle x', i' \rangle \prec \langle x, i \rangle$, $x' \in X$, s.t. there is no $\langle x'', i'' \rangle \in \mathcal{U}$, $x'' \in X$, with $\langle x'', i'' \rangle \prec \langle x', i' \rangle$.

(Writing down all details here again might make it easier to read applications of the definition later on.)

When considering the models of a language $\mathcal{L}$, $\mathcal{M}$ will be called smooth iff it is $\boldsymbol{D_\mathcal{L}}$-smooth; $\boldsymbol{D_\mathcal{L}}$ is the default.

Obviously, the richer the set $\mathcal{Y}$ is, the stronger the condition $\mathcal{Y}$-smoothness will be.

A remark for the intuition: Smoothness is perhaps best motivated through Gabbay's concept of reactive diagrams; see, e.g., [Gab04] and [Gab08], and also [GS08c], [GS08f]. In this concept, smaller, or "better", elements attack bigger, or "less good", elements. But when a attacks b, and b attacks c, then one might consider the attack of b against c weakened by the attack of a against b. In a smooth structure, for every attack against some element x, there is also an uncontested attack against x, as it originates in an element y, which is not attacked itself.

Fact 1.2.20

Let $\prec$ be an irreflexive, binary relation on X; then the following two conditions are equivalent:

(1) There is an Ω and an irreflexive, total, binary relation $\prec'$ on Ω and a function $f: X \to \Omega$ s.t. $x \prec y \Leftrightarrow f(x) \prec' f(y)$ for all $x, y \in X$.

(2) Let $x, y, z \in X$ and $x \bot y$ with respect to $\prec$ (i.e., neither $x \prec y$ nor $y \prec x$); then $z \prec x \Rightarrow z \prec y$ and $x \prec z \Rightarrow y \prec z$.

1.2 Basic Definitions and Overview

Proof

Exercise, solution in the Appendix.

Definition 1.2.13

We call an irreflexive, binary relation $\prec$ on X which satisfies (1) (equivalently (2)) of Fact 1.2.20 ranked. By abuse of language, we also call a preferential structure $\langle X, \prec \rangle$ ranked iff $\prec$ is.

Fact 1.2.21

If $\prec$ on X is ranked, and free of cycles, then $\prec$ is transitive.

Proof

Let $x \prec y \prec z$. If $x \perp z$, then $y \succ z$, resulting in a cycle of length 2. If $z \prec x$, then we have a cycle of length 3. So $x \prec z$.

□

The smoothness condition says that if $x \in X$ is not a minimal element of X, then there is $x' \in \mu(X)$ $x' \prec x$. In the finite case without copies, smoothness is a trivial consequence of transitivity and lack of cycles. But note that in the other cases infinite descending chains might still exist, even if the smoothness condition holds, they are just "short-circuited": we might have such chains, but below every element in the chain is a minimal element. In the authors' opinion, smoothness is difficult to justify as a structural property (or, in a more philosophical spirit, as a property of the world): why should we always have such minimal elements below non-minimal ones? Smoothness has, however, a justification from its consequences. Its attractiveness comes from two sides:

First, it generates a very valuable logical property, cumulativity (CUM): If $\mathcal{M}$ is smooth, and $\overline{\overline{T}}$ is the set of $\models_\mathcal{M}$-consequences, then for $T \subseteq \overline{T'} \subseteq \overline{\overline{T}} \Rightarrow \overline{\overline{T}} = \overline{\overline{T'}}$.

Second, for certain approaches, it facilitates completeness proofs, as we can look directly at "ideal" elements, without having to bother about intermediate stages. See in particular the work by Lehmann and his co-authors, [KLM90], [LM92].

"Smoothness", or, as it is also called, "stopperedness" seems - in the authors' opinion - a misnamer. We think it should better be called something like "weak transitivity": consider the case where $a \succ b \succ c$, but $c \not\prec a$, with $c \in \mu(X)$. It is then not necessarily the case that $a \succ c$, but there is c' "sufficiently close to c", i.e., in $\mu(X)$, such that $a \succ c'$. Results and proof techniques underline this idea. First, in the general case with copies, and in the smooth case (in the presence of $(\cup)!$), transitivity does not add new properties, it is "already present", second, the construction of smoothness by sequences σ (see below in Section 1.5.3.2) is very close in spirit to a transitive construction.

The second condition, rankedness, seems easier to justify already as a property of the structure. It says that, essentially, the elements are ordered in layers: If a and b are not comparable, then they are in the same layer. So, if c is above (below) a, it will also be above (below) b - like pancakes or geological strata. Apart from the triangle inequality (and leaving aside cardinality questions), this is then just a distance from some imaginary, ideal point. Again, this property has important consequences on the resulting model choice functions and consequence relations, making proof techniques for the non-ranked and the ranked case very different.

Note that, if $\mathcal{Y}$ is closed under finite intersections, in the presence of $(\mu \subseteq)$, (μPR) is equivalent to $(\mu PR')$.

Also note that $(\mu =')$ is very close to Rational Monotony: Rational Monotony says: $\alpha \hspace{0.5mm}\mid\hspace{-2mm}\sim \beta, \alpha \hspace{0.5mm}\mid\hspace{-2mm}\not\sim \neg\gamma \to \alpha \wedge \gamma \hspace{0.5mm}\mid\hspace{-2mm}\sim \beta$. Or, $\mu(A) \subseteq B, \mu(A) \cap C \neq \emptyset \to \mu(A \cap C) \subseteq B$ for all A, B, C. This is not quite, but almost: $\mu(A \cap C) \subseteq \mu(A) \cap C$ (it depends how many B there are, if $\mu(A)$ is some such B, the fit is perfect).

See Table 1.2 and Table 1.3 for the definitions.

Some Useful and Simple Results for the Minimal Version

Fact 1.2.22

Let $\mu(A) \subseteq B, \mu(B) \subseteq A$, then $\mu(A \cup B) \subseteq \mu(A) \cap \mu(B)$.

Proof

Suppose $a \in \mu(A \cup B)$. We show $a \in \mu(A)$, $a \in \mu(B)$ is analogous.

Suppose $a \in \mu(A \cup B) - \mu(A)$.

By $\mu(A \cup B) \subseteq A \cup B$, $a \in A$ or $a \in B$.

If $a \in A$, then there must be $b \in A.b \prec a$, so $a \notin \mu(A \cup B)$, a contradiction.

So suppose $a \in B - A$. If $a \notin \mu(B)$, then there is $b \in B.b \prec a$, so $a \notin \mu(A \cup B)$, a contradiction. So $a \in \mu(B) \subseteq A$, contradiction.

□

Fact 1.2.23

If μ is generated by a smooth relation, $\mu(A) \subseteq B, \mu(B) \subseteq A$, then $\mu(A) = \mu(B)$.

Proof

Let $a \in \mu(A) - \mu(B)$. As $\mu(A) \subseteq B$, $a \in B$. So there is $b \prec a$, $b \in \mu(B) \subseteq A$, contradiction.

□

1.2 Basic Definitions and Overview

Fact 1.2.24

Let $\mu(A) \subseteq B$, $\mu(B) \subseteq A$, μ generated by a smooth relation, then $\mu(A \cap B) \subseteq \mu(A) \cap \mu(B)$.

Proof

Suppose not, let $a \in \mu(A \cap B) \subseteq A \cap B \subseteq A$, but $a \notin \mu(A)$. So there must be $b \prec a$, $b \in \mu(A) \subseteq A$. By $\mu(A) \subseteq B$, $b \in B$, so $b \in A \cap B$, contradiction.

□

Many-Valued Logic

We conclude with a short remark on many-valued logics.

We can, of course, consider for a given ϕ the set of models where ϕ has maximal truth value TRUE, and then take the minimal ones as usual. The resulting logic $\mathrel|\!\sim$ then makes $\phi \mathrel|\!\sim \psi$ true iff the minimal models with value TRUE assign TRUE also to ψ.

But this does not seem to be the adequate way. So we adapt the definition of preferential structures to the many-valued situation. (For more details, see [GS10].)

Definition 1.2.14

Let $\mathcal{L}$ be given with model set M.

Let a binary relation $\prec$ be given on $\mathcal{X}$, where $\mathcal{X}$ is a set of pairs $\langle m, i \rangle$, $m \in M$, i some index as usual. (We use here the assumption that the truth value is independent of indices.)

Let $f : M \to V$ be given; we define $\mu(f)$, the minimal models of f :

$$\mu(f)(m) := \begin{cases} FALSE \; iff \; \forall \langle m, i \rangle \in \mathcal{X} \exists \langle m', i' \rangle \prec \langle m, i \rangle. f(m') \geq f(m) \\ \\ f(m) \quad otherwise \end{cases}$$

This generalizes the idea that only models of ϕ can destroy models of ϕ.

Obviously, for all $v \in V$, $v \neq FALSE$, $\{m : \mu(f)(m) = v\} \subseteq \{m : f(m) = v\}$.

A structure is called smooth iff for all f_ϕ and for all $\langle m, i \rangle$ such that there is a $\langle m', i' \rangle \prec \langle m, i \rangle$ with $f_\phi(m') \geq f_\phi(m)$, there is $\langle m'', i'' \rangle \prec \langle m, i \rangle$ with $f_\phi(m'') \geq f_\phi(m)$, and no $\langle n, j \rangle \prec \langle m'', i'' \rangle$ with $f_\phi(n) \geq f_\phi(m'')$.

A structure will be called definablity preserving iff for all f_ϕ, $\mu(f_\phi)$ is again the f_ψ for some ψ.

The Limit Version

Definition 1.2.15

(1) General preferential structures

  (1.1) The version without copies:

   Let $\mathcal{M} := \langle U, \prec \rangle$. Define

   $Y \subseteq X \subseteq U$ is a minimizing initial segment, or MISE, of X iff:

   (a) $\forall x \in X \exists x \in Y. y \preceq x$ - where $y \preceq x$ stands for $x \prec y$ or $x = y$ (i.e., Y is minimizing) and

   (b) $\forall y \in Y, \forall x \in X (x \prec y \Rightarrow x \in Y)$ (i.e., Y is downward closed or an initial part).

  (1.2) The version with copies:

   Let $\mathcal{M} := \langle \mathcal{U}, \prec \rangle$ be as above. Define for $Y \subseteq X \subseteq \mathcal{U}$

   Y is a minimizing initial segment, or MISE of X iff:

   (a) $\forall \langle x, i \rangle \in X \exists \langle y, j \rangle \in Y. \langle y, j \rangle \preceq \langle x, i \rangle$

   and

   (b) $\forall \langle y, j \rangle \in Y, \forall \langle x, i \rangle \in X (\langle x, i \rangle \prec \langle y, j \rangle \Rightarrow \langle x, i \rangle \in Y)$.

  (1.3) For $X \subseteq \mathcal{U}$, let $\Lambda(X)$ be the set of MISE of X.

  (1.4) We say that a set $\mathcal{X}$ of MISE is cofinal in another set of MISE $\mathcal{X}'$ (for the same base set X) iff for all $Y' \in \mathcal{X}'$, there is $Y \in \mathcal{X}$, $Y \subseteq Y'$.

  (1.5) A MISE X is called definable iff $\{x : \exists i. \langle x, i \rangle \in X\} \in \boldsymbol{D}_\mathcal{L}$.

(2) Ranked preferential structures:

   In the case of ranked structures, we may assume without loss of generality that the MISE sets have a particularly simple form:

   For $X \subseteq \mathcal{U}$ $A \subseteq X$ is MISE iff

   ($X \neq \emptyset$ and)

   $\forall x \in X \exists a \in A (a \prec x \text{ or } a = x)$ and

   $\forall a \in A \forall x \in X (x \prec a \vee x \bot a \Rightarrow x \in A)$.

   (A is downward and horizontally closed.)

1.2.4.2 Logical Properties

Definition 1.2.16

We define the consequence relation of a preferential structure for a given propositional language $\mathcal{L}$.

(1) Validity in a preferential structure:

Let $\mathcal{M}$ be as above.

(1.1) If m is a classical model of a language $\mathcal{L}$, we say by abuse of language

$\langle m, i \rangle \models \phi$ iff $m \models \phi$,

and if X is any set of such pairs, that

$X \models \phi$ iff for all $\langle m, i \rangle \in X$ $m \models \phi$.

(1.2) If $\mathcal{M}$ is a preferential structure, and X is a set of $\mathcal{L}$-models for a classical propositional language $\mathcal{L}$, or is a set of pairs $\langle m, i \rangle$ where the m are such models, we call $\mathcal{M}$ a classical preferential structure or model.

(2) The semantical consequence relation defined by such a structure:

(2.1) in the minimal version:

$T \models_{\mathcal{M}} \phi$ iff $\mu_{\mathcal{M}}(M(T)) \models \phi$, i.e., $\mu_{\mathcal{M}}(M(T)) \subseteq M(\phi)$.

(2.2) in the limit version:

$T \models_{\mathcal{M}} \phi$ iff there is $Y \in \Lambda(\mathcal{U} \restriction M(T))$ such that $Y \models \phi$.

($\mathcal{U} \restriction M(T) := \{\langle x, i \rangle \in \mathcal{U} : x \in M(T)\}$ - if there are no copies, we simplify in the obvious way.)

(3) $\mathcal{M}$ will be called definability preserving iff for all $X \in \mathbf{D}_{\mathcal{L}}$ $\mu_{\mathcal{M}}(X) \in \mathbf{D}_{\mathcal{L}}$.

As $\mu_{\mathcal{M}}$ is defined on $\mathbf{D}_{\mathcal{L}}$, but need by no means always result in some new definable set, this is (and reveals itself as a quite strong) additional property.

1.2.5 Algebraic and Structural Semantics

We make now a major conceptual distinction, between an "algebraic" and a "structural" semantics, which can best be illustrated by an example.

Consider nonmonotonic logics as discussed above. In preferential structures, we only consider the minimal elements, say $\mu(X)$, if X is a set of models. Abstractly, we thus have a choice function μ, defined on the power set of the model set, and μ has certain properties, e.g., $\mu(X) \subseteq X$. More important is the following property: $X \subseteq Y \to \mu(Y) \cap X \subseteq \mu(X)$. (The proof is trivial: suppose there were $x \in \mu(Y) \cap X$, $x \notin \mu(X)$. Then there must be $x' \prec x$, $x' \in X \subseteq Y$, but then x cannot be minimal in Y.)

Thus, all preferential structures generate μ functions with certain properties, and once we have a complete list, we can show that any arbitrary model choice function with these properties can be generated by an appropriate preferential structure.

Note that we do not need here the fact that we have a relation between models, just any relation on an arbitrary set suffices. It seems natural to call the complete list of properties of such μ-functions an algebraic semantics, forgetting that the function itself was created by a preferential structure, which is the structural semantics.

This distinction is very helpful, it not only incites us to separate the two semantics conceptually, but also to split completeness proof in two parts: One part, where we show correspondence between the logical side and the algebraic semantics, and a second one, where we show the correspondence between the algebraic and the structural semantics. The latter part will usually be more difficult, but any result obtained here is independent from logics itself, and can thus often be re-used in other logical contexts. On the other hand, there are often some subtle problems for the correspondence between the logics and the algebraic semantics (see definability preservation, in particular the discussion in [Sch04]), which we can then more clearly isolate, identify, and solve.

1.2.5.1 Abstract or Algebraic Semantics

In all cases, we see that the structural semantics define a set operator, and thus an algebraic semantics:

- in nonmonotonic logics (and Deontic Logic), the function chooses the minimal (morally best) models, a subset, $\mu(X) \subseteq X$
- in (distance based) theory revision, we have abinary operator, say $|$ which chooses the ϕ-models closest to the set of K-models: $M(K) \mid M(\phi)$

1.2 Basic Definitions and Overview

- in Theory Update, the operator chooses the i-th coordinate of all best sequences
- in the Logic of Counterfactual Conditionals, whave again a binary operator $m \mid M(\phi)$ which chooses the ϕ-models closest to m, or, when we consider a whole set X of models as starting points $X \mid M(\phi) = \bigcup \{m \mid M(\phi) : m \in X\}$.
- in Modal and Intuitionistic Logic, seen from some model m, we choose a subset of all the models (thus not a subset of a more restricted model set), those which can be reached from m.

Thus, in each case, the structure "sends" us to another model set, and this expresses the change from the original situation to the "most plausible", "best", "possible" etc. situations. It seems natural to call all such logics "generalized modal logics", as they all use the idea of a model choice function.

(Note again that we have neglected here the possibility that there are no best or closest models (or sequences), but only ever better ones.)

Abstract semantics are interpretations of the operators of the language (all, flat, top level or not) by functions (or relations in the case of $\mid\!\sim$), which assign to sets of models sets of models, $\mathcal{O} : \mathcal{P}(\mathcal{M}) \to \mathcal{P}(\mathcal{M})$ - $\mathcal{P}$ the power set operator, and $\mathcal{M}$ the set of basic models -, or binary functions for binary operators, etc.

These functions are determined or restricted by the laws for the corresponding operators. E.g., in classical, preferential, or modal logic, $\wedge$ is interpreted by $\cap$, etc.; in preferential logic ∇ by μ; in modal logic, we interpret $\Box$, etc.

Operators may be truth-functional or not. $\neg$ is truth-functional. It suffices to know the truth value of ϕ at some point, to know that of $\neg\phi$ at the same point. $\Box$ is not truth-functional: ϕ and ψ may hold, and $\Box\phi$, but not $\Box\psi$, all at the same point (= base model), we have to look at the full picture, not only at some model.

We consider first those operators, which have a unique possible interpretation, like $\wedge$, which is interpreted by $\cap$, $\neg$ by C, the set theoretic complement, etc. ∇ (standing for "most", "the important", etc.) e.g., has only restrictions to its interpretation, like $\mu(X) \subseteq X$, etc. Given a set of models without additional structure, we do not know its exact form, we know it only once we have fixed the additional structure (the relation in this case).

If the models contain already the operator, the function will respect it, i.e., we cannot have ϕ and $\neg\phi$ in the same model, as $\neg$ is interpreted by C. Thus, the functions can, at least in some cases, control consistency.

If, e.g., the models contain $\wedge$, then we have two ways to evaluate $\phi \wedge \psi$: we can first evaluate ϕ, then ψ, and use the function for $\wedge$ to evaluate $\phi \wedge \psi$. Alternatively, we can look directly at the model for $\phi \wedge \psi$ - provided we considered the full language in constructing the models.

As we can apply one function to the result of the other, we can evaluate complicated formulas, using the functions on the set of models. Consequently, if $\mathrel|\!\sim$ or ∇ is evaluated by μ, we can consider $\mu(\mu(X))$ etc., thus, the machinery for the flat case gives immediately an interpretation for nested formulas, too - whether we looked for it, or not.

As far as we see, our picture covers the usual presentations of classical logic, preferential, intuionist, and modal logic, but also of linear logic (where we have more structure on the set of basic models, a monoid, with a distinct set $\bot$, plus some topology for! and? - see below), and quantum logic a la Birkhoff/von Neumann.

We can introduce new truth-functional operators into the language as follows: Suppose we have a distinct truth value TRUE, then we may define $\mathcal{O}_X(\phi) = TRUE$ iff the truth-value of ϕ is an element of X. This might sometimes be helpful. Making the truth value explicit as element of the object language may facilitate the construction of an accompanying proof system - experience will tell whether this is the case. In this view, $\neg$ has now a double meaning in the classical situation: it is an operator for the truth value "false", and an operator on the model set, and corresponds to the complement. "Is true" is the identical truth functional operator, $is - true(\phi)$ and ϕ have the same truth value.

If the operators have a unique interpretation, this might be all there is to say in this abstract framework. (This does not mean that it is impossible to introduce new operators which are independent from any additional structure, and based only on the set of models for the basic language. We can, for instance, introduce a "CON" operator, saying that ϕ is consistent, and $CON(\phi)$ will hold everywhere iff ϕ is consistent, i.e., holds in at least one model. Or, for a more bizarre example, a 3 operator, which says that ϕ has at least 3 models (which is then dependent on the language). We can also provide exactly one additional structure, e.g., in the following way: Introduce a ranked order between models as follows: At the bottom, put the single model which makes all propositional variables true, on the next level those which make exactly one propositional variable true, then two, etc., with the model making all false on top. So there is room to play, if one can find many useful examples is another question.)

If the operator has no unique interpretation (like ∇, $\square$, etc., which are only restricted), the situation seems more complicated.

It is sometimes useful to consider the abstract semantics as a (somehow coherent) system of filters. For instance, in preferential structures, $\mu(X) \subseteq X$ can be seen as the basis of a principal filter. Thus, $\phi \mathrel|\!\sim \psi$ iff ψ holds in all minimal models of ϕ, iff there is a "big" subset of $M(\phi)$ where ψ holds, recalling that a filter is an abstraction of size - sets in the filter are big, their complements small, and the other sets have medium size. Thus, the "normal" elements form the smallest big subset. Rules like $X \subseteq Y \to \mu(Y) \cap X \subseteq \mu(X)$ form the coherence between the individual filters, we cannot choose them totally independently. Particularly for

1.2 Basic Definitions and Overview

preferential structures, the reasoning with small and big subsets can be made very precise and intuitively appealing, and we will come back to this point later. We can also introduce a generalized quantifier, say ∇, with the same meaning, i.e., $\phi \hspace{1pt}\vert\hspace{-3pt}\sim \psi$ iff $\nabla(\phi).\psi$, i.e., "almost everywhere", or "in the important cases" where ϕ holds, so will ψ. This is then the syntactic analogue of the semantical filter system. These aspects are discussed in detail in Chapter 3.

1.2.5.2 Structural Semantics

Structural semantics generate the abstract or algebraic semantics, i.e., the behaviour of the functions or relations (and of the operators in the language when we work with "rich" basic models). Preferences between models generate corresponding μ-functions, relations in Kripke structures generate the functions corresponding to $\Box$-operators, etc.

Ideally, structural semantics capture the essence of what we want to reason and speak about (beyond classical logic), they come, or should come, first. Next, we try to see the fundamental ingredients and laws of such structures, code them in an algebraic semantics and the language, i.e., extract the functions and operators, and their laws. In a backward movement, we make the roles of the operators (or relations) precise (should they be nested or not?, etc.), and define the basic models and the algebraic operators. This may result in minor modifications of the structural semantics (like introduction of copies), but should still be close to the point of outset. In this view, the construction of a logic is a back-and-forth movement.

1.2.6 Tables for Logical and Semantical Rules

Tables 1.2–1.5 show a summary of logical and semantical rules. The numbers in the correspondence columns of Tables 1.2 and 1.3 refer to Propositions 1.2.18 and 3.2.13. For the interdependencies of algebraic rules shown in Table 1.4 see Fact 1.2.14. For the rules listed in Table 1.5 see Proposition 1.2.18

1.2.7 Tables for Preferential Structures

Table 1.6 shows a summary of preferential structures.

Table 1.2 Logical rules, definitions and connections, Part I (the numbers in the Correspondence Columns refer to Propositions 1.2.18 and 3.2.13)

Logical rule		Correspondence			
Formula Version	Theory Version		Model set	Corr.	Size Rules
Basics					
(SC) Supraclassicality	(SC)	$\Rightarrow (4.1)$	$(\mu \subseteq)$	trivial	(Opt)
$\alpha \vdash \beta \Rightarrow \alpha \mathrel{\|\!\sim} \beta$	$\overline{T} \subseteq \overline{\overline{T}}$	$\Leftarrow (4.2)$	$f(X) \subseteq X$		
(REF) Reflexivity					
$T \cup \{\alpha\} \mathrel{\|\!\sim} \alpha$			(trivially true)		
(LLE)	(LLE)				
Left Logical Equivalence					
$\vdash \alpha \leftrightarrow \alpha', \alpha \mathrel{\|\!\sim} \beta \Rightarrow$	$\overline{T} = \overline{T'} \Rightarrow \overline{\overline{T}} = \overline{\overline{T'}}$				
$\alpha' \mathrel{\|\!\sim} \beta$					
(RW) Right Weakening	(RW)		(upward closure)	trivial	(iM)
$\alpha \mathrel{\|\!\sim} \beta, \vdash \beta \rightarrow \beta' \Rightarrow$	$T \mathrel{\|\!\sim} \beta, \vdash \beta \rightarrow \beta' \Rightarrow$				
$\alpha \mathrel{\|\!\sim} \beta'$	$T \mathrel{\|\!\sim} \beta'$				
(wOR)	(wOR)	$\Rightarrow (3.1)$	(μwOR)	$\Leftrightarrow (1)$	$(eM\mathcal{I})$
$\alpha \mathrel{\|\!\sim} \beta, \alpha' \mathrel{\|\!\sim} \beta \Rightarrow$	$\overline{\overline{T}} \cap \overline{\overline{T'}} \subseteq \overline{\overline{T \vee T'}}$	$\Leftarrow (3.2)$	$f(X \cup Y) \subseteq f(X) \cup Y$		
$\alpha \vee \alpha' \mathrel{\|\!\sim} \beta$					
$(disjOR)$	$(disjOR)$	$\Rightarrow (2.1)$	$(\mu disjOR)$	$\Leftrightarrow (4)$	$(I \cup disj)$
$\alpha \vdash \neg\alpha', \alpha \mathrel{\|\!\sim} \beta,$	$\neg Con(T \cup T') \Rightarrow$	$\Leftarrow (2.2)$	$X \cap Y = \emptyset \Rightarrow$		
$\alpha' \mathrel{\|\!\sim} \beta \Rightarrow \alpha \vee \alpha' \mathrel{\|\!\sim} \beta$	$\overline{\overline{T}} \cap \overline{\overline{T'}} \subseteq \overline{\overline{T \vee T'}}$		$f(X \cup Y) \subseteq f(X) \cup f(Y)$		
(CP)	(CP)	$\Rightarrow (5.1)$	$(\mu\emptyset)$	trivial	(I_1)
Consistency Preservation		$\Leftarrow (5.2)$			
$\alpha \mathrel{\|\!\sim} \bot \Rightarrow \alpha \vdash \bot$	$T \mathrel{\|\!\sim} \bot \Rightarrow T \vdash \bot$		$f(X) = \emptyset \Rightarrow X = \emptyset$		
			$(\mu\emptyset fin)$		(I_1)
			$X \neq \emptyset \Rightarrow f(X) \neq \emptyset$		
			for finite X		

1.2 Basic Definitions and Overview

(AND_1) $\alpha \vdash \beta \Rightarrow \alpha \not\vdash \neg\beta$				(I_2)
(AND_n) $\alpha \vdash \beta_1, \ldots, \alpha \vdash \beta_{n-1} \Rightarrow$ $\alpha \not\vdash (\neg\beta_1 \vee \ldots \vee \neg\beta_{n-1})$				(I_n)
(AND) $\alpha \vdash \beta, \alpha \vdash \beta' \Rightarrow$ $\alpha \vdash \beta \wedge \beta'$	(AND) $T \vdash \beta, T \vdash \beta' \Rightarrow$ $T \vdash \beta \wedge \beta'$	(closure under finite intersection)	trivial	(I_ω)
(CCL) Classical Closure	(CCL) $\overline{\overline{T}}$ classically closed	(trivially true)	trivial	$(iM) + (I_\omega)$
(OR) $\alpha \vdash \beta, \alpha' \vdash \beta \Rightarrow$ $\alpha \vee \alpha' \vdash \beta$	(OR) $\overline{\overline{T}} \cap \overline{\overline{T'}} \subseteq \overline{\overline{T \vee T'}}$	(μOR) $f(X \cup Y) \subseteq f(X) \cup f(Y)$	$\Rightarrow (1.1)$ $\Leftarrow (1.2)$	$(eM\mathcal{I}) + (I_\omega)$
$\overline{\alpha \wedge \alpha'} \subseteq \overline{\overline{\alpha} \cup \{\alpha'\}}$	(PR) $\overline{\overline{T \cup T'}} \subseteq \overline{\overline{T} \cup T'}$	(μPR) $X \subseteq Y \Rightarrow$ $f(Y) \cap X \subseteq f(X)$	$\Rightarrow (6.1)$ $\Leftarrow (\mu dp) + (\mu \subseteq) (6.2)$ $\not\Leftarrow -(\mu dp) (6.3)$ $\Leftarrow (\mu \subseteq) (6.4)$ $T' = \phi$	$\Leftrightarrow (3)$
		$(\mu PR')$ $f(X) \cap Y \subseteq f(X \cap Y)$	$\Leftarrow (6.5)$ $T' = \phi$	
(CUT) $T \vdash \alpha; T \cup \{\alpha\} \vdash \beta \Rightarrow$ $T \vdash \beta$	(CUT) $T \subseteq \overline{\overline{T'}} \subseteq \overline{\overline{T}} \Rightarrow$ $\overline{\overline{T'}} \subseteq \overline{\overline{T}}$	(μCUT) $f(X) \subseteq Y \subseteq X \Rightarrow$ $f(X) \subseteq f(Y)$	$\Leftarrow (8.1)$ $\not\Rightarrow (8.2)$	$(eM\mathcal{I}) + (I_\omega)$

Table 1.3 Logical rules, definitions and connections, Part II (the numbers in the Correspondence Columns refer to Propositions 1.2.18 and 3.2.13)

Logical rule		Correspondence	Model set	Corr.	Size Rules
Formula Version	Theory Version				
Cumulativity					
(wCM) $\alpha \mathrel{\|\!\sim} \beta, \alpha' \vdash \alpha, \alpha \wedge \beta \vdash \alpha' \Rightarrow$ $\alpha' \mathrel{\|\!\sim} \beta$				trivial	$(eM\mathcal{F})$
(CM_2) $\alpha \mathrel{\|\!\sim} \beta, \alpha \mathrel{\|\!\sim} \beta' \Rightarrow \alpha \wedge \beta' \not\mathrel{\|\!\sim} \neg\beta'$					(I_2)
(CM_n) $\alpha \mathrel{\|\!\sim} \beta_1, \ldots, \alpha \mathrel{\|\!\sim} \beta_n \Rightarrow$ $\alpha \wedge \beta_1 \wedge \ldots \wedge \beta_{n-1} \not\mathrel{\|\!\sim} \neg\beta_n$					(I_n)
(CM) Cautious Monotony $\alpha \mathrel{\|\!\sim} \beta, \alpha \mathrel{\|\!\sim} \beta' \Rightarrow$ $\alpha \wedge \beta \mathrel{\|\!\sim} \beta'$	(CM) $T \subseteq \overline{T'} \subseteq \overline{\overline{T}} \Rightarrow$ $\overline{\overline{T}} \subseteq \overline{\overline{T'}}$	$\Rightarrow$ (8.1) $\Leftarrow$ (8.2)	(μCM) $f(X) \subseteq Y \subseteq X \Rightarrow$ $f(Y) \subseteq f(X)$	$\Leftrightarrow$ (5)	$(\mathcal{M}_\omega^+)(4)$
or $(ResM)$ Restricted Monotony $T \mathrel{\|\!\sim} \alpha, \beta \Rightarrow T \cup \{\alpha\} \mathrel{\|\!\sim} \beta$		$\Rightarrow$ (9.1) $\Leftarrow$ (9.2)	$(\mu ResM)$ $f(X) \subseteq A \cap B \Rightarrow$ $f(X \cap A) \subseteq B$		
(CUM) Cumulativity $\alpha \mathrel{\|\!\sim} \beta \Rightarrow$ $(\alpha \mathrel{\|\!\sim} \beta' \Leftrightarrow \alpha \wedge \beta \mathrel{\|\!\sim} \beta')$	(CUM) $T \subseteq \overline{T'} \subseteq \overline{\overline{T}} \Rightarrow$ $\overline{\overline{T}} = \overline{\overline{T'}}$	$\Rightarrow$ (11.1) $\Leftarrow$ (11.2)	(μCUM) $f(X) \subseteq Y \subseteq X \Rightarrow$ $f(Y) = f(X)$	$\Leftarrow$ (9.1) $\not\Leftarrow$ (9.2)	$(eM\mathcal{I}) + (I_\omega) + (\mathcal{M}_\omega^+)(4)$
	$(\subseteq\supseteq)$ $T \subseteq \overline{T'}, T' \subseteq \overline{T} \Rightarrow$ $\overline{T'} = \overline{T}$	$\Rightarrow$ (10.1) $\Leftarrow$ (10.2)	$(\mu \subseteq\supseteq)$ $f(X) \subseteq Y, f(Y) \subseteq X \Rightarrow$ $f(X) = f(Y)$	$\Leftarrow$ (10.1) $\not\Leftarrow$ (10.2)	$(eM\mathcal{I}) + (I_\omega) + (eM\mathcal{F})$

1.2 Basic Definitions and Overview

Rationality				
$(RatM)$ Rational Monotony $\alpha \mid\!\sim \beta, \alpha \not\mid\!\sim \neg\beta' \Rightarrow$ $\alpha \wedge \beta' \mid\!\sim \beta$	$(RatM)$ $\dfrac{Con(T \cup \overline{\overline{T'}}), T \vdash T' \Rightarrow}{\overline{\overline{T}} \supseteq \overline{\overline{T'}} \cup T}$	$\Rightarrow$ (12.1) $\Leftarrow (\mu dp)$ (12.2) $\not\Leftarrow -(\mu dp)$ (12.3) $\Leftarrow T = \phi$ (12.4)	$(\mu RatM)$ $X \subseteq Y, X \cap f(Y) \neq \emptyset \Rightarrow$ $f(X) \subseteq f(Y) \cap X$	$\Leftrightarrow$ (6) $(\mathcal{M}^{++})$
	$(RatM =)$ $\dfrac{Con(T \cup \overline{\overline{T'}}), T \vdash T' \Rightarrow}{\overline{\overline{T}} = \overline{\overline{T'}} \cup T}$	$\Rightarrow$ (13.1) $\Leftarrow (\mu dp)$ (13.2) $\not\Leftarrow -(\mu dp)$ (13.3) $\Leftarrow T = \phi$ (13.4)	$(\mu =)$ $X \subseteq Y, X \cap f(Y) \neq \emptyset \Rightarrow$ $f(X) = f(Y) \cap X$	
	$(Log =')$ $\dfrac{Con(\overline{\overline{T'}} \cup T) \Rightarrow}{\overline{\overline{T \cup T'}} = \overline{\overline{T'}} \cup T}$	$\Rightarrow$ (14.1) $\Leftarrow (\mu dp)$ (14.2) $\not\Leftarrow -(\mu dp)$ (14.3) $\Leftarrow T = \phi$ (14.4)	$(\mu =')$ $f(Y) \cap X \neq \emptyset \Rightarrow$ $f(Y \cap X) = f(Y) \cap X$	
(DR) $\alpha \vee \beta \mid\!\sim \gamma \Rightarrow$ $\alpha \mid\!\sim \gamma$ or $\beta \mid\!\sim \gamma$	$(Log \parallel)$ $\overline{\overline{T \vee T'}}$ is one of $\overline{\overline{T}}$, or $\overline{\overline{T'}}$, or $\overline{\overline{T}} \cap \overline{\overline{T'}}$ (by CCL)	$\Rightarrow$ (15.1) $\Leftarrow$ (15.2)	$(\mu \parallel)$ $f(X \cup Y)$ is one of $f(X)$, $f(Y)$, or $f(X) \cup f(Y)$	
	$(Log\cup)$ $Con(\overline{\overline{T'}} \cup T), \neg Con(\overline{\overline{T'}} \cup \overline{\overline{T}}) \Rightarrow$ $\neg Con(\overline{\overline{T \vee T'}} \cup T')$	$\Rightarrow (\mu \subseteq) + (\mu =)$ (16.1) $\Leftarrow (\mu dp)$ (16.2) $\not\Leftarrow -(\mu dp)$ (16.3)	$(\mu\cup)$ $f(Y) \cap (X - f(X)) \neq \emptyset$ $\Rightarrow$ $f(X \cup Y) \cap Y = \emptyset$	
	$(Log\cup')$ $Con(\overline{\overline{T'}} \cup T), \neg Con(\overline{\overline{T'}} \cup \overline{\overline{T}}) \Rightarrow$ $\overline{\overline{T \vee T'}} = \overline{\overline{T}}$	$\Rightarrow (\mu \subseteq) + (\mu =)$ (17.1) $\Leftarrow (\mu dp)$ (17.2) $\not\Leftarrow -(\mu dp)$ (17.3)	$(\mu\cup')$ $f(Y) \cap (X - f(X)) \neq \emptyset$ $\Rightarrow$ $f(X \cup Y) = f(X)$	
			$(\mu \in)$ $a \in X - f(X) \Rightarrow$ $\exists b \in X. a \notin f(\{a,b\})$	

Table 1.4 Interdependencies of algebraic rules (see Fact 1.2.14)

		Basics	
(1.1)	(μPR)	$\Rightarrow (\cap) + (\mu \subseteq)$	$(\mu PR')$
(1.2)		$\Leftarrow$	
(2.1)	(μPR)	$\Rightarrow (\mu \subseteq)$	(μOR)
(2.2)		$\Leftarrow (\mu \subseteq) + (-)$	
(2.3)		$\Rightarrow (\mu \subseteq)$	(μwOR)
(2.4)		$\Leftarrow (\mu \subseteq) + (-)$	
(3)	(μPR)	$\Rightarrow$	(μCUT)
(4)	$(\mu \subseteq) + (\mu \subseteq \supseteq) + (\mu CUM)$ $+(\mu RatM) + (\cap)$	$\not\Rightarrow$	(μPR)
		Cumulativity	
(5.1)	(μCM)	$\Rightarrow (\cap) + (\mu \subseteq)$	$(\mu ResM)$
(5.2)		$\Leftarrow$ (infin.)	
(6)	$(\mu CM) + (\mu CUT)$	$\Leftrightarrow$	(μCUM)
(7)	$(\mu \subseteq) + (\mu \subseteq \supseteq)$	$\Rightarrow$	(μCUM)
(8)	$(\mu \subseteq) + (\mu CUM) + (\cap)$	$\Rightarrow$	$(\mu \subseteq \supseteq)$
(9)	$(\mu \subseteq) + (\mu CUM)$	$\not\Rightarrow$	$(\mu \subseteq \supseteq)$
		Rationality	
(10)	$(\mu RatM) + (\mu PR)$	$\Rightarrow$	$(\mu =)$
(11)	$(\mu =)$	$\Rightarrow$	$(\mu PR) + (\mu RatM)$
(12.1)	$(\mu =)$	$\Rightarrow (\cap) + (\mu \subseteq)$	$(\mu =')$
(12.2)		$\Leftarrow$	
(13)	$(\mu \subseteq) + (\mu =)$	$\Rightarrow (\cup)$	$(\mu \cup)$
(14)	$(\mu \subseteq) + (\mu \emptyset) + (\mu =)$	$\Rightarrow (\cup)$	$(\mu \parallel), (\mu \cup'), (\mu CUM)$
(15)	$(\mu \subseteq) + (\mu \parallel)$	$\Rightarrow (-)$ of $\mathcal{Y}$	$(\mu =)$
(16)	$(\mu \parallel) + (\mu \in) + (\mu PR) +$ $(\mu \subseteq)$	$\Rightarrow (\cup) +$ sing.	$(\mu =)$
(17)	$(\mu CUM) + (\mu =)$	$\Rightarrow (\cup) +$ sing.	$(\mu \in)$
(18)	$(\mu CUM) + (\mu =) + (\mu \subseteq)$	$\Rightarrow (\cup)$	$(\mu \parallel)$
(19)	$(\mu PR) + (\mu CUM) + (\mu \parallel)$	$\Rightarrow$ sufficient, e.g., true in $\mathbf{D}_\mathcal{L}$	$(\mu =)$.
(20)	$(\mu \subseteq) + (\mu PR) + (\mu =)$	$\not\Rightarrow$	$(\mu \parallel)$
(21)	$(\mu \subseteq) + (\mu PR) + (\mu \parallel)$	$\not\Rightarrow$ (without $(-)$)	$(\mu =)$
(22)	$(\mu \subseteq) + (\mu PR) + (\mu \parallel) +$ $(\mu =) + (\mu \cup)$	$\not\Rightarrow$	$(\mu \in)$ (thus not representable by ranked structures)

1.2 Basic Definitions and Overview

Table 1.5 Logical and algebraic rules (see Proposition 1.2.18)

		Basics	
(1.1)	(OR)	$\Rightarrow$	(μOR)
(1.2)		$\Leftarrow$	
(2.1)	$(disjOR)$	$\Rightarrow$	$(\mu disjOR)$
(2.2)		$\Leftarrow$	
(3.1)	(wOR)	$\Rightarrow$	(μwOR)
(3.2)		$\Leftarrow$	
(4.1)	(SC)	$\Rightarrow$	$(\mu \subseteq)$
(4.2)		$\Leftarrow$	
(5.1)	(CP)	$\Rightarrow$	$(\mu \emptyset)$
(5.2)		$\Leftarrow$	
(6.1)	(PR)	$\Rightarrow$	(μPR)
(6.2)		$\Leftarrow (\mu dp) + (\mu \subseteq)$	
(6.3)		$\not\Leftarrow -(\mu dp)$	
(6.4)		$\Leftarrow (\mu \subseteq)$ $T' = \phi$	
(6.5)	(PR)	$\Leftarrow$ $T' = \phi$	$(\mu PR')$
(7.1)	(CUT)	$\Rightarrow$	(μCUT)
(7.2)		$\Leftarrow$	
		Cumulativity	
(8.1)	(CM)	$\Rightarrow$	(μCM)
(8.2)		$\Leftarrow$	
(9.1)	$(ResM)$	$\Rightarrow$	$(\mu ResM)$
(9.2)		$\Leftarrow$	
(10.1)	$(\subseteq \supseteq)$	$\Rightarrow$	$(\mu \subseteq \supseteq)$
(10.2)		$\Leftarrow$	
(11.1)	(CUM)	$\Rightarrow$	(μCUM)
(11.2)		$\Leftarrow$	
		Rationality	
(12.1)	$(RatM)$	$\Rightarrow$	$(\mu RatM)$
(12.2)		$\Leftarrow (\mu dp)$	
(12.3)		$\not\Leftarrow -(\mu dp)$	
(12.4)		$\Leftarrow$ $T = \phi$	
(13.1)	$(RatM =)$	$\Rightarrow$	$(\mu =)$
(13.2)		$\Leftarrow (\mu dp)$	
(13.3)		$\not\Leftarrow -(\mu dp)$	
(13.4)		$\Leftarrow$ $T = \phi$	
(14.1)	$(Log =')$	$\Rightarrow$	$(\mu =')$
(14.2)		$\Leftarrow (\mu dp)$	
(14.3)		$\not\Leftarrow -(\mu dp)$	
(14.4)		$\Leftarrow T = \phi$	
(15.1)	$(Log \parallel)$	$\Rightarrow$	$(\mu \parallel)$
(15.2)		$\Leftarrow$	
(16.1)	$(Log\cup)$	$\Rightarrow (\mu \subseteq) + (\mu =)$	$(\mu\cup)$
(16.2)		$\Leftarrow (\mu dp)$	
(16.3)		$\not\Leftarrow -(\mu dp)$	
(17.1)	$(Log\cup')$	$\Rightarrow (\mu \subseteq) + (\mu =)$	$(\mu\cup')$
(17.2)		$\Leftarrow (\mu dp)$	
(17.3)		$\not\Leftarrow -(\mu dp)$	

Table 1.6 Preferential representation

μ-function		Pref. Structure		Logic
$(\mu \subseteq)$	$\Leftrightarrow$ Proposition 2.3.5	reactive	$\Leftrightarrow$ Proposition 2.3.14	$(LLE) + (CCL) + (SC)$
$(\mu \subseteq) + (\mu CUM)$	$\Rightarrow (\cap)$ Proposition 2.3.13	reactive + essentially smooth		
$(\mu \subseteq) + (\mu \subseteq \supseteq)$	$\Rightarrow$ Proposition 2.3.13	reactive + essentially smooth	$\Leftrightarrow$ Proposition 2.3.14	$(LLE) + (CCL) + (SC) + (\subseteq \supseteq)$
$(\mu \subseteq) + (\mu CUM) + (\mu \subseteq \supseteq)$	$\Leftarrow$ Fact 2.3.4	reactive + essentially smooth		
$(\mu \subseteq) + (\mu PR)$	$\Leftarrow$ Fact 1.3.2 $\Rightarrow$ Proposition 1.3.1 page 56	general	$\Rightarrow (\mu dp)$ $\Leftarrow$ $\not\Rightarrow$ without (μdp) Example 1.7.1 $\not\Leftarrow$ without (μdp) Proposition 5.2.15 in [Sch04] Proposition 1.7.14	$(LLE) + (RW) + (SC) + (PR)$ any "normal" characterization of any size
$(\mu \subseteq) + (\mu PR)$	$\Leftarrow$ Fact 1.3.2 $\Rightarrow$ Proposition 1.3.5	transitive	$\Rightarrow (\mu dp)$ $\Leftarrow$ $\not\Rightarrow$ without (μdp) Example 1.7.1 $\Leftrightarrow$ without (μdp) Proposition 5.2.5, 5.2.11 in [Sch04], Proposition 1.7.5 Proposition 1.7.10	$(LLE) + (RW) + (SC) + (PR)$ using "small" exception sets
$(\mu \subseteq) + (\mu PR) + (\mu CUM)$	$\Leftarrow$ Fact 1.3.10 $\Rightarrow (\cup)$ Proposition 3.3.4 in [Sch04], Proposition 1.3.14 $\not\Rightarrow$ without $(\cup)$ See [Sch04], Section 1.5.2.3 and Section 1.5.2.4	smooth	$\Rightarrow (\mu dp)$ Proposition 1.3.20 $\Leftarrow (\cup)$ Proposition 1.3.20 $\not\Rightarrow$ without (μdp) Example 1.7.1	$(LLE) + (RW) + (SC) + (PR) + (CUM)$

1.2 Basic Definitions and Overview

		smooth+transitive		
$(\mu\subseteq)+(\mu PR)+(\mu CUM)$	$\Leftarrow$ Fact 1.3.10 $\Rightarrow$ (U) Proposition 3.3.8 in [Sch04], Proposition 1.3.18		$\Rightarrow (\mu dp)$ Proposition 1.3.20 $\Leftarrow$ (U) Proposition 1.3.20	$(LLE)+(RW)+$ $(SC)+(PR)+$ (CUM)
			$\not\Rightarrow$ without (μdp) Example 1.7.1 $\not\Leftarrow$ without (μdp) Proposition 5.2.9, 5.2.11 in [Sch04], Proposition 1.7.6 Proposition 1.7.10	using "small" exception sets
$(\mu\subseteq)+(\mu=)+(\mu PR)+$ $(\mu=')+(\mu\|)+(\mu\cup)+$ $(\mu\cup')+(\mu\in)+(\mu RatM)$	$\Leftarrow$ Fact 1.4.7	ranked, ≥ 1 copy		
$(\mu\subseteq)+(\mu=)+(\mu PR)+$ $(\mu\cup)+(\mu\in)$	$\not\Rightarrow$ Example 1.4.10	ranked		
$(\mu\subseteq)+(\mu=)+(\mu\emptyset)$	$\Leftrightarrow$, (U) Proposition 3.10.11 in [Sch04], Proposition 1.4.8	ranked, 1 copy $+ (\mu\emptyset)$		
$(\mu\subseteq)+(\mu=)+(\mu\emptyset)$	$\Leftrightarrow$, (U) Proposition 3.10.11 in [Sch04], Proposition 1.4.8	ranked, smooth, 1 copy $+ (\mu\emptyset)$		
$(\mu\subseteq)+(\mu=)+(\mu\emptyset fin)+$ $(\mu\in)$	$\Leftrightarrow$, (U), singletons Proposition 3.10.12 in [Sch04], Proposition 1.4.9	ranked, smooth, ≥ 1 copy $+ (\mu\emptyset fin)$		
$(\mu\subseteq)+(\mu PR)+(\mu\|)+$ $(\mu\cup)+(\mu\in)$	$\Leftrightarrow$, (U), singletons Proposition 3.10.14 in [Sch04], Proposition 1.4.11	ranked ≥ 1 copy	$\not\Rightarrow$ without (μdp) Example 1.2.9 $\not\Leftarrow$ without (μdp) Proposition 5.2.16 in [Sch04], Proposition 1.7.15	$(RatM), (RatM=)$, $(Log\cup), (Log\cup')$ any "normal" characterization of any size

1.3 Basic Cases

1.3.1 Introduction

Nonmonotonic logics were, historically, studied from two different points of view: the syntactic side, where rules like (AND), (CUM) (see Definition 1.2.10) were postulated for their naturalness in reasoning, and from the semantic side, by the introduction of preferential structures (see Definition 1.2.11 and Definition 1.2.16). This work was done on the one hand side by Gabbay [Gab85], Makinson [Mak94], and others, and for the second approach by Shoham and others, see [Sho87b], [BS85]. Both approaches were brought together by Kraus, Lehmann, Magidor and others, see [KLM90], [LM92], in their completeness results.

A preferential structure $\mathcal{M}$ defines a logic $\mid\!\sim$ by $T \mid\!\sim \phi$ iff ϕ holds in all $\mathcal{M}$-minimal models of T. This is made precise in Definition 1.2.11 and Definition 1.2.16. At the same time, $\mathcal{M}$ defines also a model set function, by assigning to the set of models of T the set of its minimal models. As logics can speak only about definable model sets (here the model set defined by T), $\mathcal{M}$ defines a function from the definable sets of models to arbitrary model sets: $\mu_\mathcal{M} : D(\mathcal{L}) \to \mathcal{P}(M(\mathcal{L}))$. This is the general framework, within which we will work most of the time. Different logics and situations (see e.g., Plausibility Logic, Section 1.5.2, but also update situations, will force us to generalize, we then consider functions $f : \mathcal{Y} \to \mathcal{P}(W)$, where W is an arbitrary set, and $\mathcal{Y} \subseteq \mathcal{P}(W)$.

($\mathcal{Y}$ is intended to be the set of definable model sets, and we treat here the case of definability preserving functions, so we may also assume here $f : \mathcal{Y} \to \mathcal{Y}$ instead of $f : \mathcal{Y} \to \mathcal{P}(W)$.)

Example 1.3.1

This simple example illustrates the importance of copies. Such examples seem to have appeared for the first time in print in [KLM90], but can probably be attributed to folklore.

Consider the propositional language $\mathcal{L}$ of two propositional variables p, q, and the classical preferential model $\mathcal{M}$ defined by

$m \models p \wedge q$, $m' \models p \wedge q$, $m_2 \models \neg p \wedge q$, $m_3 \models \neg p \wedge \neg q$, with $m_2 \prec m$, $m_3 \prec m'$, and let $\models_\mathcal{M}$ be its consequence relation. (m and m' are logically identical.)

Obviously, $Th(m) \vee \{\neg p\} \models_\mathcal{M} \neg p$, but there is no complete theory T' s.t. $Th(m) \vee T' \models_\mathcal{M} \neg p$. (If there were one, T' would correspond to m, m_2, m_3, or the missing $m_4 \models p \wedge \neg q$, but we need two models to kill all copies of m.) On the other hand, if there were just one copy of m, then one other model, i.e., a complete theory would suffice. More formally, if we admit at most one copy of each model in a structure $\mathcal{M}$, $m \not\models T$, and $Th(m) \vee T \models_\mathcal{M} \phi$ for some ϕ s.t. $m \models \neg \phi$ - i.e., m is not

1.3 Basic Cases

minimal in the models of $Th(m) \vee T$ - then there is a complete T' with $T' \vdash T$ and $Th(m) \vee T' \models_{\mathcal{M}} \phi$, i.e., there is m'' with $m'' \models T'$ and $m'' \prec m$.

$\square$

We work in some universe W, there is a function $f : \mathcal{Y} \to \mathcal{P}(W)$, where $\mathcal{Y} \subseteq \mathcal{P}(W)$, f will have certain properties, and perhaps $\mathcal{Y}$, too, and we will try to represent f by a preferential structure $\mathcal{Z}$ of a certain type, i.e., we want $f = \mu_{\mathcal{Z}}$, with $\mu_{\mathcal{Z}}$ the μ-function or choice function of a preferential structure $\mathcal{Z}$. Note that the codomain of f is not necessarily a subset of $\mathcal{Y}$ - so we have to pay attention not to apply f twice.

1.3.2 General Preferential Structures

We discuss first general preferential structures with arbitrarily many copies. We recall the main conditions and develop the results.

There are two main possibilities: an element $x \in X - \mu(X)$ is minimized by some element in X (perhaps by x itself in an infinite descending chain or a cycle), or, we need a set of elements in X to minimize x. The first possibility works directly with elements x, the second variant needs copies: we have to destroy all copies, and for this, we need a full set of elements.

In the general case with copies, transitivity does not need new conditions, the case without copies does.

(On the other hand, as the case of smoothness shows, which is in itself a weak form of transitivity, there are more specific situations, where transitivity does not change conditions either, see the discussion of the smooth case below.)

1.3.2.1 General Minimal Preferential Structures

The following construction was already used in [Sch92], and is the basis for all other constructions for nonranked minimal preferential structures. We analyse this construction now.

Suppose we know that $x \in X - f(X)$. So there must be some $x' \prec x$, $x' \in X$. In the general case, we have no possibility to determine which x' is smaller than x. There might also be several such x'. Chosing one x' arbitrarily might be wrong. We would pretend to know something we do not know.

Working with copies of x solves the problem. For each $x' \in X$ we make a copy of x, $\{\langle x, x' \rangle : x' \in X\}$ whose minimality is destroyed by $x' : x' \prec \langle x, x' \rangle$. Now, we need all elements of X to make all copies of x non-minimal. This expresses exactly our knowledge (and ignorance): x is non-minimal in X, x' or x'', or .... is smaller than x, or several are smaller than x, but we do not know which ones. The construction codes this "or" without arbitrarily chosing one. From an intuitionist point of view, it is highly non-constructive.

Of course, any $X' \supseteq X$ will also minimize x, but any $X' \subseteq X$ need not minimize x - unless we made the same construction for X' and x. In this way, we can express that X, and X', etc. minimize x, and we will code this "and" in our construction:

$\mathcal{Y}_x := \{Y \in \mathcal{Y} : x \in Y - \mu(Y)\}$ - the set of all Y which minimize x,

$\Pi_x := \Pi \mathcal{Y}_x$ - the set of all choice functions for $\mathcal{Y}_x$,

$\mathcal{X} := \{\langle x, f \rangle : f \in \Pi_x\}$ - each such choice function provides one copy of x,

$\langle x', f' \rangle \prec \langle x, f \rangle :\leftrightarrow x' \in ran(f)$.

See Definition 1.3.1 and Construction 1.3.1.

For the transitive case, we need more control over the smaller elements, choice functions will be replaced by trees (essentially of choice functions).

For more comments, see [Sch04], section 3.2.1 there.

Proposition 1.3.1 is the basic result for non-ranked preferential structures. Most other results and techniques are variations and further developments of the same fundamental idea.

Proposition 1.3.1

An operation $\mu : \mathcal{Y} \to \mathcal{Y}$ is representable by a preferential structure iff μ satisfies $(\mu \subseteq)$ and (μPR).

We show the easy half first:

Fact 1.3.2

Every preferential structure satisfies $(\mu \subseteq)$ and (μPR).

Proof

Exercise, solution in the Appendix.

We turn to the more difficult direction.

Definition 1.3.1

For $x \in Z$, let

$\mathcal{Y}_x := \{Y \in \mathcal{Y} : x \in Y - \mu(Y)\}$,

$\Pi_x := \Pi \mathcal{Y}_x$.

Note that $\emptyset \notin \mathcal{Y}_x$, $\Pi_x \neq \emptyset$, and that $\Pi_x = \{\emptyset\}$ iff $\mathcal{Y}_x = \emptyset$.

The following Claim 1.3.3 is the core of the completeness proof.

1.3 Basic Cases

Claim 1.3.3

Let $\mu : \mathcal{Y} \to \mathcal{Y}$ satisfy $(\mu \subseteq)$ and (μPR), and let $U \in \mathcal{Y}$. Then $x \in \mu(U) \leftrightarrow x \in U \wedge \exists f \in \Pi_x . ran(f) \cap U = \emptyset$.

Proof

Case 1: $\mathcal{Y}_x = \emptyset$, thus $\Pi_x = \{\emptyset\}$. "$\to$": Take $f := \emptyset$. "$\leftarrow$": $x \in U \in \mathcal{Y}, \mathcal{Y}_x = \emptyset \to x \in \mu(U)$ by definition of $\mathcal{Y}_x$.

Case 2: $\mathcal{Y}_x \neq \emptyset$. "$\to$": Let $x \in \mu(U) \subseteq U$. It suffices to show $Y \in \mathcal{Y}_x \to Y - U \neq \emptyset$. But if $Y \subseteq U$ and $Y \in \mathcal{Y}_x$, then $x \in Y - \mu(Y)$, contradicting (μPR). "$\leftarrow$": If $x \in U - \mu(U)$, then $U \in \mathcal{Y}_x$, so $\forall f \in \Pi_x . ran(f) \cap U \neq \emptyset$.

$\square$

Proof

One direction is trivial, and was shown in Fact 1.3.2.

We turn to the other direction. The preferential structure is defined in Construction 1.3.1, Claim 1.3.4 shows representation. The construction has the same role as Proposition 1.3.1 - it is basic for much of the rest of the chapter.

Construction 1.3.1

Let

$\mathcal{X} := \{\langle x, f \rangle : x \in Z \wedge f \in \Pi_x\}$,

$\langle x', f' \rangle \prec \langle x, f \rangle :\leftrightarrow x' \in ran(f)$, and

$\mathcal{Z} := \langle \mathcal{X}, \prec \rangle$.

Claim 1.3.4

For $U \in \mathcal{Y}$, $\mu(U) = \mu_{\mathcal{Z}}(U)$.

Proof

By Claim 1.3.3, it suffices to show that for all $U \in \mathcal{Y}$ $x \in \mu_{\mathcal{Z}}(U) \leftrightarrow x \in U$ and $\exists f \in \Pi_x . ran(f) \cap U = \emptyset$. So let $U \in \mathcal{Y}$. "$\to$": If $x \in \mu_{\mathcal{Z}}(U)$, then there is $\langle x, f \rangle$ minimal in $\mathcal{X} \upharpoonright U$ (recall from Definition 1.2.1 that $\mathcal{X} \upharpoonright U := \{\langle x, i \rangle \in \mathcal{X} : x \in U\}$), so $x \in U$, and there is no $\langle x', f' \rangle \prec \langle x, f \rangle$, $x' \in U$, so by $\Pi_{x'} \neq \emptyset$ there is no $x' \in ran(f)$, $x' \in U$, but then $ran(f) \cap U = \emptyset$. "$\leftarrow$": If $x \in U$, and there is $f \in \Pi_x$, $ran(f) \cap U = \emptyset$, then $\langle x, f \rangle$ is minimal in $\mathcal{X} \upharpoonright U$.

$\square$ (Claim 1.3.4 and Proposition 1.3.1)

1.3.2.2 Transitive Minimal Preferential Structures

Discussion

To treat transitivity, trees of height $\leq \omega$ seem the right way to code successors of an element. To give us better control of successors, we define in Construction 1.3.2 that one element with its tree is a successor of another element with its tree, iff the former is an initial segment of the latter. As before, transitivity will be for free. The new construction with trees as indices respects transitivity, it "looks ahead", and not all elements $\langle y_1, t_{y_1} \rangle$ are smaller than $\langle x, t_x \rangle$, where y_1 is a child of x in t_x (or $y_1 \in ran(f)$). The old construction did not result in transitivity, as Example 1.3.2 shows.

A more detailed discussion can be found in [Sch04], Section 3.2.2.

Example 1.3.2

As we consider only one set in each case, we can index with elements, instead of with functions. So suppose $x, y_1, y_2 \in X$, $y_1, z_1, z_2 \in Y$, $x \notin \mu(X)$, $y_1 \notin \mu(Y)$, and that we need y_1 and y_2 to minimize x, so there are two copies $\langle x, y_1 \rangle$, $\langle x, y_2 \rangle$, likewise we need z_1 and z_2 to minimize y_1, thus we have $\langle x, y_1 \rangle \succ \langle y_1, z_1 \rangle$, $\langle x, y_1 \rangle \succ \langle y_1, z_2 \rangle$, $\langle x, y_2 \rangle \succ y_2$, $\langle y_1, z_1 \rangle \succ z_1$, $\langle y_1, z_2 \rangle \succ z_2$ (the z_i and y_2 are not killed). If we take the transitive closure, we have $\langle x, y_1 \rangle \succ z_k$ for any i, k, so for any z_k $\{z_k, y_2\}$ will minimize all of x, which is not intended.

□

Proposition 1.3.5 is the basic result for the transitive case.

Proposition 1.3.5

An operation $\mu : \mathcal{Y} \to \mathcal{Y}$ is representable by a transitive preferential structure iff μ satisfies $(\mu \subseteq)$ and (μPR).

Proof

The trivial direction follows from the trivial direction in Proposition 1.3.1.

We turn to the other direction.

See the proof of Proposition 3.2.4 in [Sch04] for more comments.

Construction 1.3.2

(1) For $x \in Z$, let T_x be the set of trees t_x s.t.

    (a) all nodes are elements of Z,

    (b) the root of t_x is x,

1.3 Basic Cases

(c) $height(t_x) \leq \omega$,

(d) if y is an element in t_x, then there is $f \in \Pi_y := \Pi\{Y \in \mathcal{Y}: y \in Y - \mu(Y)\}$ s.t. the set of children of y is $ran(f)$.

(2) For $x, y \in Z, t_x \in T_x, t_y \in T_y$, set $t_x \triangleright t_y$ iff y is a (direct) child of the root x in t_x, and t_y is the subtree of t_x beginning at y.

(3) Let $\mathcal{Z} := \; < \{\langle x, t_x \rangle : x \in Z, t_x \in T_x\}, \langle x, t_x \rangle \succ \langle y, t_y \rangle$ iff $t_x \triangleright t_y >$.

Fact 1.3.6

(1) The construction ends at some y iff $\mathcal{Y}_y = \emptyset$, consequently $T_x = \{x\}$ iff $\mathcal{Y}_x = \emptyset$. (We identify the tree of height 1 with its root.)

(2) If $\mathcal{Y}_x \neq \emptyset$, tc_x, the totally ordered tree of height ω, branching with $card = 1$, and with all elements equal to x is an element of T_x. Thus, with (1), $T_x \neq \emptyset$ for any x.

(3) If $f \in \Pi_x$, $f \neq \emptyset$, then the tree tf_x with root x and otherwise composed of the subtrees t_y for $y \in ran(f)$, where $t_y := y$ iff $\mathcal{Y}_y = \emptyset$, and $t_y := tc_y$ iff $\mathcal{Y}_y \neq \emptyset$, is an element of T_x. (Level 0 of tf_x has x as element, the $t'_y s$ begin at level 1.)

(4) If y is an element in t_x and t_y the subtree of t_x starting at y, then $t_y \in T_y$.

(5) $\langle x, t_x \rangle \succ \langle y, t_y \rangle$ implies $y \in ran(f)$ for some $f \in \Pi_x$.

□

Claim 1.3.7 shows basic representation.

Claim 1.3.7

$\forall U \in \mathcal{Y}. \mu(U) = \mu_{\mathcal{Z}}(U)$

Proof

Exercise, solution in the Appendix.

We consider now the transitive closure of $\mathcal{Z}$. (Recall that $\prec^*$ denotes the transitive closure of $\prec$.) Claim 1.3.8 shows that transitivity does not destroy what we have achieved. The trees tf_x will play a crucial role in the demonstration.

Claim 1.3.8

Let $\mathcal{Z}' := \; < \{\langle x, t_x \rangle : x \in Z, t_x \in T_x\}, \langle x, t_x \rangle \succ \langle y, t_y \rangle$ iff $t_x \triangleright^* t_y >$. Then $\mu_{\mathcal{Z}} = \mu_{\mathcal{Z}'}$.

Proof

Suppose there is $U \in \mathcal{Y}$, $x \in U$, $x \in \mu_{\mathcal{Z}}(U)$, $x \notin \mu_{\mathcal{Z}'}(U)$. Then there must be an element $\langle x, t_x \rangle \in \mathcal{Z}$ with no $\langle x, t_x \rangle \succ \langle y, t_y \rangle$ for any $y \in U$. Let $f \in \Pi_x$ determine the set of children of x in t_x, then $ran(f) \cap U = \emptyset$, consider tf_x. As all elements $\neq x$ of tf_x are already in $ran(f)$, no element of tf_x is in U. Thus there is no $\langle z, t_z \rangle \prec^* \langle x, tf_x \rangle$ in $\mathcal{Z}$ with $z \in U$, so $\langle x, tf_x \rangle$ is minimal in $\mathcal{Z}' \upharpoonright U$, contradiction.

□ (Claim 1.3.8 and Proposition 1.3.5)

We give now the direct proof, which we cannot adapt to the smooth case. Such easy results must be part of the folklore, but we give them for completeness' sake.

Proposition 1.3.9

In the general case, every preferential structure is equivalent to a transitive one - i.e. they have the same μ-functions.

Proof

If $\langle a, i \rangle \succ \langle b, j \rangle$, we create an infinite descending chain of new copies $\langle b, \langle j, a, i, n \rangle \rangle$, $n \in \omega$, where $\langle b, \langle j, a, i, n \rangle \rangle \succ \langle b, \langle j, a, i, n' \rangle \rangle$ if $n' > n$, and make $\langle a, i \rangle \succ \langle b, \langle j, a, i, n \rangle \rangle$ for all $n \in \omega$, but cancel the pair $\langle a, i \rangle \succ \langle b, j \rangle$ from the relation (otherwise, we would not have achieved anything), but $\langle b, j \rangle$ stays as element in the set. Now, the relation is trivially transitive, and all these $\langle b, \langle j, a, i, n \rangle \rangle$ just kill themselves, there is no need to minimize them by anything else. We just continued $\langle a, i \rangle \succ \langle b, j \rangle$ in a way it cannot bother us. For the $\langle b, j \rangle$, we do of course the same thing again. So, we have full equivalence, i.e. the μ-functions of both structures are identical (this is trivial to see).

□

1.3.3 Smooth Minimal Preferential Structures with Arbitrarily Many Copies

1.3.3.1 Discussion

We know that if $x \in X - f(X)$, then for each copy $\langle x, i \rangle$ of x, there must be $x' \prec x$, $x' \in f(X)$. We have to assure that obtaining minimization for x in X does not destroy smoothness elsewhere, or, if it does, we have to repair it.

For some given x, and a copy $\langle x, \sigma \rangle$ to be constructed, we will

- minimize x, where necessary, using again a cartesian product as in the not necessarily smooth case, choosing in $f(Y)$ for suitable $Y : \sigma_0 \in \Pi\{f(Y) : x \in Y - f(Y)\}$.

1.3 Basic Cases

- if X is such that $x \in f(X)$, and $ran(\sigma_0) \cap X \neq \emptyset$, we have destroyed minimality of the copy $\langle x, \sigma \rangle$ under construction in X, and have to put a new element minimal in this X below it, to preserve smoothness: $\sigma_1 \in \Pi\{f(X) : x \in f(X)$ and $ran(\sigma_0) \cap X \neq \emptyset\}$.
- we might have destroyed minimality in some X, this time by the new $ran(\sigma_1)$, so we repeat the procedure for σ_1, and so on, infinitely often.

We then show that for each x and U with $x \in f(U)$ there is such $\langle x, \sigma \rangle$, s.t. all $ran(\sigma_i)$ have empty intersection with U, even with $H(U)$, a sufficiently big "hull" around U, this guarantees minimality of x in U for some copy.

The hull $H(U)$ is defined as $\bigcup\{X : f(X) \subseteq U\}$. Anything inside the hull will be "sucked" into U - any element in the hull will be minimized by some element in some $f(X) \subseteq U$, and thus by U.

A more detailed discussion is in [Sch04], section 3.3.1.

First, again the easy half:

Fact 1.3.10

Every smooth preferential structure satisfies $(\mu \subseteq)$, (μPR), and (μCUM).

Proof

By Fact 1.3.2, it remains to show (μCUM). But, if $\mu(X) \subseteq Y \subseteq X$, then, by smoothness, any $y \in Y - \mu(X)$ will be minimized by an element in $\mu(X) \subseteq Y$.

$\Box$

1.3.3.2 The Constructions

Let $\mu : \mathcal{Y} \to \mathcal{Y}$, and $\mathcal{Y}$ be closed under finite finite unions and finite intersections.

Definition 1.3.2

Define $H(U) := \bigcup\{X : \mu(X) \subseteq U\}$.

The following Fact 1.3.11 contains the basic properties of μ and $H(U)$ which we will need for the representation construction.

Fact 1.3.11

Let A, U, U', Y and all A_i be in $\mathcal{Y}$.

$(\mu \subseteq)$ and (μPR) entail:

(1) $A = \bigcup\{A_i : i \in I\} \to \mu(A) \subseteq \bigcup\{\mu(A_i) : i \in I\}$,

(2) $U \subseteq H(U)$, and $U \subseteq U' \to H(U) \subseteq H(U')$,
(3) $\mu(U \cup Y) - H(U) \subseteq \mu(Y)$.

$(\mu \subseteq), (\mu PR), (\mu CUM)$ entail:

(4) $U \subseteq A, \mu(A) \subseteq H(U) \to \mu(A) \subseteq U$,
(5) $\mu(Y) \subseteq H(U) \to Y \subseteq H(U)$ and $\mu(U \cup Y) = \mu(U)$,
(6) $x \in \mu(U), x \in Y - \mu(Y) \to Y \not\subseteq H(U)$,
(7) $Y \not\subseteq H(U) \to \mu(U \cup Y) \not\subseteq H(U)$.

Proof

(1) $\mu(A) \cap A_j \subseteq \mu(A_j) \subseteq \bigcup \mu(A_i)$, so by $\mu(A) \subseteq A = \bigcup A_i$ $\mu(A) \subseteq \bigcup \mu(A_i)$.
(2) trivial.
(3) $\mu(U \cup Y) - H(U) \subseteq_{(2)} \mu(U \cup Y) - U \subseteq_{(\mu \subseteq)} \mu(U \cup Y) \cap Y \subseteq_{(\mu PR)} \mu(Y)$.
(4) Exercise, solution in the Appendix.
(5) Let $\mu(Y) \subseteq H(U)$, then by $\mu(U) \subseteq H(U)$ and (1) $\mu(U \cup Y) \subseteq \mu(U) \cup \mu(Y) \subseteq H(U)$, so by (4) $\mu(U \cup Y) \subseteq U$ and $U \cup Y \subseteq H(U)$. Moreover, $\mu(U \cup Y) \subseteq U \subseteq U \cup Y \to_{(\mu CUM)} \mu(U \cup Y) = \mu(U)$.
(6) If not, $Y \subseteq H(U)$, so $\mu(Y) \subseteq H(U)$, so $\mu(U \cup Y) = \mu(U)$ by (5), but $x \in Y - \mu(Y) \to_{(\mu PR)} x \notin \mu(U \cup Y) = \mu(U)$, contradiction.
(7) $\mu(U \cup Y) \subseteq H(U) \to_{(5)} U \cup Y \subseteq H(U)$.

□

Assume now $(\mu \subseteq), (\mu PR), (\mu CUM)$ to hold.

Definition 1.3.3

For $x \in Z$, let

$\mathcal{W}_x := \{\mu(Y): Y \in \mathcal{Y} \wedge x \in Y - \mu(Y)\}$,
$\Gamma_x := \Pi \mathcal{W}_x$, and
$K := \{x \in Z: \exists X \in \mathcal{Y}. x \in \mu(X)\}$.

Note that we consider here now $\mu(Y)$ in $\mathcal{W}_x$, and not Y as in $\mathcal{Y}_x$ in Definition 1.3.1.

Remark 1.3.12

(1) $x \in K \to \Gamma_x \neq \emptyset$,
(2) $g \in \Gamma_x \to ran(g) \subseteq K$.

1.3 Basic Cases

Proof

(1) We have to show that $Y \in \mathcal{Y}$, $x \in Y - \mu(Y) \to \mu(Y) \neq \emptyset$. By $x \in K$, there is $X \in \mathcal{Y}$ s.t. $x \in \mu(X)$. Suppose $x \in Y$, $\mu(Y) = \emptyset$. Then $x \in X \cap Y$, so by $x \in \mu(X)$ and (μPR) $x \in \mu(X \cap Y)$. But $\mu(Y) = \emptyset \subseteq X \cap Y \subseteq Y$, so by (μCUM) $\mu(X \cap Y) = \emptyset$, $contradiction$.

(2) By definition, $\mu(Y) \subseteq K$ for all $Y \in \mathcal{Y}$.

□

The following claim is the analogue of Claim 1.3.3 above.

Claim 1.3.13

Let $U \in \mathcal{Y}$, $x \in K$. Then

(1) $x \in \mu(U) \leftrightarrow x \in U \wedge \exists f \in \Gamma_x.ran(f) \cap U = \emptyset$,

(2) $x \in \mu(U) \leftrightarrow x \in U \wedge \exists f \in \Gamma_x.ran(f) \cap H(U) = \emptyset$.

Proof

(1) Case 1: $\mathcal{W}_x = \emptyset$, thus $\Gamma_x = \{\emptyset\}$. "→": Take $f := \emptyset$. "←": $x \in U \in \mathcal{Y}$, $\mathcal{W}_x = \emptyset \to x \in \mu(U)$ by definition of $\mathcal{W}_x$.

Case 2: $\mathcal{W}_x \neq \emptyset$. "→": Let $x \in \mu(U) \subseteq U$. It suffices to show $Y \in \mathcal{W}_x \to \mu(Y) - H(U) \neq \emptyset$. But $Y \in \mathcal{W}_x \to x \in Y - \mu(Y) \to$ (by Fact 1.3.11, (6)) $Y \not\subseteq H(U) \to$ (by Fact 1.3.11, (5)) $\mu(Y) \not\subseteq H(U)$. "←": If $x \in U - \mu(U)$, $U \in \mathcal{W}_x$, moreover $\Gamma_x \neq \emptyset$ by Remark 1.3.12, (1) and thus (or by the same argument) $\mu(U) \neq \emptyset$, so $\forall f \in \Gamma_x.ran(f) \cap U \neq \emptyset$.

(2) The proof is verbatim the same as for (1).

□ (Claim 1.3.13)

Proposition 1.3.14 is the basic representation result for the smooth case.

Proposition 1.3.14

Let $\mathcal{Y}$ be closed under finite unions and finite intersections, and $\mu : \mathcal{Y} \to \mathcal{Y}$. Then there is a $\mathcal{Y}$-smooth preferential structure $\mathcal{Z}$, s.t. for all $X \in \mathcal{Y}$ $\mu(X) = \mu_{\mathcal{Z}}(X)$ iff μ satisfies $(\mu \subseteq)$, (μPR), (μCUM).

Proof

Outline of "←": We first define a structure $\mathcal{Z}$ (in a way very similar to Construction 1.3.1) which represents μ, but is not necessarily $\mathcal{Y}$-smooth, refine it to $\mathcal{Z}'$ and show that $\mathcal{Z}'$ represents μ too, and that $\mathcal{Z}'$ is $\mathcal{Y}$-smooth.

In the structure $\mathcal{Z}'$, all pairs destroying smoothness in $\mathcal{Z}$ are successively repaired, by adding minimal elements: If $\langle y, j \rangle$ is not minimal, and has no minimal $\langle x, i \rangle$

below it, we just add one such $\langle x, i \rangle$. As the repair process might itself generate such "bad" pairs, the process may have to be repeated infinitely often. Of course, one has to take care that the representation property is preserved.

The proof given is close to the minimum one has to show (except that we avoid $H(U)$, instead of U - as was done in the old proof of [Sch96-1]). We could simplify further, we do not, in order to stay closer to the construction that is really needed. The reader will find the simplification as building block of the proof in Section 1.3.3.3. (In the simplified proof, we would consider for x, U s.t. $x \in \mu(U)$ the pairs $\langle x, g_U \rangle$ with $g_U \in \Pi\{\mu(U \cup Y) : x \in Y \not\subseteq H(U)\}$, giving minimal elements. For the U s.t. $x \in U - \mu(U)$, we would choose $\langle x, g \rangle$ s.t. $g \in \Pi\{\mu(Y) : x \in Y \in \mathcal{Y}\}$ with $\langle x', g'_U \rangle \prec \langle x, g \rangle$ for $\langle x', g'_U \rangle$ as above.)

Construction 1.3.3 represents μ. The structure will not yet be smooth, we will mend it afterwards in Construction 1.3.4.

Construction 1.3.3

(Construction of $\mathcal{Z}$) Let $\mathcal{X} := \{\langle x, g \rangle : x \in K, g \in \Gamma_x\}$, $\langle x', g' \rangle \prec \langle x, g \rangle :\leftrightarrow x' \in ran(g)$, $\mathcal{Z} := \langle \mathcal{X}, \prec \rangle$.

Claim 1.3.15

$\forall U \in \mathcal{Y}. \mu(U) = \mu_{\mathcal{Z}}(U)$

Proof

Case 1: $x \notin K$. Then $x \notin \mu(U)$ and $x \notin \mu_{\mathcal{Z}}(U)$.

Case 2: $x \in K$. By Claim 1.3.13, (1) it suffices to show that for all $U \in \mathcal{Y}$ $x \in \mu_{\mathcal{Z}}(U) \leftrightarrow x \in U \wedge \exists f \in \Gamma_x . ran(f) \cap U = \emptyset$. Fix $U \in \mathcal{Y}$. "$\rightarrow$": $x \in \mu_{\mathcal{Z}}(U) \rightarrow$ ex. $\langle x, f \rangle$ minimal in $\mathcal{X} \upharpoonright U$, thus $x \in U$ and there is no $\langle x', f' \rangle \prec \langle x, f \rangle$, $x' \in U$, $x' \in K$. But if $x' \in K$, then by Remark 1.3.12, (1), $\Gamma_{x'} \neq \emptyset$, so we find suitable f'. Thus, $\forall x' \in ran(f). x' \notin U$ or $x' \notin K$. But $ran(f) \subseteq K$, so $ran(f) \cap U = \emptyset$. "$\leftarrow$": If $x \in U$, $f \in \Gamma_x$ s.t. $ran(f) \cap U = \emptyset$, then $\langle x, f \rangle$ is minimal in $\mathcal{X} \upharpoonright U$.

$\square$ (Claim 1.3.15)

We now construct the refined structure $\mathcal{Z}'$.

Construction 1.3.4

(Construction of $\mathcal{Z}'$)

σ is called x-admissible sequence iff

1. σ is a sequence of length $\leq \omega$, $\sigma = \{\sigma_i : i \in \omega\}$,
2. $\sigma_o \in \Pi\{\mu(Y) : Y \in \mathcal{Y} \wedge x \in Y - \mu(Y)\}$,
3. $\sigma_{i+1} \in \Pi\{\mu(X) : X \in \mathcal{Y} \wedge x \in \mu(X) \wedge ran(\sigma_i) \cap X \neq \emptyset\}$.

1.3 Basic Cases

By 2., σ_0 minimizes x, and by 3., if $x \in \mu(X)$, and $ran(\sigma_i) \cap X \neq \emptyset$, i.e. we have destroyed minimality of x in X, x will be above some y minimal in X to preserve smoothness.

Let Σ_x be the set of x-admissible sequences, for $\sigma \in \Sigma_x$ let $\widehat{\sigma} := \bigcup \{ran(\sigma_i) : i \in \omega\}$.

Note that by the argument in the proof of Remark 1.3.12, (1), $\Sigma_x \neq \emptyset$, if $x \in K$.

Let $\mathcal{X}' := \{\langle x, \sigma \rangle \colon x \in K \wedge \sigma \in \Sigma_x\}$ and $\langle x', \sigma' \rangle \prec' \langle x, \sigma \rangle :\leftrightarrow x' \in \widehat{\sigma}$.

Finally, let $\mathcal{Z}' := \langle \mathcal{X}', \prec' \rangle$, and $\mu' := \mu_{\mathcal{Z}'}$.

It is now easy to show that $\mathcal{Z}'$ represents μ, and that $\mathcal{Z}'$ is smooth. For $x \in \mu(U)$, we construct a special x-admissible sequence $\sigma^{x,U}$ using the properties of $H(U)$ as described at the beginning of this section.

Claim 1.3.16

For all $U \in \mathcal{Y}$ $\mu(U) = \mu_{\mathcal{Z}}(U) = \mu'(U)$.

Proof

If $x \notin K$, then $x \notin \mu_{\mathcal{Z}}(U)$, and $x \notin \mu'(U)$ for any U. So assume $x \in K$. If $x \in U$ and $x \notin \mu_{\mathcal{Z}}(U)$, then for all $\langle x, f \rangle \in \mathcal{X}$, there is $< x', f' > \in \mathcal{X}$ with $\langle x', f' \rangle \prec \langle x, f \rangle$ and $x' \in U$. Let now $\langle x, \sigma \rangle \in \mathcal{X}'$, then $\langle x, \sigma_0 \rangle \in \mathcal{X}$, and let $\langle x', f' \rangle \prec \langle x, \sigma_0 \rangle$ in $\mathcal{Z}$ with $x' \in U$. As $x' \in K$, $\Sigma_{x'} \neq \emptyset$, let $\sigma' \in \Sigma_{x'}$. Then $\langle x', \sigma' \rangle \prec' \langle x, \sigma \rangle$ in $\mathcal{Z}'$. Thus $x \notin \mu'(U)$. Thus, for all $U \in \mathcal{Y}$, $\mu'(U) \subseteq \mu_{\mathcal{Z}}(U) = \mu(U)$.

It remains to show $x \in \mu(U) \rightarrow x \in \mu'(U)$.

Assume $x \in \mu(U)$ (so $x \in K$), $U \in \mathcal{Y}$, we will construct minimal σ, i.e. show that there is $\sigma^{x,U} \in \Sigma_x$ s.t. $\widehat{\sigma^{x,U}} \cap U = \emptyset$. We construct this $\sigma^{x,U}$ inductively, with the stronger property that $ran(\sigma_i^{x,U}) \cap H(U) = \emptyset$ for all $i \in \omega$.

$\sigma_0^{x,U} : x \in \mu(U), x \in Y - \mu(Y) \rightarrow \mu(Y) - H(U) \neq \emptyset$ by Fact 1.3.11, (6)+(5). Let $\sigma_0^{x,U} \in \Pi\{\mu(Y) - H(U) : Y \in \mathcal{Y}, x \in Y - \mu(Y)\}$, so $ran(\sigma_0^{x,U}) \cap H(U) = \emptyset$.

$\sigma_i^{x,U} \rightarrow \sigma_{i+1}^{x,U}$: By induction hypothesis, $ran(\sigma_i^{x,U}) \cap H(U) = \emptyset$. Let $X \in \mathcal{Y}$ be s.t. $x \in \mu(X)$, $ran(\sigma_i^{x,U}) \cap X \neq \emptyset$. Thus $X \not\subseteq H(U)$, so $\mu(U \cup X) - H(U) \neq \emptyset$ by Fact 1.3.11, (7). Let $\sigma_{i+1}^{x,U} \in \Pi\{\mu(U \cup X) - H(U) : X \in \mathcal{Y}, x \in \mu(X), ran(\sigma_i^{x,U}) \cap X \neq \emptyset\}$, so $ran(\sigma_{i+1}^{x,U}) \cap H(U) = \emptyset$. As $\mu(U \cup X) - H(U) \subseteq \mu(X)$ by Fact 1.3.11, (3), the construction satisfies the x-admissibility condition.

□

It remains to show:

Claim 1.3.17

$\mathcal{Z}'$ is $\mathcal{Y}$-smooth.

Proof

Exercise, solution in the Appendix.

(Claim 1.3.17 and Proposition 1.3.14 are thus shown.)

1.3.3.3 Smooth and Transitive Minimal Preferential Structures

Discussion

In a certain way, it is not surprising that transitivity does not impose stronger conditions in the smooth case either. Smoothness is itself a weak kind of transitivity: If an element is not minimal, then there is a minimal element below it, i.e., $x \succ y$ with y not minimal is possible, there is $z' \prec y$, but then there is z minimal with $x \succ z$. This is "almost" $x \succ z'$, transitivity.

Note that even beyond Fact 1.3.11, closure of the domain under finite unions is used in the construction of the trees. This - or something like it - is necessary, as we have to respect the hulls of all elements treated so far (the predecessors), and not only of the first element, because of transitivity. For the same reason, we need more bookkeeping, to annotate all the hulls (or the union of the respective U's) of all predecessors to be respected. One can perhaps do with a weaker operation than union - i.e. just look at the hulls of all U's separately, to obtain a transitive construction where unions are lacking, see the case of plausibility logic below - but we have not investigated this problem.

The Construction

Recall that $\mathcal{Y}$ will be closed under finite unions and finite intersections in this section, and let again $\mu : \mathcal{Y} \to \mathcal{Y}$.

Proposition 1.3.18 is the representation result for the smooth transitive case.

Proposition 1.3.18

Let $\mathcal{Y}$ be closed under finite unions and finite intersections, and $\mu : \mathcal{Y} \to \mathcal{Y}$. Then there is a $\mathcal{Y}$-smooth transitive preferential structure $\mathcal{Z}$, s.t. for all $X \in \mathcal{Y}$ $\mu(X) = \mu_{\mathcal{Z}}(X)$ iff μ satisfies $(\mu \subseteq)$, (μPR), (μCUM).

1.3 Basic Cases

Proof

The Idea

We have to adapt Construction 1.3.4 (x-admissible sequences) to the transitive situation, and to our construction with trees. If $\langle \emptyset, x \rangle$ is the root, $\sigma_0 \in \Pi\{\mu(Y) : x \in Y - \mu(Y)\}$ determines some children of the root. To preserve smoothness, we have to compensate and add other children by the $\sigma_{i+1} : \sigma_{i+1} \in \Pi\{\mu(X) : x \in \mu(X), ran(\sigma_i) \cap X \neq \emptyset\}$. On the other hand, we have to pursue the same construction for the children so constructed. Moreover, these indirect children have to be added to those children of the root, which have to be compensated (as the first children are compensated by σ_1) to preserve smoothness. Thus, we build the tree in a simultaneous vertical and horizontal induction.

This construction can be simplified, by considering immediately all $Y \in \mathcal{Y}$ s.t. $x \in Y \not\subseteq H(U)$ - independent of whether $x \notin \mu(Y)$ (as done in σ_0), or whether $x \in \mu(Y)$, and some child y constructed before is in Y (as done in the σ_{i+1}), or whether $x \in \mu(Y)$, and some indirect child y of x is in Y (to take care of transitivity, as indicated above). We make this simplified construction.

There are two ways to proceed. First, we can take as $\prec^*$ in the trees the transitive closure of $\triangleleft$. Second, we can deviate from the idea that children are chosen by selection functions f, and take nonempty subsets of elements instead, making more elements children than in the first case. We take the first alternative, as it is more in the spirit of the construction.

We will suppose for simplicity that $Z = K$ - the general case in easy to obtain by a technique similar to that in Section 1.3.3, but complicates the picture.

For each $x \in Z$, we construct trees t_x, which will be used to index different copies of x, and control the relation $\prec$.

These trees t_x will have the following form:

(a) the root of t is $\langle \emptyset, x \rangle$ or $\langle U, x \rangle$ with $U \in \mathcal{Y}$ and $x \in \mu(U)$,

(b) all other nodes are pairs $\langle Y, y \rangle$, $Y \in \mathcal{Y}$, $y \in \mu(Y)$,

(c) $ht(t) \leq \omega$,

(d) if $\langle Y, y \rangle$ is an element in t_x, then there is some $\mathcal{Y}(y) \subseteq \{W \in \mathcal{Y} : y \in W\}$, and $f \in \Pi\{\mu(W) : W \in \mathcal{Y}(y)\}$ s.t. the set of children of $\langle Y, y \rangle$ is $\{\langle Y \cup W, f(W) \rangle : W \in \mathcal{Y}(y)\}$.

The first coordinate is used for bookkeeping when constructing children, in particular for condition (d).

The relation $\prec$ will essentially be determined by the subtree relation.

We first construct the trees t_x for those sets U where $x \in \mu(U)$, and then take care of the others. In the construction for the minimal elements, at each level $n > 0$, we may have several ways to choose a selection function f_n, and each such choice leads to the construction of a different tree - we construct all these trees. (We could also construct only one tree, but then the choice would have to be made coherently for different x, U. It is simpler to construct more trees than necessary.)

We control the relation by indexing with trees, just as it was done in the not necessarily smooth case before.

Definition 1.3.4

If t is a tree with root $\langle a, b \rangle$, then t/c will be the same tree, only with the root $\langle c, b \rangle$.

Construction 1.3.5

(A) The set T_x of trees t for fixed x:

  (1) Construction of the set $T\mu_x$ of trees for those sets $U \in \mathcal{Y}$, where $x \in \mu(U)$:

    Let $U \in \mathcal{Y}, x \in \mu(U)$. The trees $t_{U,x} \in T\mu_x$ are constructed inductively, observing simultaneously:

    If $\langle U_{n+1}, x_{n+1} \rangle$ is a child of $\langle U_n, x_n \rangle$, then

    (a) $x_{n+1} \in \mu(U_{n+1}) - H(U_n)$,

    and

    (b) $U_n \subseteq U_{n+1}$.

    Set $U_0 := U$, $x_0 := x$.

    Level 0: $\langle U_0, x_0 \rangle$.

    Level $n \to n+1$: Let $\langle U_n, x_n \rangle$ be in level n. Suppose $Y_{n+1} \in \mathcal{Y}$, $x_n \in Y_{n+1}$, and $Y_{n+1} \not\subseteq H(U_n)$. Note that $\mu(U_n \cup Y_{n+1}) - H(U_n) \neq \emptyset$ by Fact 1.3.11, (7), and $\mu(U_n \cup Y_{n+1}) - H(U_n) \subseteq \mu(Y_{n+1})$ by Fact 1.3.11, (3). Choose $f_{n+1} \in \Pi\{\mu(U_n \cup Y_{n+1}) - H(U_n) : Y_{n+1} \in \mathcal{Y}, x_n \in Y_{n+1} \not\subseteq H(U_n)\}$ (for the construction of this tree, at this element), and let the set of children of $\langle U_n, x_n \rangle$ be $\{\langle U_n \cup Y_{n+1}, f_{n+1}(Y_{n+1}) \rangle : Y_{n+1} \in \mathcal{Y}, x_n \in Y_{n+1} \not\subseteq H(U_n)\}$. (If there is no such Y_{n+1}, $\langle U_n, x_n \rangle$ has no children.) Obviously, (a) and (b) hold.

    We call such trees U, x-trees.

  (2) Construction of the set T'_x of trees for the nonminimal elements. Let $x \in Z$. Construct the tree t_x as follows (here, one tree per x suffices for all U):

    Level 0: $\langle \emptyset, x \rangle$

    Level 1: Choose arbitrary $f \in \Pi\{\mu(U) : x \in U \in \mathcal{Y}\}$. Note that $U \neq \emptyset \to \mu(U) \neq \emptyset$ by $Z = K$ (by Remark 1.3.12, (1)). Let $\{\langle U, f(U) \rangle :$

$x \in U \in \mathcal{Y}\}$ be the set of children of $< \emptyset, x >$. This assures that the element will be nonminimal.

Level > 1: Let $\langle U, f(U) \rangle$ be an element of level 1, as $f(U) \in \mu(U)$, there is a $t_{U,f(U)} \in T\mu_{f(U)}$. Graft one of these trees $t_{U,f(U)} \in T\mu_{f(U)}$ at $\langle U, f(U) \rangle$ on the level 1. This assures that a minimal element will be below it to guarantee smoothness.

Finally, let $T_x := T\mu_x \cup T'_x$.

(B) The relation $\triangleleft$ between trees: For $x, y \in Z$, $t \in T_x$, $t' \in T_y$, set $t \triangleright t'$ iff for some Y $\langle Y, y \rangle$ is a child of the root $\langle X, x \rangle$ in t, and t' is the subtree of t beginning at this $\langle Y, y \rangle$.

(C) The structure $\mathcal{Z}$: Let $\mathcal{Z} := < \{\langle x, t_x \rangle : x \in Z, t_x \in T_x\}, \langle x, t_x \rangle \succ \langle y, t_y \rangle$ iff $t_x \triangleright^* t_y >$.

The rest of the proof are simple observations.

Fact 1.3.19

(1) If $t_{U,x}$ is an U, x-tree, $\langle U_n, x_n \rangle$ an element of $t_{U,x}$, $\langle U_m, x_m \rangle$ a direct or indirect child of $\langle U_n, x_n \rangle$, then $x_m \notin H(U_n)$.

(2) Let $\langle Y_n, y_n \rangle$ be an element in $t_{U,x} \in T\mu_x$, t' the subtree starting at $\langle Y_n, y_n \rangle$, then t' is a Y_n, y_n-tree.

(3) $\prec$ is free from cycles.

(4) If $t_{U,x}$ is an U, x-tree, then $\langle x, t_{U,x} \rangle$ is $\prec$-minimal in $\mathcal{Z} \upharpoonright U$.

(5) No $\langle x, t_x \rangle$, $t_x \in T'_x$ is minimal in any $\mathcal{Z} \upharpoonright U$, $U \in \mathcal{Y}$.

(6) Smoothness is respected for the elements of the form $\langle x, t_{U,x} \rangle$.

(7) Smoothness is respected for the elements of the form $\langle x, t_x \rangle$ with $t_x \in T'_x$.

(8) $\mu = \mu_{\mathcal{Z}}$.

Proof

(1) trivial by (a) and (b).

(2) trivial by (a).

(3) Note that no $\langle x, t_x \rangle$ $t_x \in T'_x$ can be smaller than any other element (smaller elements require $U \neq \emptyset$ at the root). So no cycle involves any such $\langle x, t_x \rangle$. Consider now $\langle x, t_{U,x} \rangle$, $t_{U,x} \in T\mu_x$. For any $\langle y, t_{V,y} \rangle \prec \langle x, t_{U,x} \rangle$, $y \notin H(U)$ by (1), but $x \in \mu(U) \subseteq H(U)$, so $x \neq y$.

(4) This is trivial by (1).

(5) Let $x \in U \in \mathcal{Y}$, then f as used in the construction of level 1 of t_x chooses $y \in \mu(U) \neq \emptyset$, and some $\langle y, t_{U,y} \rangle$ is in $\mathcal{Z} \upharpoonright U$ and below $\langle x, t_x \rangle$.

(6) Exercise, solution in the Appendix.

(7) Let $x \in A \in \mathcal{Y}$, $\langle x, t_x \rangle$, $t_x \in T'_x$, and consider the subtree t beginning at $\langle A, f(A) \rangle$, then t is one of the $A, f(A)$-trees, and $\langle f(A), t \rangle$ is minimal in $\mathcal{Z} \upharpoonright A$ by (4).

(8) Let $x \in \mu(U)$. Then any $\langle x, t_{U,x} \rangle$ is $\prec$-minimal in $\mathcal{Z} \upharpoonright U$ by (4), so $x \in \mu_{\mathcal{Z}}(U)$. Conversely, let $x \in U - \mu(U)$. By (5), no $\langle x, t_x \rangle$ is minimal in U. Consider now some $\langle x, t_{V,x} \rangle \in \mathcal{Z}$, so $x \in \mu(V)$. As $x \in U - \mu(U)$, $U \not\subseteq H(V)$ by Fact 1.3.11, (6). Thus U was considered in the construction of level 1 of $t_{V,x}$. Let t be the subtree of $t_{V,x}$ beginning at $\langle V \cup U, f_1(U) \rangle$, by $\mu(V \cup U) - H(V) \subseteq \mu(U)$ (Fact 1.3.11, (3)), $f_1(U) \in \mu(U) \subseteq U$, and $\langle f_1(U), t \rangle \prec \langle x, t_{V,x} \rangle$.

□ (Fact 1.3.19 and Proposition 1.3.18)

1.3.4 The Logical Characterization of General and Smooth Preferential Models

Discussion

The translations from the algebraic to the logical characterizations and conversely are usually quite straightforward - as long as the operations are definability preserving, see Section 1.7 for results without.

For more discussion, see Section 3.4 in [Sch04].

Proposition 1.3.20

Let $\mathrel{\mid\!\sim}$ be a logic for $\mathcal{L}$. Recall from Definition 2.3.2 $T^{\mathcal{M}} := Th(\mu_{\mathcal{M}}(M(T)))$, where $\mathcal{M}$ is a preferential structure.

(1) Then there is a (transitive) definability preserving classical preferential model $\mathcal{M}$ s.t. $\overline{\overline{T}} = T^{\mathcal{M}}$ iff

(LLE) $\overline{T} = \overline{T'} \to \overline{\overline{T}} = \overline{\overline{T'}}$,

(CCL) $\overline{\overline{T}}$ is classically closed,

(SC) $T \subseteq \overline{\overline{T}}$,

(PR) $\overline{\overline{T \cup T'}} \subseteq \overline{\overline{T}} \cup T'$

for all $T, T' \subseteq \mathcal{L}$.

(2) The structure can be chosen smooth, iff, in addition

(CUM) $T \subseteq T' \subseteq \overline{\overline{T}} \to \overline{\overline{T}} = \overline{\overline{T'}}$

holds.

This is an immediate consequence of Proposition 1.2.18, see also Table 1.5.

1.4 Ranked Case

For the limit versions, and the cases without definability preservation, the reader is referred to Section 1.6 and Section 1.7 respectively.

1.4.1 Ranked Preferential Structures

1.4.1.1 Introduction

We discuss the following versions:

(1) Ranked structures which preserve nonemptiness (property $(\mu\emptyset)$) $X \neq \emptyset \to \mu(X) \neq \emptyset$, they are almost equivalent to smooth ranked structures.
(2) The more general case, but without copies of elements, which is very similar to case (1), as the decisive property, $(\mu\emptyset)$, still holds for finite sets. The order itself may, however, now be nonwellfounded.
(3) The general case with copies.

This diversity leads to the confusing list of conditions in Definition 1.2.10, of positive interrelations in Fact 1.2.14, of negative results in Fact 1.2.14 and Fact 1.4.10.

The crucial property is that incomparable elements have the same behavior: $a \perp b$ (i.e. neither $a \prec b$ nor $b \prec a$) and $c \prec a$ ($c \succ a$) imply $c \prec b$ ($c \succ b$).

The main positive results for minimal ranked structures are Proposition 1.4.8 and Proposition 1.4.9 for structures without copies, Proposition 1.4.11 for the general case. Proposition 1.4.11 is the most general result we show in this context.

The condition

$(\mu =) \ X \subseteq Y, \mu(Y) \cap X \neq \emptyset \to \mu(Y) \cap X = \mu(X)$

plays a central role. It is a strengthening of the basic condition (μPR), and is a very strong property.

More discussion and details can be found in [Sch04], section 3.10.1.

1.4.1.2 The Details

Introductory Facts and Definitions

The material of Fact 1.4.1 to Fact 1.4.5 is taken from [Sch96-1], and is mostly folklore.

We first note the following trivial

Fact 1.4.1

In a ranked structure, smoothness and the condition

$(\mu\emptyset)$ $X \neq \emptyset \to \mu(X) \neq \emptyset$

are (almost) equivalent.

Proof

Suppose $(\mu\emptyset)$ holds, and let $x \in X - \mu(X)$, $x' \in \mu(X)$. Then $x' \prec x$ by rankedness. Conversely, if the structure is smooth and there is an element $x \in X$ in the structure (recall that structures may have "gaps", but this condition is a minor point, which we shall neglect here - this is the precise meaning of "almost"), then either $x \in \mu(X)$ or there is $x' \prec x$, $x' \in \mu(X)$, so $\mu(X) \neq \emptyset$.

□

Note further that if we have no copies (and there is some $x \in X$ in the structure), $(\mu\emptyset)$ holds for all finite sets, and this will be sufficient to construct the relation for representation results, as we shall see.

Fact 1.4.2

In the presence of $(\mu =)$ and $(\mu \subseteq)$, $f(Y) \cap (X - f(X)) \neq \emptyset$ is equivalent to $f(Y) \cap X \neq \emptyset$ and $f(Y) \cap f(X) = \emptyset$.

Proof

Exercise, solution in the Appendix.

Definition 1.4.1

Let $\mathcal{Z} = \langle \mathcal{X}, \prec \rangle$ be a preferential structure. Call $\mathcal{Z}$ $1 - \infty$ over Z, iff for all $x \in Z$ there are exactly one or infinitely many copies of x, i.e. for all $x \in Z$ $\{u \in \mathcal{X} : u = \langle x, i \rangle \text{ for some } i\}$ has cardinality 1 or $\geq \omega$.

Lemma 1.4.3

Let $\mathcal{Z} = \langle \mathcal{X}, \prec \rangle$ be a preferential structure and $f : \mathcal{Y} \to \mathcal{P}(Z)$ with $\mathcal{Y} \subseteq \mathcal{P}(Z)$ be represented by $\mathcal{Z}$, i.e. for $X \in \mathcal{Y}$ $f(X) = \mu_{\mathcal{Z}}(X)$, and $\mathcal{Z}$ be ranked and free of cycles. Then there is a structure $\mathcal{Z}'$, $1 - \infty$ over Z, ranked and free of cycles, which also represents f.

1.4 Ranked Case

Proof

We construct $\mathcal{Z}' = \langle \mathcal{X}', \prec' \rangle$.

Let $A := \{x \in Z:$ there is some $\langle x, i \rangle \in \mathcal{X}$, but for all $\langle x, i \rangle \in \mathcal{X}$ there is $\langle x, j \rangle \in \mathcal{X}$ with $\langle x, j \rangle \prec \langle x, i \rangle\}$,

let $B := \{x \in Z:$ there is some $\langle x, i \rangle \in \mathcal{X}$, s.t. for no $\langle x, j \rangle \in \mathcal{X}$ $\langle x, j \rangle \prec \langle x, i \rangle\}$,

let $C := \{x \in Z:$ there is no $\langle x, i \rangle \in \mathcal{X}\}$.

Let $c_i : i < \kappa$ be an enumeration of C. We introduce for each such c_i ω many copies $\langle c_i, n \rangle : n < \omega$ into $\mathcal{X}'$, put all $\langle c_i, n \rangle$ above all elements in $\mathcal{X}$, and order the $\langle c_i, n \rangle$ by $\langle c_i, n \rangle \prec' \langle c_{i'}, n' \rangle :\leftrightarrow (i = i'$ and $n > n')$ or $i > i'$. Thus, all $\langle c_i, n \rangle$ are comparable.

If $a \in A$, then there are infinitely many copies of a in $\mathcal{X}$, as $\mathcal{X}$ was cycle-free, we put them all into $\mathcal{X}'$. If $b \in B$, we choose exactly one such minimal element $\langle b, m \rangle$ (i.e. there is no $\langle b, n \rangle \prec \langle b, m \rangle$) into $\mathcal{X}'$, and omit all other elements. (For definiteness, assume in all applications $m = 0$.) For all elements from A and B, we take the restriction of the order $\prec$ of $\mathcal{X}$. This is the new structure $\mathcal{Z}'$.

Obviously, adding the $\langle c_i, n \rangle$ does not introduce cycles, irreflexivity and rankedness are preserved. Moreover, any substructure of a cycle-free, irreflexive, ranked structure also has these properties, so $\mathcal{Z}'$ is $1 - \infty$ over Z, ranked and free of cycles.

We show that $\mathcal{Z}$ and $\mathcal{Z}'$ are equivalent. Let then $X \subseteq Z$, we have to prove $\mu(X) = \mu'(X)$ ($\mu := \mu_{\mathcal{Z}}, \mu' := \mu_{\mathcal{Z}'}$).

Let $z \in X - \mu(X)$. If $z \in C$ or $z \in A$, then $z \notin \mu'(X)$. If $z \in B$, let $\langle z, m \rangle$ be the chosen element. As $z \notin \mu(X)$, there is $x \in X$ s.t. some $\langle x, j \rangle \prec \langle z, m \rangle$. x cannot be in C. If $x \in A$, then also $\langle x, j \rangle \prec' \langle z, m \rangle$. If $x \in B$, then there is some $\langle x, k \rangle$ also in $\mathcal{X}'$. $\langle x, j \rangle \prec \langle x, k \rangle$ is impossible. If $\langle x, k \rangle \prec \langle x, j \rangle$, then $\langle z, m \rangle \succ \langle x, k \rangle$ by transitivity. If $\langle x, k \rangle \bot \langle x, j \rangle$, then also $\langle z, m \rangle \succ \langle x, k \rangle$ by rankedness. In any case, $\langle z, m \rangle \succ' \langle x, k \rangle$, and thus $z \notin \mu'(X)$.

Let $z \in X - \mu'(X)$. If $z \in C$ or $z \in A$, then $z \notin \mu(X)$. Let $z \in B$, and some $\langle x, j \rangle \prec' \langle z, m \rangle$. x cannot be in C, as they were sorted on top, so $\langle x, j \rangle$ exists in $\mathcal{X}$ too and $\langle x, j \rangle \prec \langle z, m \rangle$. But if any other $\langle z, i \rangle$ is also minimal in $\mathcal{Z}$ among the $\langle z, k \rangle$, then by rankedness also $\langle x, j \rangle \prec \langle z, i \rangle$, as $\langle z, i \rangle \bot \langle z, m \rangle$, so $z \notin \mu(X)$.

□

Assume in the sequel that $\mathcal{Y}$ contains all singletons and pairs, and fix $f : \mathcal{Y} \rightarrow \mathcal{P}(Z)$. We also fix the following notation: $A := \{x \in Z : f(x) = \emptyset\}$ and $B := Z - A$ (here and in future we sometimes write $f(x)$ for $f(\{x\})$, likewise $f(x, x') = x$ for $f(\{x, x'\}) = \{x\}$, etc., when the meaning is obvious).

Corollary 1.4.4

If f can be represented by a ranked $\mathcal{Z}$ free of cycles, then there is $\mathcal{Z}'$, which is also ranked and cycle-free, all $b \in B$ occur in 1 copy, all $a \in A$ ∞ often.

Fact 1.4.5

(1) If $\mathcal{Z}'$ is as in Corollary 1.4.4, $b \in B$, $a \in A$, $f(a,b) = b$, then for all $\langle a,i \rangle \in \mathcal{X}'$ $\langle a,i \rangle \succ' \langle b,0 \rangle$.

(2) If f can be represented by a cycle-free ranked $\mathcal{Z}$, then it has the "singleton property": If $x \in X$, then $x \notin f(X) \leftrightarrow \exists x' \in X. x \notin f(x,x')$.

(3) If f is as in 2), $b, b' \in B$, then $f(b,b') \neq \emptyset$.

Proof

(1) For no $\langle a,i \rangle$ $\langle b,0 \rangle \succ' \langle a,i \rangle$, since otherwise $f(a,b) = \emptyset$. If $\langle b,0 \rangle \perp \langle a,i \rangle$, then as there is $\langle a,j \rangle \prec \langle a,i \rangle$, $\langle a,j \rangle \prec' \langle b,0 \rangle$ by rankedness, contradiction.

(2) "$\leftarrow$" holds for all preferential structures. "$\rightarrow$": If $x \in A$, then $x \notin f(x,x)$. Let $x \in B$, $\mathcal{Z}$ a $1 - \infty$ over Z structure representing f as above. So there is just one copy of x in $\mathcal{X}$, $\langle x,0 \rangle$, and there is some $\langle y,j \rangle \prec \langle x,0 \rangle$, $y \in X$, thus $x \notin f(x,y)$.

(3) In any $1 - \infty$ over Z representation of f, $\langle b,0 \rangle \perp \langle b',0 \rangle$, or $\langle b,0 \rangle \prec \langle b',0 \rangle$, or $\langle b',0 \rangle \prec \langle b,0 \rangle$. $\langle b,0 \rangle \prec \langle b',0 \rangle \prec \langle b,0 \rangle$ cannot be, as this is a cycle.

□

We summarize in the following Lemma 1.4.6 some results for the general ranked case, many of them trivial.

Lemma 1.4.6

We assume here for simplicity that all elements occur in the structure.

(1) If $\mu(X) = \emptyset$, then each element $x \in X$ either has infinitely many copies, or below each copy of each x, there is an infinite descending chain of other elements.

(2) If there is no X such that $x \in \mu(X)$, then we can make infinitely many copies of x.

(3) There is no simple way to detect whether there is for all x some X such that $x \in \mu(X)$. More precisely: there is no normal finite characterization of ranked structures, in which each x in the domain occurs in at least one $\mu(X)$.

Suppose in the sequel that for each x there is some X such that $x \in \mu(X)$. (This is the hard case.)

(4) If the language is finite, then $X \neq \emptyset$ implies $\mu(X) \neq \emptyset$.

Suppose now the language to be infinite.

(5) If we admit all theories, then $\mu(M(T)) = M(T)$ for all complete theories.

1.4 Ranked Case

(6) It is possible to have $\mu(M(\phi)) = \emptyset$ for all formulas ϕ, even though all models occur in exactly one copy.

(7) If the domain is sufficiently rich, then we cannot have $\mu(X) = \emptyset$ for "many" X.

(8) We see that a small domain (see Case (6)) can have many X with $\mu(X) = \emptyset$, but if the domain is too dense (see Case (7)), then we cannot have many $\mu(X) = \emptyset$. (We do not know any criterion to distinguish poor from rich domains.)

(9) If we have all pairs in the domain, we can easily construct the ranking.

Proof

(1), (2), (4), (5), (9) are trivial, there is nothing to show for (8).

(3) Suppose there is a normal characterization Φ of such structures, where each element x occurs at least once in a set X such that $x \in \mu(X)$. Such a characterization will be a finite boolean combination of set expressions Φ, universally quantified, in the spirit of (AND), (RM) etc.

We consider a realistic counterexample - an infinite propositional language and the sets definable by formulas. We do not necessarily assume definability preservation, and work with full equality of results.

Take an infinite propositional language $p_i : i < \omega$. Choose an arbitrary model m, say $m \models p_i : i < \omega$.

Now, determine the height of any model m' as follows: $ht(m') := $ *the* first p_i such that $m(p_i) \neq m'(p_i)$, in our example then the first p_i such that $m' \models \neg p_i$. Thus, only m has infinite height, essentially, the more different m' is from m (in an alphabetical order), the lower it is.

Make now ω many copies of m, in infinite descending order, which you put on top of the rest.

Φ has to fail for some instantiation, as $\mathcal{X}$ does not have the desired property. Write this instantiation of Φ without loss of generality as a disjunction of conjunctions: $\bigvee(\bigwedge \phi_{i,j})$.

Each (consistent, or non-empty) component $\phi_{i,j}$ has finite height, more precisely: the minimum of all heigts of its models (which is a finite height). Thus, $\mathrel{\mid\!\sim} (\phi_{i,j})$ will be just the minimally high models of $\phi_{i,j}$ in this order.

Modify now $\mathcal{X}$ such that m has only 1 copy, and is just ($+1$ suffices) above the minimum of all the finitely many $\phi_{i,j}$. Then none of the $\mathrel{\mid\!\sim} (\phi_{i,j})$ is affected, and m has now finite height, say h, and is a minimum in any $M(\phi')$ where $\phi' = $ the conjunction of the first h values of m.

(Remark: Obviously, there are two easy generalizations for this ranking: First, we can go beyond ω (but also stay below ω), second, instead of taking just one m as a scale, and which has maximal height, we can take a set M of models: $ht(m')$ is then the first p_i where $m'(p_i)$ is different from *all* $m \in M$. Note that in this case, in general, not all levels need to be filled. If e.g., $m_0, m_1 \in M$, and $m_0(p_0) = false$, $m_1(p_0) = true$, then level 0 will be empty.)

(6) Let the p_i again define an infinite language. Denote by p_i^+ the set of all $+p_j$, where $j > i$. Let T be the usual tree of models (each model is a branch) for the p_i, with an artificial root $*$. Let the first model (= branch) be $*^+$, i.e., the leftest branch in the obvious way of drawing it. Next, we choose $\neg p_0^+$, i.e., we go right, and then all the way left. Next, we consider the 4 sequences of $+/- p_0$, $+/- p_1$, two of them were done already, both ending in p_1^+, and choose the remaining two, both ending in $\neg p_1^+$, i.e., the positive prolongations of $p_0, \neg p_1$ and $\neg p_0, \neg p_1$. Thus, at each level, we take all possible prolongations, the positive ones were done already, and we count those, which begin negatively, and then continue positively. Each formula has in this counting arbitrarily big models.

This is not yet a full enumeration of all models, e.g., the branch with all models negative will never be enumerated. But it suffices for our purposes.

Reverse the order so far constructed, and put the models not enumerated on top. Then all models are considered, and each formula has arbitrarily small models, thus $\mu(\phi) = \emptyset$ for all ϕ.

(7) Let the domain contain all singletons, and let the structure be without copies. The latter can be seen by considering singletons. Suppose now there is a set X in the domain such that $\mu(X) = \emptyset$. Thus, each $x \in X$ must have infinitely many $x' \in X$ $x' \prec x$. Suppose $\mathcal{P}(X)$ is a subset of the domain. Then there must be infinite $Y \in \mathcal{P}(X)$ such that $\mu(Y) \neq \emptyset$: Suppose not. Let $\prec$ be the ranking order. Choose arbitrary $x \in X$. Consider $X' := \{x' \in X : x \prec x'\}$, then $x \in \mu(X')$, and not all such X' can be finite - assuming X is big enough, e.g., uncountable.

□

Representation

Fact 1.4.7

In all ranked structures, $(\mu \subseteq)$, $(\mu =)$, (μPR), $(\mu =')$, $(\mu \parallel)$, $(\mu \cup)$, $(\mu \cup')$, $(\mu \in)$, $(\mu RatM)$ will hold, if the corresponding closure conditions are satisfied.

Proof

Exercise, solution in [Sch04], Fact 3.10.8.

We show results for the case without copies, (Proposition 1.4.8 and Proposition 1.4.9), then negative results for the general case (Fact 1.4.10), and conclude with a characterization of the general case (Proposition 1.4.11).

More details can be found in [Sch04], section 3.10.2.2.

1.4 Ranked Case

Proposition 1.4.8

Let $\mathcal{Y} \subseteq \mathcal{P}(U)$ be closed under finite unions. Then $(\mu \subseteq), (\mu\emptyset), (\mu =)$ characterize ranked structures for which for all $X \in \mathcal{Y}$ $X \neq \emptyset \to \mu_<(X) \neq \emptyset$ hold, i.e. $(\mu \subseteq)$, $(\mu\emptyset), (\mu =)$ hold in such structures for $\mu_<$, and if they hold for some μ, we can find a ranked relation $<$ on U s.t. $\mu = \mu_<$. Moreover, the structure can be choosen $\mathcal{Y}$-smooth.

Proof

Exercise, solution see [Sch04], Proposition 3.10.11.

For the following representation result, we assume only $(\mu\emptyset fin)$, but the domain has to contain singletons.

Proposition 1.4.9

Let $\mathcal{Y} \subseteq \mathcal{P}(U)$ be closed under finite unions, and contain singletons. Then $(\mu \subseteq)$, $(\mu\emptyset fin), (\mu =), (\mu \in)$ characterize ranked structures for which for all finite $X \in \mathcal{Y}$ $X \neq \emptyset \to \mu_<(X) \neq \emptyset$ hold, i.e. $(\mu \subseteq), (\mu\emptyset fin), (\mu =), (\mu \in)$ hold in such structures for $\mu_<$, and if they hold for some μ, we can find a ranked relation $<$ on U s.t. $\mu = \mu_<$.

Proof

Exercise, solution see [Sch04], Proposition 3.10.12.

Note that the prerequisites of Proposition 1.4.9 hold in particular in the case of ranked structures without copies, where all elements of U are present in the structure - we need infinite descending chains to have $\mu(X) = \emptyset$ for $X \neq \emptyset$.

We turn now to the general case, where every element may occur in several copies.

Fact 1.4.10

(1) $(\mu \subseteq) + (\mu PR) + (\mu =) + (\mu \cup) + (\mu \in)$ do not imply representation by a ranked structure.

(2) The infinitary version of $(\mu \parallel)$:

$(\mu \parallel \infty)$ $\mu(\bigcup\{A_i : i \in I\}) = \bigcup\{\mu(A_i) : i \in I'\}$ for some $I' \subseteq I$.

will not always hold in ranked structures.

Proof

Exercise, solution see [Sch04], Fact 3.10.13.

We assume again the existence of singletons for the following representation result.

Proposition 1.4.11

Let $\mathcal{Y}$ be closed under finite unions and contain singletons. Then $(\mu \subseteq) + (\mu PR) + (\mu \parallel) + (\mu \cup) + (\mu \in)$ characterize ranked structures.

Proof

Exercise, solution see [Sch04], Proposition 3.10.14.

Smooth Ranked Structures

We assume that all elements occur in the structure, so smoothness and $\mu(X) \neq \emptyset$ for $X \neq \emptyset$ coincide.

The abstract Definition 1.4.3 is motivated by the distinction

Definition 1.4.2

$\approx (u)$, the set of $u' \in W$ which have same rank as u,

$\prec (u)$, the set of $u' \in W$ which have lower rank than u,

$\succ (u)$, the set of $u' \in W$ which have higher rank than u.

(All other $u' \in W$ will by default have unknown rank in comparison.)

We can diagnose e.g., $u' \in \approx (u)$ if $u, u' \in \mu(X)$ for some X, and $u' \in \succ (u)$ if $u \in \mu(X)$ and $u' \in X - \mu(X)$ for some X.

If we sometimes do not know more, we will have to consider also $\preceq (u)$ and $\succeq (u)$ - this will be needed in Section 4.3.5.2, where we will have only incomplete information, due to hidden dimensions.

All other $u' \in W$ will by default have unknown rank in comparison.

Definition 1.4.3

(1) Define for each $u \in W$ three subsets of W $\approx (u)$, $\prec (u)$, and $\succ (u)$. Let $\mathcal{O}$ be the set of all these subsets, i.e., $\mathcal{O} := \{\approx (u), \prec (u), \succ (u) : u \in W\}$

(2) We say that $\mathcal{O}$ is generated by a choice function f iff

 (1) $\forall U \in \mathcal{Y} \forall x, x' \in f(U)\ x' \in \approx (x)$,

 (2) $\forall U \in \mathcal{Y} \forall x \in f(U) \forall x' \in U - f(U)\ x' \in \succ (x)$

(3) $\mathcal{O}$ is said to be representable by a ranking iff there is a function $f : W \to \langle O, \blacktriangleleft \rangle$ into a total order $\langle O, \blacktriangleleft \rangle$ such that

 (1) $u' \in \approx (u) \Rightarrow f(u') = f(u)$

 (2) $u' \in \prec (u) \Rightarrow f(u') \blacktriangleleft f(u)$

 (3) $u' \in \succ (u) \Rightarrow f(u') \blacktriangleright f(u)$

(4) Let $\mathcal{C}(\mathcal{O})$ be the closure of $\mathcal{O}$ under the following operations:

1.4 Ranked Case

- $u \in \approx (u)$,
- if $u' \in \approx (u)$, then $\approx (u) = \approx (u')$, $\prec (u) = \prec (u')$, $\succ (u) = \succ (u')$,
- $u' \in \prec (u)$ iff $u \in \succ (u')$,
- $u \in \prec (u')$, $u' \in \prec (u'') \Rightarrow u \in \prec (u'')$,

  or, equivalently,

- $u \in \prec (u') \Rightarrow \prec (u') \subseteq \prec (u)$.

Note that we will generally loose much ignorance in applying the next two Facts.

Fact 1.4.12

A partial (strict) order on W can be extended to a total (strict) order.

Proof

Take an arbitrary enumeration of all pairs a, b of W : $\langle a, b \rangle_i : i \in \kappa$. Suppose all $\langle a, b \rangle_j$ for $j < i$ have been ordered, and we have no information if $a \prec b$ or $a \approx b$ or $a \succ b$. Choose arbitrarily $a \prec b$. A contradiction would be a (finite) cycle involving $\prec$. But then we would have known already that $b \preceq a$.

□

Fact 1.4.13

$\mathcal{O}$ can be represented by a ranking iff in $\mathcal{C}(\mathcal{O})$ the sets $\approx (u)$, $\prec (u)$, $\succ (u)$ are pairwise disjoint.

Proof

(Outline) By the construction of $\mathcal{C}(\mathcal{O})$ and disjointness, there are no cycles involving $\prec$. Extend the relation by Lemma 1.2.3. Let the $\approx (u)$ be the equivalence classes. Define $\approx (u) \blacktriangleleft \approx (u')$ iff $u \in \prec (u')$.

□

Proposition 1.4.14

Let $f : \mathcal{Y} \to \mathcal{P}(W)$. f is representable by a smooth ranked structure iff in $\mathcal{C}(\mathcal{O})$ the sets $\approx (u)$, $\prec (u)$, $\succ (u)$ are pairwise disjoint, where $\mathcal{O}$ is the system generated by f, as in Definition 1.4.3.

Proof

If the sets are not pairwise disjoint, we have a cycle. If not, use Fact 1.4.13.

□

1.4.2 $\mathcal{A}$-Ranked Structures

1.4.2.1 Introduction

We do now the completeness proofs for the preferential part of hierarchical conditionals. All motivation etc. will be found in Section 7.5.

First the basic semantical definition:

Definition 1.4.4

Let A be a fixed set, and $\mathcal{A}$ a finite, totally ordered (by $<$) disjoint cover by non-empty subsets of A.

- For $x \in A$, let $rg(x)$ the unique $A \in \mathcal{A}$ such that $x \in A$, so $rg(x) < rg(y)$ is defined in the natural way.
- A preferential structure $\langle \mathcal{X}, \prec \rangle$ ($\mathcal{X}$ a set of pairs $\langle x, i \rangle$) is called $\mathcal{A}$-ranked iff for all x, x' $rg(x) < rg(x')$ implies $\langle x, i \rangle \prec \langle x', i' \rangle$ for all $\langle x, i \rangle, \langle x', i' \rangle \in \mathcal{X}$.

1.4.2.2 Representation Results for $\mathcal{A}$-Ranked Structures

Discussion

The not necessarily smooth and the smooth case will be treated differently.

Strangely, the smooth case is simpler, as an added new layer in the proof settles it. Yet, this is not surprising when looking closer, as minimal elements never have higher rank, and we know from (μCUM) that minimizing by minimal elements suffices. All we have to add that any element in the minimal layer minimizes any element higher up.

In the simple, not necessarily smooth, case, we have to go deeper into the original proof to obtain the result.

The following idea, inspired by the treatment of the smooth case, will not work: Instead of minimizing by arbitrary elements, minimize only by elements of minimal rank, as the following example shows. If it worked, we might add just another layer to the original proof without $(\mu \mathcal{A})$, (see Definition 1.4.5), as in the smooth case.

Example 1.4.1

Consider the base set $\{a, b, c\}$, $\mu(\{a, b, c\}) = \{b\}$, $\mu(\{a, b\}) = \{a, b\}$, $\mu(\{a, c\}) = \emptyset$, $\mu(\{b, c\}) = \{b\}$, $\mathcal{A}$ defined by $\{a, b\} < \{c\}$.

Obviously, $(\mu \mathcal{A})$ (see Definition 1.4.5) is satisfied. μ can be represented by the (not transitive!) relation $a \prec c \prec a$, $b \prec c$, which is $\mathcal{A}$-ranked.

1.4 Ranked Case

But trying to minimize a in $\{a, b, c\}$ in the minimal layer will lead to $b \prec a$, and thus $a \notin \mu(\{a, b\})$, which is wrong.

$\square$

The proofs of the general and transitive general case are (minor) adaptations of the proofs in Section 1.5.3.2. For the smooth case, we only have to add a supplementary layer in the end (Fact 1.4.22), which will make the construction $\mathcal{A}$-ranked.

In the following, we will assume the partition $\mathcal{A}$ to be given. We could also construct it from the properties of μ, but this would need stronger closure properties of the domain. The construction of $\mathcal{A}$ is more difficult than the construction of the ranking in fully ranked structures, as $x \in \mu(X), y \in X - \mu(X)$ will guarantee only $rg(x) \leq rg(y)$, and not $rg(x) < rg(y)$, as is the case in the latter situation. This corresponds to the separate treatment of the α and other formulas in the logical version, discussed in Section 1.4.2.2.

$\mathcal{A}$-Ranked General and Transitive Structures

We will show here the following representation result:

Let $\mathcal{A}$ be given.

An operation $\mu : \mathcal{Y} \to \mathcal{P}(Z)$ is representable by an $\mathcal{A}$−ranked preferential structure iff μ satisfies $(\mu \subseteq), (\mu PR), (\mu \mathcal{A})$ (Proposition 1.4.16), and, moreover, the structure can be chosen transitive (Proposition 1.4.18).

Note that we carefully avoid any unnecessary assumptions about the domain $\mathcal{Y} \subseteq \mathcal{P}(Z)$ of the function μ.

Definition 1.4.5

We define a new condition:

Let $\mathcal{A}$ be given as defined in Definition 1.4.4.

$(\mu \mathcal{A})$ If $X \in \mathcal{Y}, A, A' \in \mathcal{A}, A < A', X \cap A \neq \emptyset, X \cap A' \neq \emptyset$ then $\mu(X) \cap A' = \emptyset$.

This new condition will be central for the modified representation.

The Basic, Not Necessarily Transitive, Case

Corollary 1.4.15

Let $\mu : \mathcal{Y} \to \mathcal{P}(Z)$ satisfy $(\mu \subseteq), (\mu PR), (\mu \mathcal{A})$, and let $U \in \mathcal{Y}$.
If $x \in U$ and $\exists x' \in U.rg(x') < rg(x)$, then $\forall f \in \Pi_x.ran(f) \cap U \neq \emptyset$.

Proof

Exercise, solution in the Appendix.

Proposition 1.4.16

Let $\mathcal{A}$ be given.

An operation $\mu : \mathcal{Y} \to \mathcal{P}(Z)$ is representable by an $\mathcal{A}$-ranked preferential structure iff μ satisfies $(\mu \subseteq)$, (μPR), $(\mu \mathcal{A})$.

Proof

One direction is trivial. The central argument is: If $a \prec b$ in X, and $X \subseteq Y$, then $a \prec b$ in Y, too.

We turn to the other direction. The preferential structure is defined in Construction 1.4.1, Claim 1.4.17 shows representation.

Construction 1.4.1

Let $\mathcal{X} := \{\langle x, f \rangle : x \in Z \wedge f \in \Pi_x\}$, and $\langle x', f' \rangle \prec \langle x, f \rangle :\leftrightarrow x' \in ran(f)$ or $rg(x') < rg(x)$.

Note that, as $\mathcal{A}$ is given, we also know $rg(x)$.

Let $\mathcal{Z} := \langle \mathcal{X}, \prec \rangle$.

Obviously, $\mathcal{Z}$ is $\mathcal{A}$-ranked.

Claim 1.4.17

For $U \in \mathcal{Y}$, $\mu(U) = \mu_{\mathcal{Z}}(U)$.

Proof

By Claim 1.3.3, it suffices to show that for all $U \in \mathcal{Y}$ $x \in \mu_{\mathcal{Z}}(U) \leftrightarrow x \in U$ and $\exists f \in \Pi_x.ran(f) \cap U = \emptyset$. So let $U \in \mathcal{Y}$.

- "$\Rightarrow$": If $x \in \mu_{\mathcal{Z}}(U)$, then there is $\langle x, f \rangle$ minimal in $\mathcal{X} \upharpoonright U$ - where $\mathcal{X} \upharpoonright U := \{\langle x, i \rangle \in \mathcal{X} : x \in U\}$), so $x \in U$, and there is no $\langle x', f' \rangle \prec \langle x, f \rangle$, $x' \in U$, so by $\Pi_{x'} \neq \emptyset$ there is no $x' \in ran(f)$, $x' \in U$, but then $ran(f) \cap U = \emptyset$.
- "$\Leftarrow$": If $x \in U$, and there is $f \in \Pi_x$, $ran(f) \cap U = \emptyset$, then by Corollary 1.4.15, there is no $x' \in U$, $rg(x') < rg(x)$, so $\langle x, f \rangle$ is minimal in $\mathcal{X} \upharpoonright U$.

$\Box$ (Claim 1.4.17 and Proposition 1.4.16)

The Transitive Case

Proposition 1.4.18

Let $\mathcal{A}$ be given.

1.4 Ranked Case

An operation $\mu : \mathcal{Y} \to \mathcal{P}(Z)$ is representable by an $\mathcal{A}$–ranked transitive preferential structure iff μ satisfies $(\mu \subseteq)$, (μPR), $(\mu \mathcal{A})$.

Construction 1.4.2

(1) For $x \in Z$, let T_x be the set of trees t_x s.t.

    (a) all nodes are elements of Z,

    (b) the root of t_x is x,

    (c) $height(t_x) \leq \omega$,

    (d) if y is an element in t_x, then there is $f \in \Pi_y := \Pi\{Y \in \mathcal{Y} : y \in Y - \mu(Y)\}$ s.t. the set of children of y is $ran(f) \cup \{y' \in Z : rg(y') < rg(y)\}$.

(2) For $x, y \in Z$, $t_x \in T_x$, $t_y \in T_y$, set $t_x \triangleright t_y$ iff y is a (direct) child of the root x in t_x, and t_y is the subtree of t_x beginning at y.

(3) Let $\mathcal{Z} := \langle \{\langle x, t_x \rangle : x \in Z, t_x \in T_x\}, \langle x, t_x \rangle \succ \langle y, t_y \rangle$ iff $t_x \triangleright t_y \rangle$.

Fact 1.4.19

(1) The construction ends at some y iff $\mathcal{Y}_y = \emptyset$ and there is no y' s.t. $rg(y') < rg(y)$, consequently $T_x = \{x\}$ iff $\mathcal{Y}_x = \emptyset$ and there are no x' with lesser rang. (We identify the tree of height 1 with its root.)

(2) We define a special tree tc_x for all x : For all nodes y in tc_x, the successors are as follows:

if $\mathcal{Y}_y \neq \emptyset$, then z is an successor iff $z = y$ or $rg(z) < rg(y)$

if $\mathcal{Y}_y = \emptyset$, then z is an successor iff $rg(z) < rg(y)$.

(In the first case, we make $f \in \mathcal{Y}_y$ always choose y itself.) tc_x is an element of T_x. Thus, with (1), $T_x \neq \emptyset$ for any x. Note: $tc_x = x$ iff $\mathcal{Y}_x = \emptyset$ and x has minimal rang.

(3) If $f \in \Pi_x$, then the tree tf_x with root x and otherwise composed of the subtrees tc_y for $y \in ran(f) \cup \{y' : rg(y') < rg(y)\}$ is an element of T_x. (Level 0 of tf_x has x as element, the $t'_y s$ begin at level 1.)

(4) If y is an element in t_x and t_y the subtree of t_x starting at y, then $t_y \in T_y$.

(5) $\langle x, t_x \rangle \succ \langle y, t_y \rangle$ implies $y \in ran(f) \cup \{x' : rg(x') < rg(x)\}$ for some $f \in \Pi_x$.

$\Box$

Claim 1.4.20 shows basic representation.

Claim 1.4.20

$\forall U \in \mathcal{Y}. \mu(U) = \mu_{\mathcal{Z}}(U)$

Proof

By Claim 1.3.3, it suffices to show that for all $U \in \mathcal{Y}$ $x \in \mu_{\mathcal{Z}}(U) \leftrightarrow x \in U \land \exists f \in \Pi_x.ran(f) \cap U = \emptyset$.

Fix $U \in \mathcal{Y}$.

"$\Rightarrow$": $x \in \mu_{\mathcal{Z}}(U) \to$ ex. $\langle x, t_x \rangle$ minimal in $\mathcal{Z} \upharpoonright U$, thus $x \in U$ and there is no $\langle y, t_y \rangle \in \mathcal{Z}$, $\langle y, t_y \rangle \prec \langle x, t_x \rangle$, $y \in U$. Let f define the first part of the set of children of the root x in t_x. If $ran(f) \cap U \neq \emptyset$, if $y \in U$ is a child of x in t_x, and if t_y is the subtree of t_x starting at y, then $t_y \in T_y$ and $\langle y, t_y \rangle \prec \langle x, t_x \rangle$, contradicting minimality of $\langle x, t_x \rangle$ in $\mathcal{Z} \upharpoonright U$. So $ran(f) \cap U = \emptyset$.

"$\Leftarrow$": Let $x \in U$, and $\exists f \in \Pi_x.ran(f) \cap U = \emptyset$. By Corollary 1.4.15, there is no $x' \in U$, $rg(x') < rg(x)$. If $\mathcal{Y}_x = \emptyset$, then the tree tc_x has no $\triangleright$-successors in U, and $\langle x, tc_x \rangle$ is $\succ$-minimal in $\mathcal{Z} \upharpoonright U$. If $\mathcal{Y}_x \neq \emptyset$ and $f \in \Pi_x$ s.t. $ran(f) \cap U = \emptyset$, then $\langle x, tf_x \rangle$ is again $\succ$-minimal in $\mathcal{Z} \upharpoonright U$.

$\square$

We consider now the transitive closure of $\mathcal{Z}$. (Recall that $\prec^*$ denotes the transitive closure of $\prec$.) Claim 1.4.21 shows that transitivity does not destroy what we have achieved. The trees tf_x play a crucial role in the demonstration.

Claim 1.4.21

Let $\mathcal{Z}' := \langle \{\langle x, t_x \rangle : x \in Z, t_x \in T_x\}, \langle x, t_x \rangle \succ \langle y, t_y \rangle$ iff $t_x \triangleright^* t_y \rangle$. Then $\mu_{\mathcal{Z}} = \mu_{\mathcal{Z}'}$.

Proof

Suppose there is $U \in \mathcal{Y}$, $x \in U$, $x \in \mu_{\mathcal{Z}}(U)$, $x \notin \mu_{\mathcal{Z}'}(U)$. Then there must be an element $\langle x, t_x \rangle \in \mathcal{Z}$ with no $\langle x, t_x \rangle \succ \langle y, t_y \rangle$ for any $y \in U$. Let $f \in \Pi_x$ determine the first part of the set of children of x in t_x, then $ran(f) \cap U = \emptyset$, consider tf_x. All elements $w \neq x$ of tf_x are already in $ran(f)$, or $rg(w) < rg(x)$ holds. (Note that the elements chosen by rang in tf_x continue by themselves or by another element of even smaller rang, but the rang order is transitive.) But all w s.t. $rg(w) < rg(x)$ were already successors at level 1 of x in tf_x. By Corollary 1.4.15, there is no $w \in U$, $rg(w) < rg(x)$. Thus, no element $\neq x$ of tf_x is in U. Thus there is no $\langle z, t_z \rangle \prec^* \langle x, tf_x \rangle$ in $\mathcal{Z}$ with $z \in U$, so $\langle x, tf_x \rangle$ is minimal in $\mathcal{Z}' \upharpoonright U$, contradiction.

$\square$ (Claim 1.4.21 and Proposition 1.4.18)

$\mathcal{A}$-Ranked Smooth Structures

All smooth cases have a simple solution. We use one of our existing proofs for the not necessarily $\mathcal{A}$-ranked case, and add one litte result:

1.4 Ranked Case

Fact 1.4.22

Let $(\mu\mathcal{A})$ hold, and let $\mathcal{Z} = \langle \mathcal{X}, \prec \rangle$ be a smooth preferential structure representing μ, i.e. $\mu = \mu_\mathcal{Z}$.

Suppose that

$\langle x, i \rangle \prec \langle y, j \rangle$ implies $rg(x) \leq rg(y)$.

Define $\mathcal{Z}' := \langle \mathcal{X}, \sqsubset \rangle$ where $\langle x, i \rangle \sqsubset \langle y, j \rangle$ iff $\langle x, i \rangle \prec \langle y, j \rangle$ or $rg(x) < rg(y)$.

Then $\mathcal{Z}'$ is $\mathcal{A}$-ranked.

$\mathcal{Z}'$ is smooth, too, and $\mu_\mathcal{Z} = \mu_{\mathcal{Z}'} =: \mu'$.

In addition, if $\prec$ is free from cycles, so is $\sqsubset$, if $\prec$ is transitive, so is $\sqsubset$.

Proof

$\mathcal{A}$-rankedness is trivial.

Suppose $\langle x, i \rangle$ is $\prec$-minimal, but not $\sqsubset$-minimal. Then there must be $\langle y, j \rangle \sqsubset \langle x, i \rangle$, $\langle y, j \rangle \not\prec \langle x, i \rangle$, $y \in X$, so $rg(y) < rg(x)$. By $(\mu\mathcal{A})$, all $x \in \mu(X)$ have minimal $\mathcal{A}$-rang among the elements of X, so this is impossible. Thus, μ-minimal elements stay μ'-minimal, so smoothness will also be preserved - remember that we increased the relation.

By prerequisite, there cannot be any cycle involving only $\prec$, but the rang order is free from cycles, too, and $\prec$ respects the rang order, so $\sqsubset$ is free from cycles.

Let $\prec$ be transitive, so is the rang order. But if $\langle x, i \rangle \prec \langle y, j \rangle$ and $rg(y) < rg(z)$ for some $\langle z, k \rangle$, then by prerequisite $rg(x) \leq rg(y)$, so $rg(x) < rg(z)$, so $\langle x, i \rangle \sqsubset \langle z, k \rangle$ by definition. Likewise for $rg(x) < rg(y)$ and $\langle y, j \rangle \prec \langle z, k \rangle$.

$\square$

All that remains to show then is that our constructions of smooth and of smooth and transitive structures satisfy the condition

$\langle x, i \rangle \prec \langle y, j \rangle$ implies $rg(x) \leq rg(y)$.

Proposition 1.4.23

Let $\mathcal{A}$ be given.

Let - for simplicity - $\mathcal{Y}$ be closed under finite unions, and $\mu : \mathcal{Y} \to \mathcal{P}(Z)$. Then there is a $\mathcal{Y}$-smooth $\mathcal{A}$-ranked preferential structure $\mathcal{Z}$, s.t. for all $X \in \mathcal{Y}$ $\mu(X) = \mu_\mathcal{Z}(X)$ iff μ satisfies $(\mu \subseteq)$, (μPR), (μCUM), $(\mu\mathcal{A})$.

Proof

Exercise, solution in the Appendix.

Proposition 1.4.24

Let $\mathcal{A}$ be given.

Let - for simplicity - $\mathcal{Y}$ be closed under finite unions, and $\mu : \mathcal{Y} \to \mathcal{P}(Z)$. Then there is a $\mathcal{Y}$-smooth $\mathcal{A}$-ranked transitive preferential structure $\mathcal{Z}$, s.t. for all $X \in \mathcal{Y}$ $\mu(X) = \mu_{\mathcal{Z}}(X)$ iff μ satisfies $(\mu \subseteq)$, (μPR), (μCUM), $(\mu \mathcal{A})$.

Proof

Consider the construction in the proof of Proposition 3.3.8 in [Sch04].

Thus, we only have to show that in $\mathcal{Z}$ defined by

$\mathcal{Z} := \langle \{\langle x, t_x \rangle : x \in Z, t_x \in T_x\}, \langle x, t_x \rangle \succ \langle y, t_y \rangle$ iff $t_x \triangleright^* t_y \rangle$, $t_x \triangleright t_y$ implies $rg(y) \leq rg(x)$.

But by construction of the trees, $x_n \in Y_{n+1}$, and $x_{n+1} \in \mu(U_n \cup Y_{n+1})$, so $rg(x_{n+1}) \leq rg(x_n)$.

□ (Proposition 1.4.24)

The Logical Properties with Definability Preservation

First, a small fact about the $\mathcal{A}$.

Fact 1.4.25

Let $\mathcal{A}$ be as above (and thus finite). Then each A_i is equivalent to a formula α_i.

Proof

We use the standard topology and its compactness. By definition, each $M(A_i)$ is closed, by finiteness all unions of such $M(A_i)$ are closed, too, so $C(M(A_i))$ is closed. By compactness, each open cover $X_j : j \in J$ of the clopen $M(A_i)$ contains a finite subcover, so also $\bigcup \{M(A_j) : j \neq i\}$ has a finite open cover. But the $M(\phi)$, ϕ a formula form a basis of the closed sets, so we are done.

□

Proposition 1.4.26

Let $\mathrel{|\!\sim}$ be a logic for $\mathcal{L}$. Set $T^{\mathcal{M}} := Th(\mu_{\mathcal{M}}(M(T)))$, where $\mathcal{M}$ is a preferential structure.

(1) Then there is a (transitive) definability preserving classical preferential model $\mathcal{M}$ s.t. $\overline{\overline{T}} = T^{\mathcal{M}}$ iff

(LLE), (CCL), (SC), (PR) hold for all $T, T' \subseteq \mathcal{L}$.

(2) The structure can be chosen smooth, iff, in addition

(CUM) holds.

1.4 Ranked Case

(3) The structure can be chosen $\mathcal{A}$-ranked, iff, in addition

($\mathcal{A}$-min) $T \not\vdash \neg\alpha_i$ and $T \not\vdash \neg\alpha_j$, $i < j$ implies $\overline{\overline{T}} \vdash \neg\alpha_j$

holds.

The proof is an immediate consequence of Proposition 1.4.27 and the respective above results. This proposition (or its analogue) was mostly already shown in [Sch92] and [Sch96-1] and is repeated here for completeness' sake, but with a new and partly stronger proof.

Proposition 1.4.27

Consider for a logic $\vdash\!\sim$ on $\mathcal{L}$ the properties

$(LLE), (CCL), (SC), (PR), (CUM), (\mathcal{A}\text{-min})$ hold for all $T, T' \subseteq \mathcal{L}$.

and for a function $\mu : \mathbf{D}_{\mathcal{L}} \to \mathcal{P}(M_{\mathcal{L}})$ the properties

(μdp) μ is definability preserving, i.e. $\mu(M(T)) = M(T')$ for some T'

$(\mu \subseteq), (\mu PR), (\mu CUM), (\mu\mathcal{A})$

for all $X, Y \in \mathbf{D}_{\mathcal{L}}$.

It then holds:

(a) If μ satisfies $(\mu dp), (\mu \subseteq), (\mu PR)$, then $\vdash\!\sim$ defined by $\overline{\overline{T}} := T^\mu := Th(\mu(M(T)))$ satisfies $(LLE), (CCL), (SC), (PR)$. If μ satisfies in addition (μCUM), then (CUM) will hold, too. If μ satisfies in addition $(\mu\mathcal{A})$, then ($\mathcal{A}$-min) will hold, too.

(b) If $\vdash\!\sim$ satisfies $(LLE), (CCL), (SC), (PR)$, then there is $\mu : \mathbf{D}_{\mathcal{L}} \to \mathcal{P}(M_{\mathcal{L}})$ s.t. $\overline{\overline{T}} = T^\mu$ for all $T \subseteq \mathcal{L}$ and μ satisfies $(\mu dp), (\mu \subseteq), (\mu PR)$. If, in addition, (CUM) holds, then (μCUM) will hold, too. If, in addition, ($\mathcal{A}$-min) holds, then $(\mu\mathcal{A})$ will hold, too.

Proof

All properties except ($\mathcal{A}$-min) and $(\mu\mathcal{A})$ are shown in Proposition 1.2.18. But the remaining two are trivial.

□

1.5 The Smooth Case Without Domain Closure

1.5.1 Introduction

We begin with two sequent calculi which do not have closure of the (semantical) domain under finite unions, and which cannot be represented by a smooth model, despite the usual Cumulativity property.

This leads to an analysis of the situation, and a replacement of Cumulativity by an infinity of conditions, which guarantee representation by a smooth structure.

The present proof follows closely the proof in Section 1.3.3, using more complicated hulls, which now depend on the element, too: $H(U, x)$. We replace the old Fact 1.3.11 by sufficiently strong conditions on $H(U, x)$, see Definition 1.5.8 of the property (HU).

It seems to be an open problem to characterize transitive smooth structures without closure of the domain under finite unions - our proof of the transitive case uses unions.

See [Sch04], section 3.7.1 for more comments.

1.5.2 Problems Without Closure Under Finite Union - two Sequent Calculi

1.5.2.1 Introduction

We discuss here two sequent calculi, which have no "or" on the left hand side, and thus, semantically speaking, no closure of the domain under finite unions. This leads to incompleteness of certain characterisations, and led the author to elaborate the importance of domain closure conditions.

More discussion can be found in [Sch04].

1.5.2.2 Introduction to Plausibility Logic

In Hilbert style axiomatisations, the axioms describe the universe U, the set of all models. E.g. the axiom $\phi \to (\psi \to \phi)$ says that $M(\phi \to (\psi \to \phi)) = U$. The rules, e.g. MP say: $X \cap Y \subseteq Z$, if $x \in M(\phi)$ and $x \in M(\phi \to \psi)$, then $x \in M(\psi)$, they describe inclusion. In sequent calculi, we try to find all X, Y, s.t. $X \subseteq Y$, i.e. we characterize inclusion.

1.5 The Smooth Case Without Domain Closure

Plausibility logic was introduced by D. Lehmann [Leh92a], [Leh92b] as a sequent calculus in a propositional language without connectives. Thus, a plausibility logic language $\mathcal{L}$ is just a set, whose elements correspond to propositional variables, and a sequent has the form $X \mathrel{|\!\sim} Y$, where X, Y are finite subsets of $\mathcal{L}$, thus, in the intuitive reading, $\bigwedge X \mathrel{|\!\sim} \bigvee Y$.

We show here

- a proof that the weak system is complete for general preferential structures,
- a counterexample which shows that the full system is not complete for smooth preferential structures,

More details can be found in [Sch04], section 3.7.1.

The Details

We abuse notation, and write $X \mathrel{|\!\sim} a$ for $X \vdash \{a\}$, $X, a \mathrel{|\!\sim} Y$ for $X \cup \{a\} \mathrel{|\!\sim} Y$, $ab \mathrel{|\!\sim} Y$ for $\{a,b\} \mathrel{|\!\sim} Y$, etc. When discussing plausibility logic, X, Y, etc. will denote finite subsets of $\mathcal{L}$, a, b, etc. elements of $\mathcal{L}$.

We first define the logical properties we will examine.

Definition 1.5.1

X and Y will be finite subsets of $\mathcal{L}$, a, etc. elements of $\mathcal{L}$. The base axiom and rules of plausibility logic are (we use the prefix "Pl" to differentiate them from the usual ones):

(PlI) (Inclusion): $X \mathrel{|\!\sim} a$ for all $a \in X$,

(PlRM) (Right Monotony): $X \mathrel{|\!\sim} Y \Rightarrow X \mathrel{|\!\sim} a, Y$,

(PlCLM) (Cautious Left Monotony): $X \mathrel{|\!\sim} a, X \mathrel{|\!\sim} Y \Rightarrow X, a \mathrel{|\!\sim} Y$,

(PlCC) (Cautious Cut): $X, a_1 \ldots a_n \mathrel{|\!\sim} Y$, and for all $1 \leq i \leq n$ $X \mathrel{|\!\sim} a_i, Y \Rightarrow X \mathrel{|\!\sim} Y$,

and as a special case of (PlCC):

(PlUCC) (Unit Cautious Cut): $X, a \mathrel{|\!\sim} Y, X \mathrel{|\!\sim} a, Y \Rightarrow X \mathrel{|\!\sim} Y$.

and we denote by PL, for plausibility logic, the full system, i.e. $(PlI) + (PlRM) + (PlCLM) + (PlCC)$.

We will show in this section that:

(1) The system $(PlI) + (PlRM) + (PlCC)$ is sound and complete for minimal preferential models (as adapted to plausibility logic).

(2) The system $(PlI) + (PlRM) + (PlCC) + (PlCLM)$ is not complete for smooth minimal preferential models, due to the absence of an "or" on the left hand side of $\mathrel{|\!\sim}$, so the sets of models of plausibility formulas are not closed under finite union, violating one of the prerequisites of Proposition 1.3.14.

(3) We show how to mend the representation result for smooth structures to work without closure under finite unions.

Before we turn to the main results, we show a considerable simplification of plausibility logic (which will be used later), and introduce the main definition.

Fact 1.5.1

We note that (partially by finiteness of all sets involved)

(1) In the presence of (PlRM), (PlI) is equivalent to (PlI'): $X \mathrel{|\!\sim} Y$ for $X \cap Y \neq \emptyset$.
(2) (PlRM) is equivalent to $(PlRM')$: $X \mathrel{|\!\sim} Y \to X \mathrel{|\!\sim} Z, Y$ for all Z.
(3) (PlCC) is equivalent to $(PlCC')$: $X, Z \mathrel{|\!\sim} Y$ and $X \mathrel{|\!\sim} z, Y$ for all $z \in Z \to X \mathrel{|\!\sim} Y$ for all X, Y, Z s.t. $(X \cup Y) \cap Z = \emptyset$.

□

Definition 1.5.2

Let PL' denote $(PlCLM) + (PlCC')$.

PL and PL' are equivalent in the following sense, i.e. (PlRM) can essentially be omitted as rule:

Fact 1.5.2

Let $\mathcal{A}$ be a set of $\mathcal{L}$-sequences, $\mathcal{A}'$ be the closure of $\mathcal{A}$ under PL, and $\mathcal{A}''$ be the closure of $\{X \mathrel{|\!\sim} Y \colon X \cap Y \neq \emptyset\} \cup \{X \mathrel{|\!\sim} Y \colon X \mathrel{|\!\sim} Y' \in \mathcal{A} \text{ for some } Y' \subseteq Y\}$ under PL'. Then $\mathcal{A}' = \mathcal{A}''$.

Proof

Exercise, solution see [Sch04], Fact 3.7.2.

This remark allows to reduce considerably the number of possible proofs of a sequence.

We now adapt the definition of a preferential model to plausibility logic. This is the central definition on the semantic side.

1.5 The Smooth Case Without Domain Closure

Definition 1.5.3

Fix a plausibility logic language $\mathcal{L}$. A model for $\mathcal{L}$ is then just an arbitrary subset of $\mathcal{L}$.

If $\mathcal{M} := \langle M, \prec \rangle$ is a preferential model s.t. M is a set of (indexed) $\mathcal{L}$-models, then for a finite set $X \subseteq \mathcal{L}$ (to be imagined on the left hand side of $\mid\!\sim$!), we define

(a) $m \models X$ iff $X \subseteq m$

(b) $M(X) := \{m: \langle m, i \rangle \in M \text{ for some } i \text{ and } m \models X\}$

(c) $\mu(X) := \{m \in M(X): \exists \langle m, i \rangle \in M. \neg \exists <m', i'> \in M \; (m' \in M(X) \wedge \langle m', i' \rangle \prec \langle m, i \rangle)\}$

(d) $X \models_{\mathcal{M}} Y$ iff $\forall m \in \mu(X). m \cap Y \neq \emptyset$.

□

(a) reflects the intuitive reading of X as $\bigwedge X$, and (d) that of Y as $\bigvee Y$ in $X \mid\!\sim Y$. Note that X is a set of "formulas", and $\mu(X) = \mu_{\mathcal{M}}(M(X))$.

We note as trivial consequences of the definition.

Fact 1.5.3

(a) $a \models_{\mathcal{M}} b$ iff for all $m \in \mu(a). b \in m$

(b) $X \models_{\mathcal{M}} Y$ iff $\mu(X) \subseteq \bigcup\{M(b) : b \in Y\}$

(c) $m \in \mu(X) \wedge X \subseteq X' \wedge m \in M(X') \rightarrow m \in \mu(X')$.

□

1.5.2.3 Completeness and Incompleteness Results for Plausibility Logic

We have taken an old proof of ours (with a correction of Fact 1.5.6, due to D. Lehmann), and not the one published in [Sch96-3], as this (somewhat more complicated) proof fits better into the general proof strategy.

$(PlI) + (PlRM) + (PlCC)$ Is Complete (and Sound) for Preferential Models

$((PlI) + (PlRM) + (PlCC)$ is the system of [Leh92b].)

Fact 1.5.4

If $X_i : i \in I$ are all finite, and Y is s.t. $Y \cap X_i \neq \emptyset$ for all $i \in I$, then there is $f \in \Pi X_i$ s.t.

(1) $ran(f) \subseteq Y$,

(2) $\neg \exists g \in \Pi X_i.ran(g) \subset ran(f)$.

Proof

Exercise, solution see [Sch04], Fact 3.7.4.

Definition 1.5.4

For $T \subseteq \mathcal{L}$ let $\overline{\overline{T}} := \{Y : T \mathrel{|\!\sim} Y\}$.

Let $\mathcal{P}_f(X)$ denote the set of finite subsets of X.

Let $\Pi'\overline{\overline{T}} := \{f \in \Pi\overline{\overline{T}} : \text{there is no } g \in \Pi\overline{\overline{T}} \text{ s.t. } ran(g) \subset ran(f)\}$.

Given a function $\mu : \mathcal{P}_f(\mathcal{L}) \to \mathcal{PP}(\mathcal{L})$, define for $T, Y \in \mathcal{P}_f(\mathcal{L})$ $T \models_\mu Y :\leftrightarrow \forall m \in \mu(T).m \cap Y \neq \emptyset$.

Lemma 1.5.5

Let $\mathrel{|\!\sim}$ satisfy (PlRM), and μ be s.t.

1. If $T \mathrel{|\!\sim} \emptyset$, then $\mu(T) = \emptyset$.
2. If $T \mathrel{|\!\not\sim} \emptyset$, then $\mu(T)$ has the properties

    (a) $\{ran(f) : f \in \Pi'\overline{\overline{T}}\} \subseteq \mu(T)$,

    (b) if $m \in \mu(T)$, then ex. $f \in \Pi'\overline{\overline{T}}$ s.t. $ran(f) \subseteq m$.

Then $\models_\mu = \mathrel{|\!\sim}$.

Proof

Exercise, solution see [Sch04], Lemma 3.7.5.

Thus, if (PlRM) holds for $\mathrel{|\!\sim}$, then for μ defined by

$$\mu(T) := \begin{cases} \emptyset & \text{iff } T \vdash \emptyset \\ \{ran(f) : f \in \Pi'\overline{\overline{T}}\} & \text{otherwise} \end{cases}$$

$\models_\mu = \mathrel{|\!\sim}$.

Definition 1.5.5

For $\mathcal{M} := \langle M, \mu \rangle$ with $M \subseteq \mathcal{P}(\mathcal{L})$, $\mu : \mathcal{P}_f(\mathcal{L}) \to \mathcal{P}(M)$, let $\models_\mathcal{M} := \models_\mu$.

Let $\mathrel{|\!\sim}$ be given, define $\mathcal{M} := \langle M, \mu \rangle$ by:

1.5 The Smooth Case Without Domain Closure

1.) $M := \mathcal{P}(\mathcal{L})$,

2.)

$$\mu(T) := \begin{cases} \emptyset & \text{iff } T \vdash \emptyset \\ \{m \subseteq \mathcal{L}: 1.\ m \cap Y \neq \emptyset \text{ for all } Y \in \overline{\overline{T}} \\ \quad 2.\ \text{ex.}\ T' \subseteq T \text{ s.t. } m = ran(f) \text{ for some } f \in \Pi'\overline{\overline{T'}}\} & \text{otherwise} \end{cases}$$

By Fact 1.5.4, μ satisfies the prerequisites of Lemma 1.5.5, so if $\mathrel|\!\sim$ satisfies (PlRM), then $\models_{\mathcal{M}} = \mathrel|\!\sim$.

For $T \subseteq \mathcal{L}$, let $M(T) := \{X \subseteq \mathcal{L} : T \subseteq X\}$ and $\mathcal{Y} := \{M(T) : T \subseteq \mathcal{L}\}$, thus $\mathcal{Y} \subseteq \mathcal{P}(M)$. Define $F : \mathcal{Y} \to \mathcal{P}(M)$ by $F(M(T)) := \mu(T)$, this is well-defined, as $T = \bigcap M(T)$.

We work now with μ and F as just defined, and first note two auxiliary facts:

Fact 1.5.6

Let $\mathrel|\!\sim$ satisfy (PlRM), $f \in \Pi'\overline{\overline{S}}$, $R \subseteq ran(f)$ be finite, then there is $Y_{S,f,R}$ s.t.

1. $S \mathrel|\!\sim r, Y_{S,f,R}$ for all $r \in R$,
2. $Y_{S,f,R} \cap ran(f) = \emptyset$.

Proof

By minimality of $ran(f)$, for each $q \in ran(f)$ exists $Z_q \in \overline{\overline{S}}$ with $Z_q \cap ran(f) = \{q\}$. Let $Y_q := Z_q - \{q\}$, so $Y_q \cap ran(f) = \emptyset$ and $S \mathrel|\!\sim q, Y_q$. Let $Y_{S,f,R} := \bigcup\{Y_r : r \in R\}$, so $Y_{S,f,R} \cap ran(f) = \emptyset$ and for $r \in R$ $S \mathrel|\!\sim r, Y_{S,f,R}$ (by (PlRM)).

□

Fact 1.5.7

Let $\mathrel|\!\sim$ satisfy $(PlRM) + (PlCC)$, $f \in \Pi'\overline{\overline{S}}$, $S \subseteq S' \subseteq ran(f)$, $S' \mathrel|\!\sim Y$, then $ran(f) \cap Y \neq \emptyset$.

Proof

Exercise, solution see [Sch04], Fact 3.7.7.

Lemma 1.5.8

(a) If $\mathrel|\!\sim$ satisfies (PlI), then $F(M(T)) \subseteq M(T)$.

(b) If $\mathrel|\!\sim$ satisfies $(PlRM) + (PlCC)$, then $M(T') \subseteq M(T) \to F(M(T)) \cap M(T') \subseteq F(M(T'))$.

Proof

Exercise, solution see [Sch04], Lemma 3.7.8.

We have shown in Section 1.3.2.2 above, Proposition 1.3.5, that such F can be represented by a (transitive, irreflexive) preferential structure.

☐ (Completeness)

Incompleteness of Full Plausibility Logic for Smooth Structures

We work in PL and construct a counterexample, a set of formulas which satisfies the axiom and rules of plausibility logic, but violates Fact 1.5.9 below, and thus cannot be represented by a smooth preferential model.

We note the following fact for smooth preferential models:

Fact 1.5.9

Let U, X, Y be any sets, $\mathcal{M}$ be smooth for at least $\{Y, X\}$ and let $\mu(Y) \subseteq U \cup X$, $\mu(X) \subseteq U$, then $X \cap Y \cap \mu(U) \subseteq \mu(Y)$.

(This is, of course, a special case of $(\mu Cum1)$, see Definition 1.5.7.)

Proof

Exercise, solution see [Sch04], Fact 3.7.9.

Example 1.5.1

Let $\mathcal{L} := \{a, b, c, d, e, f\}$, and $\mathcal{X} := \{a \mathrel{|\!\sim} b, b \mathrel{|\!\sim} a, a \mathrel{|\!\sim} c, a \mathrel{|\!\sim} fd, dc \mathrel{|\!\sim} ba, dc \mathrel{|\!\sim} e, fcba \mathrel{|\!\sim} e\}$. We show that $\mathcal{X}$ does not have a smooth representation.

Fact 1.5.10

$\mathcal{X}$ does not entail $a \mathrel{|\!\sim} e$.

Proof

Let $\mathcal{A} := \{a \mathrel{|\!\sim} b, a \mathrel{|\!\sim} c, a \mathrel{|\!\sim} ed, a \mathrel{|\!\sim} fd, b \mathrel{|\!\sim} a, b \mathrel{|\!\sim} c, b \mathrel{|\!\sim} ed, b \mathrel{|\!\sim} fd, ba \mathrel{|\!\sim} c, ba \mathrel{|\!\sim} ed, ba \mathrel{|\!\sim} fd, ca \mathrel{|\!\sim} b, ca \mathrel{|\!\sim} ed, ca \mathrel{|\!\sim} fd, cb \mathrel{|\!\sim} a, cb \mathrel{|\!\sim} ed, cb \mathrel{|\!\sim} fd, cba \mathrel{|\!\sim} ed, cba \mathrel{|\!\sim} fd, dc \mathrel{|\!\sim} ba, dc \mathrel{|\!\sim} e, edc \mathrel{|\!\sim} ba, fcba \mathrel{|\!\sim} e\}$.

Set $\mathcal{A}_0 := \{X \mathrel{|\!\sim} Y : X \cap Y \neq \emptyset\}$, $\mathcal{A}_1 := \{X \mathrel{|\!\sim} Y :$ there is $Y' \subseteq Y$ s.t. $X \mathrel{|\!\sim} Y' \in \mathcal{A}\}$, and $\mathcal{A}'' := \mathcal{A}_0 \cup \mathcal{A}_1$.

As $\mathcal{A}''$ contains $\mathcal{X}$, but not $a \mathrel{|\!\sim} e$, it suffices to show that $\mathcal{A}''$ is a plausibility logic, i.e. is closed under PL. By Fact 1.5.2, this is equivalent to showing that $\mathcal{A}''$ is closed under (PlCLM) + $(PlCC')$. We note

1.5 The Smooth Case Without Domain Closure   95

Remark 1.5.11

(a) For $X \in \{a, b, ba, ca, cb, cba\}$ and $Y \in \{a, b, c, ed, fd\}$ $X \mathrel{\mkern-2mu\vert\mkern-5mu\sim} Y \in \mathcal{A}''$,

(b) for $X \in \{dc, edc, fcba, fecba\}$, $Y \in \{e, ba\}$ $X \mathrel{\mkern-2mu\vert\mkern-5mu\sim} Y \in \mathcal{A}''$

□ (Remark 1.5.11)

Note also that all cases of $\mathcal{A}$ occur as cases of (a) or (b).

We first show closure of $\mathcal{A}''$ under (PlCLM): $X' \mathrel{\mkern-2mu\vert\mkern-5mu\sim} a'$, $X' \mathrel{\mkern-2mu\vert\mkern-5mu\sim} Y' \to X', a' \mathrel{\mkern-2mu\vert\mkern-5mu\sim} Y'$ ($a' \notin X'$):

Thus, $X' \mathrel{\mkern-2mu\vert\mkern-5mu\sim} a' \in \mathcal{A}$, and $X' = a$ and $a' = b$ or $a' = c$, $X' = b$ and $a' = a$ or $a' = c$, $X' = ba$ and $a' = c$, $X' = ca$ and $a' = b$, $X' = cb$ and $a' = a$, $X' = dc$ and $a' = e$, $X' = fcba$ and $a' = e$.

The case $X' \mathrel{\mkern-2mu\vert\mkern-5mu\sim} Y' \in \mathcal{A}_0$ is trivial. Suppose $X' \mathrel{\mkern-2mu\vert\mkern-5mu\sim} Y' \in \mathcal{A}_1$, so there is $Y \subseteq Y'$ and $X' \mathrel{\mkern-2mu\vert\mkern-5mu\sim} Y \in \mathcal{A}$. It suffices to show that then $X', a' \mathrel{\mkern-2mu\vert\mkern-5mu\sim} Y \in \mathcal{A}''$, as $\mathcal{A}''$ is obviously closed under (PlRM). But all cases are handled by Remark 1.5.11 (a) or (b): If $X' \mathrel{\mkern-2mu\vert\mkern-5mu\sim} a' \in \mathcal{A}$ and $X' \mathrel{\mkern-2mu\vert\mkern-5mu\sim} Y \in \mathcal{A}$, then X' is one of the X and Y is one of the Y in (a) or (b). But then X', a' is also one of the X in (a) or (b).

We turn to closure under ($PlCC'$). We have to show for all X', Y', Z' with $Z' \cap (X' \cup Y') = \emptyset$, $Z' \neq \emptyset$: $X', Z' \mathrel{\mkern-2mu\vert\mkern-5mu\sim} Y', X' \mathrel{\mkern-2mu\vert\mkern-5mu\sim} z', Y'$ for all $z' \in Z' \to X' \mathrel{\mkern-2mu\vert\mkern-5mu\sim} Y'$. As for no Y $\emptyset \mathrel{\mkern-2mu\vert\mkern-5mu\sim} Y \in \mathcal{A}''$, $X' \neq \emptyset$.

The case $X', Z' \mathrel{\mkern-2mu\vert\mkern-5mu\sim} Y' \in \mathcal{A}_0$ is again trivial, as $Z' \cap Y' = \emptyset$, likewise the case $X' \mathrel{\mkern-2mu\vert\mkern-5mu\sim} z', Y' \in \mathcal{A}_0$ for some $z' \in Z'$, as $Z' \cap X' = \emptyset$.

So assume without loss of generality $X' \neq \emptyset$, $Z' \neq \emptyset$, $X' \cap Z' = \emptyset$, $Y' \cap Z' = \emptyset$, $X', Z' \mathrel{\mkern-2mu\vert\mkern-5mu\sim} Y' \in \mathcal{A}_1$, and for all $z' \in Z'$ $X' \mathrel{\mkern-2mu\vert\mkern-5mu\sim} z', Y' \in \mathcal{A}_1$. We have to show $X' \mathrel{\mkern-2mu\vert\mkern-5mu\sim} Y' \in \mathcal{A}''$. Note that by definition of $\mathcal{A}_1$, and $X' \mathrel{\mkern-2mu\vert\mkern-5mu\sim} z', Y' \in \mathcal{A}_1$, X' has to occur on the left hand side in $\mathcal{A}$, so $X' \in \{a, b, ba, ca, cb, cba, dc, edc, fcba\}$. As $X', Z' \mathrel{\mkern-2mu\vert\mkern-5mu\sim} Y' \in \mathcal{A}_1$, there is some $Y \subseteq Y'$ with $X', Z' \mathrel{\mkern-2mu\vert\mkern-5mu\sim} Y \in \mathcal{A}$. Moreover, $X' \cup Z'$ has at least two elements.

Case 1: $X' \cup Z' \in \{ba, ca, cb, cba\}$. X' is a proper, nonempty subset of $X' \cup Z'$. As $X' \neq c$, Remark 1.5.11 (a) shows that $X' \mathrel{\mkern-2mu\vert\mkern-5mu\sim} Y$ too, and thus $X' \mathrel{\mkern-2mu\vert\mkern-5mu\sim} Y'$.

Case 2: $X' \cup Z' \in \{dc, edc, fcba\}$. The possible cases are:

(1) $X' \cup Z' = edc$, $X' = dc$ and $Z' = e$,

(2) $X' \cup Z' = fcba$, $X' \in \{a, b, ba, ca, cb, cba\}$, $Z' = fcba - X'$, so $f \in Z'$.

In (1), we are done by Remark 1.5.11 (b).

(2): As $X' \cup Z' \mathrel{\mkern-2mu\vert\mkern-5mu\sim} Y' \in \mathcal{A}_1$, Y' has to contain e. Moreover, by $f \in Z'$, $X' \mathrel{\mkern-2mu\vert\mkern-5mu\sim} f, Y' \in \mathcal{A}_1$, so there must be some $Y'' \subseteq f, Y'$ with $X' \mathrel{\mkern-2mu\vert\mkern-5mu\sim} Y'' \in \mathcal{A}$. If $Y'' \subseteq Y'$, we are done, as then $X' \mathrel{\mkern-2mu\vert\mkern-5mu\sim} Y' \in \mathcal{A}_1 \subseteq \mathcal{A}''$. But if $f \in Y''$, then $Y'' = fd$, so $d \in Y'$. Thus, $d, e \in Y'$. But by Remark 1.5.11 (a), $X' \mathrel{|\!\sim} ed \in \mathcal{A}''$, so $X' \mathrel{|\!\sim} Y' \in \mathcal{A}''$.

□ ($\mathcal{X}$ does not entail $a \mathrel{|\!\sim} e$, Fact 1.5.10)

Suppose now that there is a smooth preferential model $\mathcal{M} = \langle M, \prec \rangle$ for plausibility logic which represents $\mathrel{|\!\sim}$, i.e. for all X, Y finite subsets of $\mathcal{L}$ $X \mathrel{|\!\sim} Y$ iff $X \models_\mathcal{M} Y$. (See Definition 1.5.3 and Fact 1.5.3 (page 91).)

$a \mathrel{|\!\sim} a$, $a \mathrel{|\!\sim} b$, $a \mathrel{|\!\sim} c$ implies for $m \in \mu(a)$ $a, b, c \in m$. Moreover, as $a \mathrel{|\!\sim} df$, then also $d \in m$ or $f \in m$. As $a \mathrel{|\!\not\sim} e$, there must be $m \in \mu(a)$ s.t. $e \notin m$. Suppose now $m \in \mu(a)$ with $f \in m$. So $a, b, c, f \in m$, thus by $m \in \mu(a)$ and Fact 1.5.3, $m \in \mu(a, b, c, f)$. But $fcba \mathrel{|\!\sim} e$, so $e \in m$. We thus have shown that $m \in \mu(a)$ and $f \in m$ implies $e \in m$. Consequently, there must be $m \in \mu(a)$ s.t. $d \in m$, $e \notin m$. Thus, in particular, as $cd \mathrel{|\!\sim} e$, there is $m \in \mu(a)$, $a, b, c, d \in m$, $m \notin \mu(cd)$. But by $cd \mathrel{|\!\sim} ab$, and $b \mathrel{|\!\sim} a$, $\mu(cd) \subseteq M(a) \cup M(b)$ and $\mu(b) \subseteq M(a)$ by Fact 1.5.3. Let now $T := M(cd)$, $R := M(a)$, $S := M(b)$, and $\mu_\mathcal{M}$ be the choice function of the minimal elements in the structure $\mathcal{M}$, we then have by $\mu(S) = \mu_\mathcal{M}(M(S))$:

1. $\mu_\mathcal{M}(T) \subseteq R \cup S$,
2. $\mu_\mathcal{M}(S) \subseteq R$,
3. there is $m \in S \cap T \cap \mu_\mathcal{M}(R)$, but $m \notin \mu_\mathcal{M}(T)$,

but this contradicts above Fact 1.5.9.

□ (Example 1.5.1)

1.5.2.4 A Comment on the Work by Arieli and Avron

We turn to a similar case, published in [AA00]. Definitions are due to [AA00], for motivation the reader is referred there.

Definition 1.5.6

(1) A Scott consequence relation, abbreviated (scr), is a binary relation $\vdash$ between sets of formulae, that satisfies the following conditions:

(s-R) if $\Gamma \cap \Delta \neq \emptyset$, the $\Gamma \vdash \Delta$,

(M) if $\Gamma \vdash \Delta$ and $\Gamma \subseteq \Gamma'$, $\Delta \subseteq \Delta'$, then $\Gamma' \vdash \Delta'$,

(C) if $\Gamma \vdash \psi, \Delta$ and $\Gamma', \psi \vdash \Delta'$, then $\Gamma, \Gamma' \vdash \Delta, \Delta'$.

(2) A Scott cautious consequence relation, abbreviated (sccr), is a binary relation $\mathrel{|\!\sim}$ between nonempty sets of formulae, that satisfies the following conditions:

(s-R) if $\Gamma \cap \Delta \neq \emptyset$, the $\Gamma \mathrel{|\!\sim} \Delta$,

1.5 The Smooth Case Without Domain Closure

(CM) if $\Gamma \mathrel{|\!\sim} \Delta$ and $\Gamma \mathrel{|\!\sim} \psi$, then $\Gamma, \psi \mathrel{|\!\sim} \Delta$,
(CC) if $\Gamma \mathrel{|\!\sim} \psi$ and $\Gamma, \psi \mathrel{|\!\sim} \Delta$, then $\Gamma \mathrel{|\!\sim} \Delta$.

Example 1.5.2

We have two consequence relations, $\vdash$ and $\mathrel{|\!\sim}$.

The rules to consider are

(LCC^n) $\dfrac{\Gamma \mathrel{|\!\sim} \psi_1, \Delta \ldots \Gamma \mathrel{|\!\sim} \psi_n, \Delta \Gamma, \psi_1, \ldots, \psi_n \mathrel{|\!\sim} \Delta}{\Gamma \mathrel{|\!\sim} \Delta}$

(RW^n) $\dfrac{\Gamma \mathrel{|\!\sim} \psi_i, \Delta i=1 \ldots n \, \Gamma, \psi_1, \ldots, \psi_n \vdash \phi}{\Gamma \mathrel{|\!\sim} \phi, \Delta}$

(Cum) $\Gamma, \Delta \neq \emptyset, \Gamma \vdash \Delta \Rightarrow \Gamma \mathrel{|\!\sim} \Delta$

(RM) $\Gamma \mathrel{|\!\sim} \Delta \Rightarrow \Gamma \mathrel{|\!\sim} \psi, \Delta$

(CM) $\dfrac{\Gamma \mathrel{|\!\sim} \psi \, \Gamma \mathrel{|\!\sim} \Delta}{\Gamma, \psi \mathrel{|\!\sim} \Delta}$

$(s\text{-}R)$ $\Gamma \cap \Delta \neq \emptyset \Rightarrow \Gamma \mathrel{|\!\sim} \Delta$

(M) $\Gamma \vdash \Delta, \Gamma \subseteq \Gamma', \Delta \subseteq \Delta' \Rightarrow \Gamma' \vdash \Delta'$

(C) $\dfrac{\Gamma_1 \vdash \psi, \Delta_1 \, \Gamma_2, \psi \vdash \Delta_2}{\Gamma_1, \Gamma_2 \vdash \Delta_1, \Delta_2}$

Let $\mathcal{L}$ be any set. Define now $\Gamma \vdash \Delta$ iff $\Gamma \cap \Delta \neq \emptyset$. Then $(s\text{-}R)$ and (M) for $\vdash$ are trivial. For (C): If $\Gamma_1 \cap \Delta_1 \neq \emptyset$ or $\Gamma_1 \cap \Delta_1 \neq \emptyset$, the result is trivial. If not, $\psi \in \Gamma_1$ and $\psi \in \Delta_2$, which implies the result. So $\vdash$ is a (scr).

Consider now the rules for a (sccr) which is $\vdash$-plausible for this $\vdash$. (Cum) is equivalent to $(s\text{-}R)$, which is essentially (PlI) of Plausibility Logic. Consider (RW^n). If ϕ is one of the ψ_i, then the consequence $\Gamma \mathrel{|\!\sim} \phi, \Delta$ is a case of one of the other hypotheses. If not, $\phi \in \Gamma$, so $\Gamma \mathrel{|\!\sim} \phi$ by $(s\text{-}R)$, so $\Gamma \mathrel{|\!\sim} \phi, \Delta$ by (RM) (if Δ is finite). So, for this $\vdash$, (RW^n) is a consequence of $(s\text{-}R) + (RM)$.

We are left with (LCC^n), (RM), (CM), $(s\text{-}R)$, it was shown in [Sch04] and [Sch96-3] that this does not suffice to guarantee smooth representability, by failure of $(\mu Cum1)$.

1.5.3 Smooth Preferential Structures Without Domain Closure Conditions

1.5.3.1 Introduction

Smooth Structures

We will consider choice functions $g \in \Pi \{f(X) : x \in X - f(X)\}$.

(In the final construction, we will construct simultaneously for all u, U such that $u \in f(U)$ a U-minimal copy, so in the following intuitive discussion, it will suffice to find minimal u, x, etc. with the required properties. This remark is destined for readers who wonder how this will all fit together. We should also note that we will again be in the dilemma which copy to make smaller, and will do so for all candidates - violating our principle of preserving ignorance. Yet, as before, as long as we are aware of it, it will do no harm.)

To see the new problem arising now, we start with U, and suppose that $u \in f(U)$. Let now $u \in X - f(X)$, then we have to find $x \in f(X)$ below u. First, x must not be in U, as we would have destroyed minimality of u in U, this is analogous to Case (1), so we need $f(X) - U \neq \emptyset$. But let now $u \in f(Y)$, $x \in Y$. In Case (1), it was sufficient to find another copy of u, which is minimal in Y. Now, we have to do more: to find an $y \in f(Y)$, y below u, so smoothness will hold. We will call the following process the "repairing process for u, x, and Y". Suppose then that $u \in f(Y)$, and $x \in f(Y)$ for $Y \in \mathcal{Y}$. Then we have destroyed minimality of u in Y, but have repaired smoothness immediately again by finding the minimal x. The situation is different if $x \in Y - f(Y)$ (and there was no $x' \prec u\, x' \in f(Y)$ chosen at the same time). Then we have destroyed minimality of u in Y, without repairing smoothness, and we have to repair it now by finding suitable $y \prec u$, $y \in f(Y)$. Of course, y must not be in U, as this would destroy minimality of u in U.

Thus, we have to find for all Y with $u \in f(Y)$, $x \in Y - f(Y)$ some $y \in f(Y)$, $y \prec u$, $y \notin U$. Note that this repair process is individual, i.e., we do not have to find one universal y which repairs lost minimality for ALL such Y at the same time, but it suffices to do it one by one, individually for every single such Y.

But now, the solutions y for such Y may have introduced new problems: Not only x is below u, but also y is below u. If there is now $Z \in \mathcal{Y}$ such that $u \in f(Z)$, and $y \in Z - f(Z)$, then we have to do the same repairing process for u, y, Z: find suitable $z \in f(Z)$ below u, $z \notin U$, etc. So we will have an infinite repairing process, where each step may introduce new problems, which will be repaired in the next step.

To illustrate that the problem is still a bit more complicated, we make a definition, and see that we have to avoid in above situation not only U, but $H(U, u)$, to be defined now.

$H(U, u)_0 := U$,
$H(U, u)_{\alpha+1} := H(U, u)_\alpha \cup \bigcup \{X : u \in X \wedge \mu(X) \subseteq H(U, u)_\alpha\}$,
$H(U, u)_\lambda := \bigcup \{H(U, u)_\alpha : \alpha < \lambda\}$ for $limit(\lambda)$,
$H(U, u) := \bigcup \{H(U, u)_\alpha : \alpha < \kappa\}$ for κ sufficiently big

($card(Z)$ suffices, as the procedure trivializes, when we cannot add any new elements).

(HU, u) is the property:

$u \in \mu(U)$, $u \in Y - \mu(Y) \Rightarrow \mu(Y) \nsubseteq H(U, u)$ - of course for all u and U.

$(U, Y \in \mathcal{Y})$.

1.5 The Smooth Case Without Domain Closure

Fact 1.5.12

(HU, u) holds in smooth structures.

The proof is given in Fact 1.5.16 (2).

We note now that we have to consider our principle of preserving ignorance again: We can choose first arbitrary $y \in f(Y)$ to repair for u, x, Y. So which one we choose is - a priori - an arbitrary choice. Yet, this choice might have repercussions later, as different y and y' chosen to repair for u, x, Y might force different repairs for u, y, Z or u, y', Z', as Z might be such that $u \in f(Z)$, $y \in Z - f(Z)$, and Z' such that $u \in f(Z)$, $y' \in Z - f(Z')$, and it might be possible to find suitable $z \in f(Z)$, $z \notin H(U, u)$, but no suitable z', etc. So the, at first sight, arbitrary choice might reveal an impasse later on. We will see that we can easily solve the problem in the not necessarily transitive case, but we do not see at the time of writing any easy solution in the transitive case, if the domain is not necessarily closed under finite unions.

Transitive Smooth Structures

The basic, now more complicated, situation to consider is the following:

Let again $u \in f(U)$, $u \in X - f(X)$, we have to find $x \in f(X)$ - outside $H(U, u)$ as in Case (3). Thus, we need again $f(X) - H(U, u) \neq \emptyset$. Again, we have to repair all damage done, i.e., for all u, x, Y as discussed in Case (3), the infinite repair process discussed there.

Suppose now that $x \in Y - f(Y)$, so we have to find $y \in f(Y)$, outside $H(U, u)$ by transitivity of the relation, as $y \prec x \prec u$, and, in addition outside $H(X, x)$, as in Case (3), now for X and x. Thus, we need $f(Y) - (H(U, u) \cup H(X, x)) \neq \emptyset$. Moreover, we have to do the same for all elements y introduced by the above repairing process. Again, we have to do repairing: $y \prec u$, and $y \prec x$, so for all Y' such that $u \in f(Y')$, $y \in Y' - f(Y')$ we have to repair for u, y, Y', and if $x \in f(Y')$, $y \in Y' - f(Y')$ we have to repair for x, y, Y', creating new smaller elements, etc.

If $y \in Z - f(Z)$, we have to find $z \in f(Z)$, outside $H(U, u)$, $H(X, x)$, $H(Y, y)$, etc., so the further we go down, the longer the condition will be. Thus, we need $f(Z) - (H(U, u) \cup H(X, x) \cup H(Y, y)) \neq \emptyset$. And, again we have to repair, for u, z, x, z, and y, z.

And so on.

Note again the arbitrariness of choice, when there is not a unique solution, i.e., no unique x, y, z etc. This has to be considered when we want to respect preservation of ignorance, but also an early wrong choice might lead to an impasse, leading to backtracking to this early wrong choice.

We will see that the closure of the domain under ($\cup$) makes all this easily possible, but the authors do not see an easy solution in the absence of ($\cup$) at the time of writing - the problem is an initial potentially wrong choice, which we do not see how to avoid other than by trying.

So we will give here only a formal negative result by an example, see Example 1.5.4.

1.5.3.2 Detailed Discussion

Smooth Structures Without Domain Closure Conditions

We show here that, without sufficient closure properties, there is an infinity of versions of cumulativity, which collapse to usual cumulativity when the domain is closed under finite unions. Closure properties thus reveal themselves as a powerful tool to show independence of properties. We then show positive results for the smooth and the transitive smooth case.

We work in some fixed arbitrary set Z, all sets considered will be subsets of Z. Unless said otherwise, we use without further mentioning (μPR) and ($\mu \subseteq$).

Note that (μPR) and ($\mu \subseteq$) entail $\mu(A \cup B) \subseteq \mu(A) \cup \mu(B)$ whenever μ is defined for A, B, $A \cup B$. ($\mu(A \cup B) \cap A \subseteq \mu(A)$, $\mu(A \cup B) \cap B \subseteq \mu(B)$, by ($\mu PR$), but $\mu(A \cup B) \subseteq A \cup B$ by ($\mu \subseteq$).)

The important point in the counterexample in Section 1.5.2 was that the condition

$\mu(T) \subseteq R \cup S$ and $\mu(S) \subseteq R$ imply $S \cap T \cap \mu(R) \subseteq \mu(T)$

holds in all smooth models, but not in the example.

Thus, we need new conditions, which take care of the "semi-transitivity" of smoothness, coding it directly and not by a simple condition, which uses finite union. For this purpose, we modify the definition of $H(U)$, and replace it by $H(U, x)$:

Definition 1.5.7

For any ordinal α, we define

($\mu Cum\alpha$):

If for all $\beta \leq \alpha$ $\mu(X_\beta) \subseteq U \cup \bigcup \{X_\gamma : \gamma < \beta\}$ hold, then so does $\bigcap \{X_\gamma : \gamma \leq \alpha\} \cap \mu(U) \subseteq \mu(X_\alpha)$.

($\mu Cumt\alpha$):

If for all $\beta \leq \alpha$ $\mu(X_\beta) \subseteq U \cup \bigcup \{X_\gamma : \gamma < \beta\}$ hold, then so does $X_\alpha \cap \mu(U) \subseteq \mu(X_\alpha)$.

("t" stands for transitive, see Fact 1.5.13, (2.2) below.)

1.5 The Smooth Case Without Domain Closure

$(\mu Cum\infty)$ and $(\mu Cumt\infty)$ will be the class of all $(\mu Cum\alpha)$ or $(\mu Cumt\alpha)$ - read their "conjunction", i.e., if we say that $(\mu Cum\infty)$ holds, we mean that all $(\mu Cum\alpha)$ hold.

Note:

The first conditions thus have the form:

$(\mu Cum0)$ $\mu(X_0) \subseteq U \Rightarrow X_0 \cap \mu(U) \subseteq \mu(X_0)$,

$(\mu Cum1)$ $\mu(X_0) \subseteq U, \mu(X_1) \subseteq U \cup X_0 \Rightarrow X_0 \cap X_1 \cap \mu(U) \subseteq \mu(X_1)$,

$(\mu Cum2)$ $\mu(X_0) \subseteq U, \mu(X_1) \subseteq U \cup X_0, \mu(X_2) \subseteq U \cup X_0 \cup X_1 \Rightarrow$
$X_0 \cap X_1 \cap X_2 \cap \mu(U) \subseteq \mu(X_2)$.

$(\mu Cumt\alpha)$ differs from $(\mu Cum\alpha)$ only in the consequence, the intersection contains only the last X_α - in particular, $(\mu Cum0)$ and $(\mu Cumt0)$ coincide.

Recall that condition $(\mu Cum1)$ is the crucial condition in [Leh92a], which failed, despite (μCUM), but which has to hold in all smooth models. This condition $(\mu Cum1)$ was the starting point of the investigation.

We briefly mention some major results on above conditions, taken from Fact 1.5.13 and shown there - we use the same numbering:

(1.1) $(\mu Cum\alpha) \Rightarrow (\mu Cum\beta)$ for all $\beta \leq \alpha$

(1.2) $(\mu Cumt\alpha) \Rightarrow (\mu Cumt\beta)$ for all $\beta \leq \alpha$

(2.1) All $(\mu Cum\alpha)$ hold in smooth preferential structures

(2.2) All $(\mu Cumt\alpha)$ hold in transitive smooth preferential structures

(3.1) $(\mu Cum\beta) + (\cup) \Rightarrow (\mu Cum\alpha)$ for all $\beta \leq \alpha$

(3.2) $(\mu Cumt\beta) + (\cup) \Rightarrow (\mu Cumt\alpha)$ for all $\beta \leq \alpha$

(5.2) $(\mu Cum\alpha) \Rightarrow (\mu CUM)$ for all α

(5.3) $(\mu CUM) + (\cup) \Rightarrow (\mu Cum\alpha)$ for all α

The following inductive definition of $H(U, u)$ and of the property (HU, u) concerns closure under $(\mu Cum\infty)$, its main property is formulated in Fact 1.5.15, its main interest is its use in the proof of Proposition 1.5.19.

Definition 1.5.8

$(H(U, u)_\alpha, H(U)_\alpha, (HU, u), (HU).)$

$H(U, u)_0 := U$,

$H(U, u)_{\alpha+1} := H(U, u)_\alpha \cup \bigcup \{X : u \in X \wedge \mu(X) \subseteq H(U, u)_\alpha\}$,

$H(U, u)_\lambda := \bigcup \{H(U, u)_\alpha : \alpha < \lambda\}$ for $limit(\lambda)$,

$H(U, u) := \bigcup \{H(U, u)_\alpha : \alpha < \kappa\}$ for κ sufficiently big $(card(Z))$ suffices, as the procedure trivializes, when we cannot add any new elements).

(HU, u) is the property:

$u \in \mu(U), u \in Y - \mu(Y) \Rightarrow \mu(Y) \nsubseteq H(U, u)$ - of course for all u and U. $(U, Y \in \mathcal{Y})$.

Thus, (HU, u) entails $\mu(U) \subseteq H(U, u), u \in \mu(U) \cap Y \Rightarrow u \in \mu(Y)$.

For the case with $(\cup)$, we further define, independent of u,

$H(U)_0 := U$,
$H(U)_{\alpha+1} := H(U)_\alpha \cup \bigcup\{X : \mu(X) \subseteq H(U)_\alpha\}$,
$H(U)_\lambda := \bigcup\{H(U)_\alpha : \alpha < \lambda\}$ for $limit(\lambda)$,
$H(U) := \bigcup\{H(U)_\alpha : \alpha < \kappa\}$ again for κ sufficiently big

(HU) is the property:

$u \in \mu(U), u \in Y - \mu(Y) \Rightarrow \mu(Y) \nsubseteq H(U)$ - of course for all U. $(U, Y \in \mathcal{Y})$.

Thus, (HU) entails $\mu(Y) \subseteq H(U) \Rightarrow \mu(U) \cap Y \subseteq \mu(Y)$.

Obviously, $H(U, u) \subseteq H(U)$, so $(HU) \Rightarrow (HU, u)$.

Example 1.5.3

This important example shows that the conditions $(\mu Cum\alpha)$ and $(\mu Cumt\alpha)$ defined in Definition 1.5.7 are all different in the absence of $(\cup)$, in its presence they all collapse (see Fact 1.5.13 below). More precisely, the following (class of) examples shows that the $(\mu Cum\alpha)$ increase in strength. For any finite or infinite ordinal $\kappa > 0$ we construct an example such that

(a) (μPR) and $(\mu \subseteq)$ hold
(b) (μCUM) holds
(c) $(\cap)$ holds
(d) $(\mu Cumt\alpha)$ holds for $\alpha < \kappa$
(e) $(\mu Cum\kappa)$ fails.

Proof

We define a suitable base set and a non-transitive binary relation $\prec$ on this set, as well as a suitable set $\mathcal{X}$ of subsets, closed under arbitrary intersections, but not under finite unions, and define μ on these subsets as usual in preferential structures by $\prec$. Thus, (μPR) and $(\mu \subseteq)$ will hold. It will be immediate that $(\mu Cum\kappa)$ fails, and we will show that (μCUM) and $(\mu Cumt\alpha)$ for $\alpha < \kappa$ hold by examining the cases.

1.5 The Smooth Case Without Domain Closure

For simplicity, we first define a set of generators for $\mathcal{X}$, and close under ($\bigcap$) afterwards. The set U will have a special position, it is the "useful" starting point to construct chains corresponding to above definitions of ($\mu Cum\alpha$) and ($\mu Cumt\alpha$).

In the sequel, i, j will be successor ordinals, λ etc. limit ordinals, α, β, κ any ordinals, thus e.g., $\lambda \leq \kappa$ will imply that λ is a limit ordinal $\leq \kappa$, etc.

The base set and the relation $\prec$:

$\kappa > 0$ is fixed, but arbitrary. We go up to $\kappa > 0$.

The base set is $\{a, b, c\} \cup \{d_\lambda : \lambda \leq \kappa\} \cup \{x_\alpha : \alpha \leq \kappa + 1\} \cup \{x'_\alpha : \alpha \leq \kappa\}$.
$a \prec b \prec c$, $x_\alpha \prec x_{\alpha+1}$, $x_\alpha \prec x'_\alpha$, $x'_0 \prec x_\lambda$ (for any λ) - $\prec$ is NOT transitive.

The generators:

$U := \{a, c, x_0\} \cup \{d_\lambda : \lambda \leq \kappa\}$ - i.e., ...$\{d_\lambda : lim(\lambda) \wedge \lambda \leq \kappa\}$,
$X_i := \{c, x_i, x'_i, x_{i+1}\}$ ($i < \kappa$),
$X_\lambda := \{c, d_\lambda, x_\lambda, x'_\lambda, x_{\lambda+1}\} \cup \{x'_\alpha : \alpha < \lambda\}$ ($\lambda < \kappa$),
$X'_\kappa := \{a, b, c, x_\kappa, x'_\kappa, x_{\kappa+1}\}$ if κ is a successor,
$X'_\kappa := \{a, b, c, d_\kappa, x_\kappa, x'_\kappa, x_{\kappa+1}\} \cup \{x'_\alpha : \alpha < \kappa\}$ if κ is a limit.

Thus, $X'_\kappa = X_\kappa \cup \{a, b\}$ if X_κ were defined.

Note that there is only one X'_κ, and X_α is defined only for $\alpha < \kappa$, so we will not have X_α and X'_α at the same time.

Thus, the values of the generators under μ are:

$\mu(U) = U$,
$\mu(X_i) = \{c, x_i\}$,
$\mu(X_\lambda) = \{c, d_\lambda\} \cup \{x'_\alpha : \alpha < \lambda\}$,
$\mu(X'_i) = \{a, x_i\}$ ($i > 0$, i has to be a successor),
$\mu(X'_\lambda) = \{a, d_\lambda\} \cup \{x'_\alpha : \alpha < \lambda\}$.

(We do not assume that the domain is closed under μ.)

Intersections:

We consider first pairwise intersections:

(1) $U \cap X_0 = \{c, x_0\}$,
(2) $U \cap X_i = \{c\}$, $i > 0$,
(3) $U \cap X_\lambda = \{c, d_\lambda\}$,
(4) $U \cap X'_i = \{a, c\}$ ($i > 0$),
(5) $U \cap X'_\lambda = \{a, c, d_\lambda\}$,
(6) $X_i \cap X_j$:

(6.1) $j = i+1$ $\{c, x_{i+1}\}$,

(6.2) else $\{c\}$,

(7) $X_i \cap X_\lambda$:

(7.1) $i < \lambda$ $\{c, x'_i\}$,

(7.2) $i = \lambda + 1$ $\{c, x_{\lambda+1}\}$,

(7.3) $i > \lambda + 1$ $\{c\}$,

(8) $X_\lambda \cap X_{\lambda'}$: $\{c\} \cup \{x'_\alpha : \alpha \leq min(\lambda, \lambda')\}$.

As X'_κ occurs only once, $X_\alpha \cap X'_\kappa$ etc. give no new results.

Note that μ is constant on all these pairwise intersections.

Iterated intersections:

As c is an element of all sets, sets of the type $\{c, z\}$ do not give any new results. The possible subsets of $\{a, c, d_\lambda\}$: $\{c\}, \{a, c\}, \{c, d_\lambda\}$ exist already. Thus, the only source of new sets via iterated intersections is $X_\lambda \cap X_{\lambda'} = \{c\} \cup \{x'_\alpha : \alpha \leq min(\lambda, \lambda')\}$. But, to intersect them, or with some old sets, will not generate any new sets either. Consequently, the example satisfies $(\bigcap)$ for $\mathcal{X}$ defined by U, X_i $(i < \kappa)$, X_λ $(\lambda < \kappa)$, X'_κ, and above paiwise intersections.

We will now verify the positive properties. This is tedious, but straightforward, we have to check the different cases.

Validity of (μCUM) :

Consider the prerequisite $\mu(X) \subseteq Y \subseteq X$. If $\mu(X) = X$ or if $X - \mu(X)$ is a singleton, X cannot give a violation of (μCUM). So we are left with the following candidates for X :

(1) $X_i := \{c, x_i, x'_i, x_{i+1}\}$, $\mu(X_i) = \{c, x_i\}$

Interesting candidates for Y will have 3 elements, but they will all contain a. (If $\kappa < \omega : U = \{a, c, x_0\}$.)

(2) $X_\lambda := \{c, d_\lambda, x_\lambda, x'_\lambda, x_{\lambda+1}\} \cup \{x'_\alpha : \alpha < \lambda\}$, $\mu(X_\lambda) = \{c, d_\lambda\} \cup \{x'_\alpha : \alpha < \lambda\}$

The only sets to contain d_λ are X_λ, U, $U \cap X_\lambda$. But $a \in U$, and $U \cap X_\lambda$ is finite. (X_λ and X'_λ cannot be present at the same time.)

(3) $X'_i := \{a, b, c, x_i, x'_i, x_{i+1}\}$, $\mu(X'_i) = \{a, x_i\}$

a is only in U, X'_i, $U \cap X'_i = \{a, c\}$, but $x_i \notin U$, as $i > 0$.

(4) $X'_\lambda := \{a, b, c, d_\lambda, x_\lambda, x'_\lambda, x_{\lambda+1}\} \cup \{x'_\alpha : \alpha < \lambda\}$, $\mu(X'_\lambda) = \{a, d_\lambda\} \cup \{x'_\alpha : \alpha < \lambda\}$

d_λ is only in X'_λ and U, but U contains no x'_α.

Thus, (μCUM) holds trivially.

1.5 The Smooth Case Without Domain Closure

$(\mu Cumt\alpha)$ hold for $\alpha < \kappa$:

To simplify language, we say that we reach Y from X iff $X \neq Y$ and there is a sequence X_β, $\beta \leq \alpha$ and $\mu(X_\beta) \subseteq X \cup \bigcup\{X_\gamma : \gamma < \beta\}$, and $X_\alpha = Y$, $X_0 = X$. Failure of $(\mu Cumt\alpha)$ would then mean that there are X and Y, we can reach Y from X, and $x \in (\mu(X) \cap Y) - \mu(Y)$. Thus, in a counterexample, $Y = \mu(Y)$ is impossible, so none of the intersections can be such Y.

To reach Y from X, we have to get started from X, i.e., there must be Z such that $\mu(Z) \subseteq X$, $Z \not\subseteq X$ (so $\mu(Z) \neq Z$). Inspection of the different cases shows that we cannot reach any set Y from any case of the intersections, except from (1), (6.1), (7.2).

If Y contains a globally minimal element (i.e., there is no smaller element in any set), it can only be reached from any X which already contains this element. The globally minimal elements are a, x_0, and the d_λ, $\lambda \leq \kappa$.

By these observations, we see that X_λ and X'_κ can only be reached from U. From no X_α U can be reached, as the globally minimal a is missing. But U cannot be reached from X'_κ either, as the globally minimal x_0 is missing.

When we look at the relation $\prec$ defining μ, we see that we can reach Y from X only by going upwards, adding bigger elements. Thus, from X_α, we cannot reach any X_β, $\beta < \alpha$, the same holds for X'_κ and X_β, $\beta < \kappa$. Thus, from X'_κ, we cannot go anywhere interesting (recall that the intersections are not candidates for a Y giving a contradiction).

Consider now X_α. We can go up to any $X_{\alpha+n}$, but not to any X_λ, $\alpha < \lambda$, as d_λ is missing, neither to X'_κ, as a is missing. And we will be stopped by the first $\lambda > \alpha$, as x_λ will be missing to go beyond X_λ. Analogous observations hold for the remaining intersections (1), (6.1), (7.2). But in all these sets we can reach, we will not destroy minimality of any element of X_α (or of the intersections).

Consequently, the only candidates for failure will all start with U. As the only element of U not globally minimal is c, such failure has to have $c \in Y - \mu(Y)$, so Y has to be X'_κ. Suppose we omit one of the X_α in the sequence going up to X'_κ. If $\kappa \geq \lambda > \alpha$, we cannot reach X_λ and beyond, as x'_α will be missing. But we cannot go to $X_{\alpha+n}$ either, as $x_{\alpha+1}$ is missing. So we will be stopped at X_α. Thus, to see failure, we need the full sequence $U = X_0$, $X'_\kappa = Y_\kappa$, $Y_\alpha = X_\alpha$ for $0 < \alpha < \kappa$.

$(\mu Cum\kappa)$ fails:

The full sequence $U = X_0$, $X'_\kappa = Y_\kappa$, $Y_\alpha = X_\alpha$ for $0 < \alpha < \kappa$ shows this, as $c \in \mu(U) \cap X'_\kappa$, but $c \notin \mu(X'_\kappa)$.

Consequently, the example satisfies $(\bigcap)$, (μCUM), $(\mu Cumt\alpha)$ for $\alpha < \kappa$, and $(\mu Cum\kappa)$ fails.

□

Fact 1.5.13

We summarize some properties of $(\mu Cum\alpha)$ and $(\mu Cumt\alpha)$ - sometimes with some redundancy. Unless said otherwise, α, β etc. will be arbitrary ordinals. For (1) to (6) (μPR) and $(\mu \subseteq)$ are assumed to hold, for (7) only $(\mu \subseteq)$.

(1) Downward:

(1.1) $(\mu Cum\alpha) \Rightarrow (\mu Cum\beta)$ for all $\beta \leq \alpha$

(1.2) $(\mu Cumt\alpha) \Rightarrow (\mu Cumt\beta)$ for all $\beta \leq \alpha$

(2) Validity of $(\mu Cum\alpha)$ and $(\mu Cumt\alpha)$:

(2.1) All $(\mu Cum\alpha)$ hold in smooth preferential structures

(2.2) All $(\mu Cumt\alpha)$ hold in transitive smooth preferential structures

(2.3) $(\mu Cumt\alpha)$ for $0 < \alpha$ do not necessarily hold in smooth structures without transitivity, even in the presence of $(\bigcap)$

(3) Upward:

(3.1) $(\mu Cum\beta) + (\cup) \Rightarrow (\mu Cum\alpha)$ for all $\beta \leq \alpha$

(3.2) $(\mu Cumt\beta) + (\cup) \Rightarrow (\mu Cumt\alpha)$ for all $\beta \leq \alpha$

(3.3) $\{(\mu Cumt\beta) : \beta < \alpha\} + (\mu CUM) + (\bigcap) \not\Rightarrow (\mu Cum\alpha)$ for $\alpha > 0$.

(4) Connection $(\mu Cum\alpha)/(\mu Cumt\alpha)$:

(4.1) $(\mu Cumt\alpha) \Rightarrow (\mu Cum\alpha)$

(4.2) $(\mu Cum\alpha) + (\bigcap) \not\Rightarrow (\mu Cumt\alpha)$

(4.3) $(\mu Cum\alpha) + (\cup) \Rightarrow (\mu Cumt\alpha)$

(5) (μCUM) and $(\mu Cumi)$:

(5.1) $(\mu CUM) + (\cup)$ entail:

(5.1.1) $\mu(A) \subseteq B \Rightarrow \mu(A \cup B) = \mu(B)$

(5.1.2) $\mu(X) \subseteq U, U \subseteq Y \Rightarrow \mu(Y \cup X) = \mu(Y)$

(5.1.3) $\mu(X) \subseteq U, U \subseteq Y \Rightarrow \mu(Y) \cap X \subseteq \mu(U)$

(5.2) $(\mu Cum\alpha) \Rightarrow (\mu CUM)$ for all α

(5.3) $(\mu CUM) + (\cup) \Rightarrow (\mu Cum\alpha)$ for all α

(5.4) $(\mu CUM) + (\cap) \Rightarrow (\mu Cum0)$

(6) (μCUM) and $(\mu Cumt\alpha)$:

1.5 The Smooth Case Without Domain Closure

(6.1) $(\mu Cumt\alpha) \Rightarrow (\mu CUM)$ for all α

(6.2) $(\mu CUM) + (\cup) \Rightarrow (\mu Cumt\alpha)$ for all α

(6.3) $(\mu CUM) \not\Rightarrow (\mu Cumt\alpha)$ for all $\alpha > 0$

(7) $(\mu Cum0) \Rightarrow (\mu PR)$

Proof

We prove these facts in a different order: (1), (2), (5.1), (5.2), (4.1), (6.1), (6.2), (5.3), (3.1), (3.2), (4.2), (4.3), (5.4), (3.3), (6.3), (7).

(1.1)

For $\beta < \gamma \leq \alpha$ set $X_\gamma := X_\beta$. Let the prerequisites of $(\mu Cum\beta)$ hold. Then for γ with $\beta < \gamma \leq \alpha$ $\mu(X_\gamma) \subseteq X_\beta$ by $(\mu \subseteq)$, so the prerequisites of $(\mu Cum\alpha)$ hold, too, so by $(\mu Cum\alpha)$ $\bigcap\{X_\delta : \delta \leq \beta\} \cap \mu(U) = \bigcap\{X_\delta : \delta \leq \alpha\} \cap \mu(U) \subseteq \mu(X_\alpha) = \mu(X_\beta)$.

(1.2)

Analogous.

(2.1)

Exercise, solution in the Appendix.

(2.2)

We use the following Fact: Let, in a smooth transitive structure, $\mu(X_\beta) \subseteq U \cup \bigcup\{X_\gamma : \gamma < \beta\}$ for all $\beta \leq \alpha$, and let $x \in \mu(U)$. Then there is no $y \prec x$, $y \in U \cup \bigcup\{X_\gamma : \gamma \leq \alpha\}$.

Proof of the Fact by induction: Suppose such $y \in U \cup X_0$ exists. $y \in U$ is impossible. Let $y \in X_0$, by $\mu(X_0) \subseteq U$, $y \in X_0 - \mu(X_0)$, so there is $z \in \mu(X_0)$, $z \prec y$, so $z \prec x$ by transitivity, but $\mu(X_0) \subseteq U$. Let the result hold for all $\beta < \alpha$, but fail for α, so $\neg \exists y \prec x.y \in U \cup \bigcup\{X_\gamma : \gamma < \alpha\}$, but $\exists y \prec x.y \in U \cup \bigcup\{X_\gamma : \gamma \leq \alpha\}$, so $y \in X_\alpha$. If $y \in \mu(X_\alpha)$, then $y \in U \cup \bigcup\{X_\gamma : \gamma < \alpha\}$, but this is impossible, so $y \in X_\alpha - \mu(X_\alpha)$, let by smoothness $z \prec y$, $z \in \mu(X_\alpha)$, so by transitivity $z \prec x$, *contradiction*. The result is easily modified for the case with copies.

Let the prerequisites of $(\mu Cumt\alpha)$ hold, then those of the Fact will hold, too. Let now $x \in \mu(U) \cap (X_\alpha - \mu(X_\alpha))$, by smoothness, there must be $y \prec x$, $y \in \mu(X_\alpha) \subseteq U \cup \bigcup\{X_\gamma : \gamma < \alpha\}$, contradicting the Fact.

(2.3)

Let $\alpha > 0$, and consider the following structure over $\{a, b, c\} : U := \{a, c\}$, $X_0 := \{b, c\}$, $X_\alpha := \ldots := X_1 := \{a, b\}$, and their intersections, $\{a\}$, $\{b\}$, $\{c\}$, $\emptyset$ with the order $c \prec b \prec a$ (without transitivity). This is preferential, so (μPR) and $(\mu \subseteq)$ hold. The structure is smooth for U, all X_β, and their intersections. We have $\mu(X_0) \subseteq U$, $\mu(X_\beta) \subseteq U \cup X_0$ for all $\beta \leq \alpha$, so

$\mu(X_\beta) \subseteq U \cup \bigcup\{X_\gamma : \gamma < \beta\}$ for all $\beta \leq \alpha$ but $X_\alpha \cap \mu(U) = \{a\} \not\subseteq \{b\} = \mu(X_\alpha)$ for $\alpha > 0$.

(5.1)

(5.1.1) $\mu(A) \subseteq B \Rightarrow \mu(A \cup B) \subseteq \mu(A) \cup \mu(B) \subseteq B \Rightarrow_{(\mu CUM)} \mu(B) = \mu(A \cup B)$.

(5.1.2) $\mu(X) \subseteq U \subseteq Y \Rightarrow$ (by (5.1.1)) $\mu(Y \cup X) = \mu(Y)$.

(5.1.3) $\mu(Y) \cap X =$ (by (5.1.2)) $\mu(Y \cup X) \cap X \subseteq \mu(Y \cup X) \cap (X \cup U) \subseteq$ (by (μPR)) $\mu(X \cup U) =$ (by (5.1.1)) $\mu(U)$.

(5.2)

Using (1.1), it suffices to show $(\mu Cum0) \Rightarrow (\mu CUM)$. Let $\mu(X) \subseteq U \subseteq X$. By $(\mu Cum0)$ $X \cap \mu(U) \subseteq \mu(X)$, so by $\mu(U) \subseteq U \subseteq X \Rightarrow \mu(U) \subseteq \mu(X)$. $U \subseteq X \Rightarrow \mu(X) \cap U \subseteq \mu(U)$ by (μPR), but also $\mu(X) \subseteq U$, so $\mu(X) \subseteq \mu(U)$.

(4.1)

Trivial.

(6.1)

Follows from (4.1) and (5.2).

(6.2)

Let the prerequisites of $(\mu Cumt\alpha)$ hold.

We first show by induction $\mu(X_\alpha \cup U) \subseteq \mu(U)$.

Proof:

$\alpha = 0 : \mu(X_0) \subseteq U \Rightarrow \mu(X_0 \cup U) = \mu(U)$ by (5.1.1). Let for all $\beta < \alpha$ $\mu(X_\beta \cup U) \subseteq \mu(U) \subseteq U$. By prerequisite, $\mu(X_\alpha) \subseteq U \cup \bigcup\{X_\beta : \beta < \alpha\}$, thus $\mu(X_\alpha \cup U) \subseteq \mu(X_\alpha) \cup \mu(U) \subseteq \bigcup\{U \cup X_\beta : \beta < \alpha\}$,

moreover for all $\beta < \alpha$ $\mu(X_\beta \cup U) \subseteq U \subseteq X_\alpha \cup U$, so $\mu(X_\alpha \cup U) \cap (U \cup X_\beta) \subseteq \mu(U)$ by (5.1.3), thus $\mu(X_\alpha \cup U) \subseteq \mu(U)$.

Consequently, under the above prerequisites, we have $\mu(X_\alpha \cup U) \subseteq \mu(U) \subseteq U \subseteq U \cup X_\alpha$, so by (μCUM) $\mu(U) = \mu(X_\alpha \cup U)$, and, finally, $\mu(U) \cap X_\alpha = \mu(X_\alpha \cup U) \cap X_\alpha \subseteq \mu(X_\alpha)$ by (μPR).

Note that finite unions take us over the limit step, essentially, as all steps collapse, and $\mu(X_\alpha \cup U)$ will always be $\mu(U)$, so there are no real changes.

(5.3)

Follows from (6.2) and (4.1).

(3.1)

Follows from (5.2) and (5.3).

(3.2)

Follows from (6.1) and (6.2).

1.5 The Smooth Case Without Domain Closure

(4.2)

Follows from (2.3) and (2.1).

(4.3)

Follows from (5.2) and (6.2).

(5.4)

$\mu(X) \subseteq U \Rightarrow \mu(X) \subseteq U \cap X \subseteq X \Rightarrow \mu(X \cap U) = \mu(X) \Rightarrow X \cap \mu(U) = (X \cap U) \cap \mu(U) \subseteq \mu(X \cap U) = \mu(X)$

(3.3)

See Example 1.5.3.

(6.3)

See Example 1.5.3.

(7)

Trivial. Let $X \subseteq Y$, so by $(\mu \subseteq)$ $\mu(X) \subseteq X \subseteq Y$, so by $(\mu Cum0)$ $X \cap \mu(Y) \subseteq \mu(X)$.

□

Fact 1.5.14

Assume $(\mu \subseteq)$.

We have for $(\mu Cum\infty)$ and (HU, u):

(1) $x \in \mu(Y), \mu(Y) \subseteq H(U, x) \Rightarrow Y \subseteq H(U, x)$

(2) $(\mu Cum\infty) \Rightarrow (HU, u)$

(3) $(HU, u) \Rightarrow (\mu Cum\infty)$

Proof

(1) Trivial by definition of $H(U, x)$.

(2) Exercise, solution in the Appendix.

(3) Suppose $(\mu Cum\alpha)$ fails, we show that then so does (HU, u) for $u = x$. As $(\mu Cum\alpha)$ fails, for all $\beta \leq \alpha$ $\mu(X_\beta) \subseteq U \cup \bigcup \{X_\gamma : \gamma < \beta\}$, but there is $x \in \bigcap \{X_\gamma : \gamma \leq \alpha\} \cap \mu(U), x \notin \mu(X_\alpha)$. Thus for all $\beta \leq \alpha$ $\mu(X_\beta) \subseteq X_\beta \subseteq H(U, x)$, moreover $x \in \mu(U), x \in X_\alpha - \mu(X_\alpha)$, but $\mu(X_\alpha) \subseteq H(U, x)$, so (HU, u) fails for $u = x$.

□

Fact 1.5.15

We continue to show results for $H(U)$ and $H(U, u)$.

Let A, X, U, U', Y and all A_i be in $\mathcal{Y}$.

(1) $H(U)$ and $H(U, u)$

   (1.1) $H(U, u) \subseteq H(U)$,
   (1.2) $(HU) \Rightarrow (HU, u)$,
   (1.3) $(\cup) + (\mu PR)$ entail $H(U) \subseteq H(U, u)$ for $u \in \mu(U)$,
   (1.4) $(\cup) + (\mu PR)$ entail $(HU, u) \Rightarrow (HU)$,

(2) $(\mu \subseteq)$ and (HU) entail:

   (2.1) (μPR),
   (2.2) (μCUM),

(3) $(\mu \subseteq)$ and (μPR) entail:

   (3.1) $A = \bigcup\{A_i : i \in I\} \Rightarrow \mu(A) \subseteq \bigcup\{\mu(A_i) : i \in I\}$,
   (3.2) $U \subseteq H(U)$, and $U \subseteq U' \Rightarrow H(U) \subseteq H(U')$,
   (3.3) $\mu(U \cup Y) - H(U) \subseteq \mu(Y)$ - if $\mu(U \cup Y)$ is defined, in particular, if $(\cup)$ holds.

(4) $(\cup)$, $(\mu \subseteq)$, (μPR), (μCUM) entail:

   (4.1) $H(U) = H_1(U)$,
   (4.2) $U \subseteq A$, $\mu(A) \subseteq H(U) \Rightarrow \mu(A) \subseteq U$,
   (4.3) $\mu(Y) \subseteq H(U) \Rightarrow Y \subseteq H(U)$ and $\mu(U \cup Y) = \mu(U)$,
   (4.4) $x \in \mu(U)$, $x \in Y - \mu(Y) \Rightarrow Y \not\subseteq H(U)$ (and thus (HU)),
   (4.5) $Y \not\subseteq H(U) \Rightarrow \mu(U \cup Y) \not\subseteq H(U)$.

(5) $(\cup)$, $(\mu \subseteq)$, (HU) entail

   (5.1) $H(U) = H_1(U)$,
   (5.2) $U \subseteq A$, $\mu(A) \subseteq H(U) \Rightarrow \mu(A) \subseteq U$,
   (5.3) $\mu(Y) \subseteq H(U) \Rightarrow Y \subseteq H(U)$ and $\mu(U \cup Y) = \mu(U)$,
   (5.4) $x \in \mu(U)$, $x \in Y - \mu(Y) \Rightarrow Y \not\subseteq H(U)$,
   (5.5) $Y \not\subseteq H(U) \Rightarrow \mu(U \cup Y) \not\subseteq H(U)$.

1.5 The Smooth Case Without Domain Closure

Proof

(1.1) and (1.2) trivial by definition.

(1.3) Proof by induction. $H(U)_0 = H(U, u)_0$ is trivial. Suppose $H(U)_\beta = H(U, u)_\beta$ has been shown for $\beta < \alpha$. The limit step is trivial, so suppose $\alpha = \beta + 1$. Let X be such that $\mu(X) \subseteq H(U)_\beta = H(U, u)_\beta$, so $X \subseteq H(U)_\alpha$. Consider $X \cup U$, so $u \in X \cup U$, $\mu(X \cup U)$ is defined and by (μPR) and $(\mu \subseteq)$ $\mu(X \cup U) \subseteq \mu(X) \cup \mu(U) \subseteq H(U)_\beta = H(U, u)_\beta$, so $X \cup U \subseteq H(U, u)_\alpha$.

(1.4) Immediate by (1.3).

(2.1) By (HU), if $\mu(Y) \subseteq H(U)$, then $\mu(U) \cap Y \subseteq \mu(Y)$. But, if $Y \subseteq U$, then $\mu(Y) \subseteq H(U)$ by $(\mu \subseteq)$.

(2.2)

Exercise, solution in the Appendix.

(3.1) $\mu(A) \cap A_j \subseteq \mu(A_j) \subseteq \bigcup \mu(A_i)$, so by $\mu(A) \subseteq A = \bigcup A_i$ $\mu(A) \subseteq \bigcup \mu(A_i)$.

(3.2) trivial.

(3.3) $\mu(U \cup Y) - H(U) \subseteq_{(3.2)} \mu(U \cup Y) - U \subseteq$ (by $(\mu \subseteq)$) $\mu(U \cup Y) \cap Y \subseteq_{(\mu PR)} \mu(Y)$.

(4.1) We show that, if $X \subseteq H_2(U)$, then $X \subseteq H_1(U)$, more precisely, if $\mu(X) \subseteq H_1(U)$, then already $X \subseteq H_1(U)$, so the construction stops already at $H_1(U)$. Suppose then $\mu(X) \subseteq \bigcup\{Y : \mu(Y) \subseteq U\}$, and let $A := X \cup U$. We show that $\mu(A) \subseteq U$, so $X \subseteq A \subseteq H_1(U)$. Let $a \in \mu(A)$. By (μPR), $(\mu \subseteq)$, $\mu(A) \subseteq \mu(X) \cup \mu(U)$. If $a \in \mu(U) \subseteq U$, we are done. If $a \in \mu(X)$, there is Y such that $\mu(Y) \subseteq U$ and $a \in Y$, so $a \in \mu(A) \cap Y$. By Fact 1.5.13, (5.1.3), we have for Y such that $\mu(Y) \subseteq U$ and $U \subseteq A$ $\mu(A) \cap Y \subseteq \mu(U)$. Thus $a \in \mu(U)$, and we are done again.

(4.2) Let $U \subseteq A$, $\mu(A) \subseteq H(U) = H_1(U)$ by (4.1). So $\mu(A) = \bigcup\{\mu(A) \cap Y : \mu(Y) \subseteq U\} \subseteq \mu(U) \subseteq U$, again by Fact 1.5.13, (5.1.3).

(4.3) Let $\mu(Y) \subseteq H(U)$, then by $\mu(U) \subseteq H(U)$ and (μPR), $(\mu \subseteq)$, $\mu(U \cup Y) \subseteq \mu(U) \cup \mu(Y) \subseteq H(U)$, so by (4.2) $\mu(U \cup Y) \subseteq U$ and $U \cup Y \subseteq H(U)$. Moreover, $\mu(U \cup Y) \subseteq U \subseteq U \cup Y \Rightarrow_{(\mu CUM)} \mu(U \cup Y) = \mu(U)$.

(4.4) If not, $Y \subseteq H(U)$, so $\mu(Y) \subseteq H(U)$, so $\mu(U \cup Y) = \mu(U)$ by (4.3), but $x \in Y - \mu(Y) \Rightarrow_{(\mu PR)} x \notin \mu(U \cup Y) = \mu(U)$, $contradiction$.

(4.5) $\mu(U \cup Y) \subseteq H(U) \Rightarrow_{(4.3)} U \cup Y \subseteq H(U)$.

(5) Trivial by (1) and (4).

□

The Representation Result

We turn now to the representation result and its proof.

We adapt Proposition 3.7.15 in [Sch04] and its proof. All we need is (HU, u) and $(\mu \subseteq)$. We modify the proof of Remark 3.7.13 (1) in [Sch04] (now Remark 1.5.17) so we will not need $(\cap)$ any more. We will give the full proof, although its essential elements have already been published, for three reasons: First, the new version will need less prerequisites than the old proof does (closure under finite intersections is not needed any more, and replaced by (HU, u)). Second, we will more clearly separate the requirements to do the construction from the construction itself, thus splitting the proof neatly into two parts.

We show how to work with $(\mu \subseteq)$ and (HU, u) only. Thus, once we have shown $(\mu \subseteq)$ and (HU, u), we have finished the substantial side, and enter the administrative part, which will not use any prerequisites about domain closure any more. At the same time, this gives a uniform proof of the difficult part for the case with and without $(\cup)$, in the former case we can even work with the stronger $H(U)$. The easy direction of the former parts needs a proof of the stronger $H(U)$, but this is easy.

Note that, in the presence of $(\mu \subseteq)$, $(HU, u) \Rightarrow (\mu Cum\infty)$ and $(\mu Cum0) \Rightarrow (\mu PR)$, by Fact 1.5.14, (3) and Fact 1.5.13, (7), so (HU, u) entails (μPR), so we can use it in our context, where (HU, u) will be the central property.

Fact 1.5.16

(HU, u) holds in all smooth models.

Proof

Suppose not. So let $x \in \mu(U)$, $x \in Y - \mu(Y)$, $\mu(Y) \subseteq H(U, x)$. By smoothness, there is $x_1 \in \mu(Y)$, $x \succ x_1$, and let κ_1 be the least κ such that $x_1 \in H(U, x)_{\kappa_1}$. κ_1 is not a limit, and $x_1 \in U'_{x_1} - \mu(U'_{x_1})$ with $x \in U'_{x_1}$ by definition of $H(U, x)$ for some U'_{x_1}, so as $x_1 \notin \mu(U'_{x_1})$, there must be (by smoothness) some other $x_2 \in \mu(U'_{x_1}) \subseteq H(U, x)_{\kappa_1 - 1}$ with $x \succ x_2$. Continue with x_2, we thus construct a descending chain of ordinals, which cannot be infinite, so there must be $x_n \in \mu(U'_{x_n}) \subseteq U$, $x \succ x_n$, contradicting minimality of x in U. (More precisely, this works for all copies of x.)

□

We first show two basic facts and then turn to the main result, Proposition 1.5.19.

Recall Definition 1.3.3.

The following remark is the same as Remark 1.3.12, but we now give two proofs.

Remark 1.5.17

(1) $x \in K \Rightarrow \Gamma_x \neq \emptyset$,

(2) $g \in \Gamma_x \Rightarrow ran(g) \subseteq K$.

1.5 The Smooth Case Without Domain Closure

Proof

(1) We give two proofs, the first uses $(\mu Cum0)$, the second the stronger (by Fact 1.5.14 (3)) (HU, u).

(a) We have to show that $Y \in \mathcal{Y}, x \in Y - \mu(Y) \Rightarrow \mu(Y) \neq \emptyset$. Suppose then $x \in \mu(X)$, this exists, as $x \in K$, and $\mu(Y) = \emptyset$, so $\mu(Y) \subseteq X$, $x \in Y$, so by $(\mu Cum0)$ $x \in \mu(Y)$.

(b) Consider $H(X, x)$, suppose $\mu(Y) = \emptyset$, $x \in Y$, so $Y \subseteq H(X, x)$, so $x \in \mu(Y)$ by (HU, u).

(2) By definition, $\mu(Y) \subseteq K$ for all $Y \in \mathcal{Y}$.

□

Claim 1.5.18

Let $U \in \mathcal{Y}, x \in K$. Then

(1) $x \in \mu(U) \Leftrightarrow x \in U \wedge \exists f \in \Gamma_x.ran(f) \cap U = \emptyset$,

(2) $x \in \mu(U) \Leftrightarrow x \in U \wedge \exists f \in \Gamma_x.ran(f) \cap H(U, x) = \emptyset$.

Proof

Exercise, solution see [Sch04], Claim 3.7.14.

The following Proposition 1.5.19 is the main positive result of Section 1.5.3.2 and shows how to characterize smooth structures in the absence of closure under finite unions. The strategy of the proof follows closely the proof of Proposition 3.3.4 in [Sch04].

Proposition 1.5.19

Let $\mu : \mathcal{Y} \to \mathcal{P}(Z)$. Then there is a $\mathcal{Y}$-smooth preferential structure $\mathcal{Z}$, such that for all $X \in \mathcal{Y}$ $\mu(X) = \mu_{\mathcal{Z}}(X)$ iff μ satisfies $(\mu \subseteq)$ and (HU, u) above.

In particular, we need no prerequisites about domain closure.

Proof

"$\Rightarrow$" (HU, u) was shown in Fact 1.5.16.

Outline of "$\Leftarrow$": We first define a structure $\mathcal{Z}$ which represents μ, but is not necessarily $\mathcal{Y}$-smooth, refine it to $\mathcal{Z}'$ and show that $\mathcal{Z}'$ represents μ too, and that $\mathcal{Z}'$ is $\mathcal{Y}$-smooth.

In the structure $\mathcal{Z}'$, all pairs destroying smoothness in $\mathcal{Z}$ are successively repaired, by adding minimal elements: If $\langle y, j \rangle$ is not minimal, and has no minimal $\langle x, i \rangle$ below it, we just add one such $\langle x, i \rangle$. As the repair process might itself generate such "bad" pairs, the process may have to be repeated infinitely often. Of course, one has to take care that the representation property is preserved.

Construction 1.5.1

(Construction of $\mathcal{Z}$)

Let $\mathcal{X} := \{\langle x, g \rangle : x \in K, g \in \Gamma_x\}$, $\langle x', g' \rangle \prec \langle x, g \rangle :\Leftrightarrow x' \in ran(g)$, $\mathcal{Z} := \langle \mathcal{X}, \prec \rangle$.

Claim 1.5.20

$\forall U \in \mathcal{Y}. \mu(U) = \mu_{\mathcal{Z}}(U)$

Proof

Exercise, solution see [Sch04], Claim 3.7.16.

The construction of the refined structure $\mathcal{Z}'$ is (almost) identical to Construction 1.3.4. The only difference is that we use here Remark 1.5.17, (1) to show $\Sigma_x \neq \emptyset$ if $x \in K$.

It is now easy to show that $\mathcal{Z}'$ represents μ, and that $\mathcal{Z}'$ is smooth. For $x \in \mu(U)$, we construct a special x-admissible sequence $\sigma^{x,U}$ using the properties of $H(U, x)$ as described at the beginning of this section.

Claim 1.5.21

For all $U \in \mathcal{Y}$ $\mu(U) = \mu_{\mathcal{Z}}(U) = \mu'(U)$.

Proof

Exercise.

Hint:

either replace in the proof of Claim 1.3.16

$H(U)$ by $H(U, x)$,

Fact 1.3.11, (6) + (5) by (HU, u) for $H(U, x)$,

$\mu(U \cup X)$ by $\mu(X)$, and

Fact 1.3.11, (7) by Fact 1.5.14, (1)

or replace in the proof of Claim 3.7.17 in [Sch04]

(HU) by (HU, x) for $H(U, x)$, and

[Sch04], Fact 3.7.12 (1) by Fact 1.5.14, (1).

It remains to show:

Claim 1.5.22

$\mathcal{Z}'$ is $\mathcal{Y}$-smooth.

1.5 The Smooth Case Without Domain Closure

Proof

Exercise, solution see [Sch04], Claim 3.7.18.

We conclude this section by showing that we cannot improve substantially.

Proposition 1.5.23

There is no fixed size characterization of μ-functions which are representable by smooth structures, if the domain is not closed under finite unions.

Proof

Suppose we have a fixed size characterization, which allows to distinguish μ-functions on domains which are not necessarily closed under finite unions, and which can be represented by smooth structures, from those which cannot be represented in this way. Let the characterization have α parameters for sets, and consider Example 1.5.3 with $\kappa = \beta + 1$, $\beta > \alpha$ (as a cardinal). This structure cannot be represented, as $(\mu Cum\kappa)$ fails - see Fact 1.5.13, (2.1). As we have only α parameters, at least one of the X_γ is not mentioned, say X_δ. Without loss of generality, we may assume that $\delta = \delta' + 1$. We change now the structure, and erase one pair of the relation, $x_\delta \prec x_{\delta+1}$. Thus, $\mu(X_\delta) = \{c, x_\delta, x_{\delta+1}\}$. But now we cannot go any more from $X_{\delta'}$ to $X_{\delta'+1} = X_\delta$, as $\mu(X_\delta) \not\subseteq X_{\delta'}$. Consequently, the only chain showing that $(\mu Cum\infty)$ fails is interrupted - and we have added no new possibilities, as inspection of cases shows. ($x_{\delta+1}$ is now globally minimal, and increasing $\mu(X)$ cannot introduce new chains, only interrupt chains.) Thus, $(\mu Cum\infty)$ holds in the modified example, and it is thus representable by a smooth structure, as above proposition shows. As we did not touch any of the parameters, the truth value of the characterization is unchanged, which was negative. So the "characterization" cannot be correct.

$\square$

The Transitive Smooth Case

Unfortunately, $(\mu Cumt\infty)$ is a necessary but not sufficient condition for smooth transitive structures, as can be seen in the following example.

Example 1.5.4

We assume no closure whatever.

$U := \{u_1, u_2, u_3, u_4\}$, $\mu(U) := \{u_3, u_4\}$

$Y_1 := \{u_4, v_1, v_2, v_3, v_4\}$, $\mu(Y_1) := \{v_3, v_4\}$

$Y_{2,1} := \{u_2, v_2, v_4\}$, $\mu(Y_{2,1}) := \{u_2, v_2\}$

$Y_{2,2} := \{u_1, v_1, v_3\}$, $\mu(Y_{2,2}) := \{u_1, v_1\}$

For no A, B $\mu(A) \subseteq B$ ($A \neq B$), so the prerequisite of ($\mu Cumt\alpha$) is false, and ($\mu Cumt\alpha$) holds, but there is no smooth transitive representation possible: Consider Y_1. If $u_4 \succ v_3$, then $Y_{2,2}$ makes this impossible, if $u_4 \succ v_4$, then $Y_{2,1}$ makes this impossible.

□

Remark 1.5.24

(1) The situation does not change when we have copies, the same argument will still work: There is a U-minimal copy $\langle u_4, i \rangle$, by smoothness and Y_1, there must be a Y_1-minimal copy, e.g., $\langle v_3, j \rangle \prec \langle u_4, i \rangle$. By smoothness and $Y_{2,2}$, there must be a $Y_{2,2}$-minimal $\langle u_1, k \rangle$ or $\langle v_1, l \rangle$ below $\langle v_3, j \rangle$. But v_1 is in Y_1, contradicting minimality of $\langle v_3, j \rangle$, u_1 is in U, contradicting minimality of $\langle u_4, i \rangle$ by transitivity. If we choose $\langle v_4, j \rangle$ minimal below $\langle u_4, i \rangle$, we will work with $Y_{2,1}$ instead of $Y_{2,2}$.

(2) We can also close under arbitrary intersections, and the example will still work: We have to consider $U \cap Y_1, U \cap Y_{2,1}, U \cap Y_{2,2}, Y_{2,1} \cap Y_{2,2}, Y_1 \cap Y_{2,1}, Y_1 \cap Y_{2,2}$, there are no further intersections to consider. We may assume $\mu(A) = A$ for all these intersections (working with copies). But then $\mu(A) \subseteq B$ implies $\mu(A) = A$ for all sets, and all ($\mu Cumt\alpha$) hold again trivially.

(3) If we had finite unions, we could form $A := U \cup Y_1 \cup Y_{2,1} \cup Y_{2,2}$, then $\mu(A)$ would have to be a subset of $\{u_3\}$ by (μPR), so by (μCUM) $u_4 \notin \mu(U)$, a contradiction. Finite unions allow us to "look ahead", without ($\cup$), we see desaster only at the end - and have to backtrack, i.e., try in our example $Y_{2,1}$, once we have seen impossibility via $Y_{2,2}$, and discover impossibility again at the end.

□

1.6 The Limit Variant

1.6.1 Introduction

We will introduce the concepts of the structural, the algebraic, and the logical limit, and will see that this allows us to separate problems in this usually quite difficult case. Some problems are simply due to the fact that a seemingly nice structural limit does not have nice algebraic properties any more, so it should not be considered. So, to have a "good" limit, the limit should not only capture the idea of a structural limit, but its algebraic counterpart should also capture the essential algebraic properties

1.6 The Limit Variant

of the minimal choice functions. Other problems are due to the fact that the nice algebraic limit does not translate to nice logical properties, and we will see that this is often due to the same problems we saw in the absence of definability preservation.

Likewise, for the limit situation, we have:

- structural limits - they are again the foundation,
- resulting abstract behaviour, which, again, has to be an abstract or algebraic limit, resulting from the structural limit,
- a logical limit, which reflects the abstract limit, and may be plagued by definability preservation problems etc. when going from the model to the logics side.

Distance based semantics give perhaps the clearest motivation for the limit variant. For instance, the Stalnaker/Lewis semantics for counterfactual conditionals defines $\phi > \psi$ to hold in a (classical) model m iff in those models of ϕ, which are closest to m, ψ holds. For this to make sense, we need, of course, a distance d on the model set. We call this approach the minimal variant. Usually, one makes a limit assumption: The set of ϕ-models closest to m is not empty if ϕ is consistent - i.e., the ϕ-models are not arranged around m in a way that they come closer and closer, without a minimal distance. This is, of course, a very strong assumption, and which is probably difficult to justify philosophically. It seems to have its only justification in the fact that it avoids degenerate cases, where, in above example, for consistent ϕ $m \models \phi > FALSE$ holds. As such, this assumption is unsatisfactory.

Our aim here is to analyze the limit version more closely, in particular, to see criteria whether the much more complex limit version can be reduced to the simpler minimal variant. In the limit version, roughly, ψ is a consequence of ϕ, if ψ holds "in the limit" in all ϕ-models. That is, iff, "going sufficiently far down", ψ will become and stay true.

The problem is not simple, as there are two sides which come into play, and sometimes we need both to cooperate to achieve a satisfactory translation.

The first component is what we call the "algebraic limit", i.e., we stipulate that the limit version should have properties which correspond to the algebraic properties of the minimal variant. An exact correspondence cannot always be achieved, and we give a translation which seems reasonable.

But once the translation is done, even if it is exact, there might still be problems linked to translation to logic.

(1) The structural limit: It is a natural and much more convincing solution to the problem described above to modify the basic definition, and work without the rather artificial assumption that the closest world exists. We adopt what we call a "limit approach", and define $m \models \phi > \psi$ iff there is a distance d' such that for all $m' \models \phi$ and $d(m, m') \leq d'$ $m' \models \psi$. Thus, from a certain point onward,

ψ becomes and stays true. We will call this definition the structural limit, as it is based directly on the structure (the distance on the model set).

(2) The algebraic limit: The model sets to consider are spheres around m, $S := \{m' \in M(\phi) : d(m, m') \leq d'\}$ for some d', such that $S \neq \emptyset$. The system of such S is nested, i.e., totally ordered by inclusion; and if $m \models \phi$, it has a smallest element $\{m\}$, etc. When we forget the underlying structure, and consider just the properties of these systems of spheres around different m, and for different ϕ, we obtain what we call the algebraic limit.

(3) The logical limit: The logical limit speaks about the logical properties which hold "in the limit", i.e., finally in all such sphere systems.

The interest to investigate this algebraic limit is twofold: first, we shall see (for other kinds of structures) that there are reasonable and not so reasonable algebraic limits. Second, this distinction permits us to separate algebraic from logical problems, which have to do with definability of model sets, in short definability problems. We will see that we find common definability problems and also common solutions in the usual minimal, and the limit variant.

In particular, the decomposition into three layers on both sides (minimal and limit version) can reveal that a (seemingly) natural notion of structural limit results in algebraic properties which have not much to do any more with the minimal variant. So, to speak about a limit variant, we will demand that this variant is not only a natural structural limit, but results in a natural abstract limit, too. Conversely, if the algebraic limit preserves the properties of the minimal variant, there is hope that it preserves the logical properties, too - not more than hope, however, due to definability problems.

1.6.2 The Algebraic Limit

There are basic problems with the algebraic limit in general preferential structures.

Example 1.6.1

Let $a \prec b$, $a \prec c$, $b \prec d$, $c \prec d$ (but $\prec$ not transitive!), then $\{a, b\}$ and $\{a, c\}$ are such S and S', but there is no $S'' \subseteq S \cap S'$ which is an initial segment. If, for instance, in a and b ψ holds, in a and c ψ', then "in the limit" ψ and ψ' will hold, but not $\psi \wedge \psi'$. This does not seem right. We should not be obliged to give up ψ to obtain ψ'.

□

Recall Definition 1.2.15.

1.6 The Limit Variant

When we look at the system of such S generated by a preferential structure and its algebraic properties, we will therefore require it to be closed under finite intersections, or at least, that if S, S' are such segments, then there must be $S'' \subseteq S \cap S'$ which is also such a segment.

We make this official. Let $\Lambda(X)$ be the set of initial segments of X, then we require:

Definition 1.6.1

$(\Lambda\cap)$ If $A, B \in \Lambda(X)$ then there is $C \subseteq A \cap B, C \in \Lambda(X)$.

More precisely, a limit should be a structural limit in a reasonable sense - whatever the underlying structure is -, and the resulting algebraic limit should respect $(\Lambda\cap)$.

We should not demand too much, either. It would be wrong to demand closure under arbitrary intersections, as this would mean that there is an initial segment which makes all consequences true - trivializing the very idea of a limit.

But we can make our requirements more precise, and bind the limit variant closely to the minimal variant, by looking at the algebraic version of both.

Before we look at deeper problems, we show some basic facts about the algebraic limit.

Fact 1.6.1

(Taken from [Sch04], Fact 3.4.3 there.)

Let the relation $\prec$ be transitive. The following hold in the limit variant of general preferential structures:

(1) If $A \in \Lambda(Y)$, and $A \subseteq X \subseteq Y$, then $A \in \Lambda(X)$.

(2) If $A \in \Lambda(Y)$, and $A \subseteq X \subseteq Y$, and $B \in \Lambda(X)$, then $A \cap B \in \Lambda(Y)$.

(3) If $A \in \Lambda(Y), B \in \Lambda(X)$, then there is $Z \subseteq A \cup B$ $Z \in \Lambda(Y \cup X)$.

(Taken from [Sch04], Proposition 3.10.16 there:)

The following hold in the limit variant of ranked structures without copies, where the domain is closed under finite unions and contains all finite sets.

(4) $A, B \in \Lambda(X) \Rightarrow A \subseteq B$ or $B \subseteq A$,

(5) $A \in \Lambda(X), Y \subseteq X, Y \cap A \neq \emptyset \Rightarrow Y \cap A \in \Lambda(Y)$,

(6) $\Lambda' \subseteq \Lambda(X), \bigcap \Lambda' \neq \emptyset \Rightarrow \bigcap \Lambda' \in \Lambda(X)$.

(7) $X \subseteq Y, A \in \Lambda(X) \Rightarrow \exists B \in \Lambda(Y).B \cap X = A$

Proof

Exercise, solution see [Sch04], Fact 3.4.3 and Proposition 3.10.16.

We have as immediate logical consequence:

Fact 1.6.2

If $\prec$ is transitive, then

(1) (AND) holds,

(2) (OR) holds,

(3) $\overline{\overline{\phi \wedge \phi'}} \subseteq \overline{\overline{\phi}} \cup \{\phi'\}$

Proof

Exercise, solution see [Sch04], Fact 3.4.4.

1.6.3 The Logical Limit

1.6.3.1 Translation Between the Minimal and the Limit Variant

A good example for problems linked to the translation from the algebraic limit to the logical limit is the property $(\mu =)$ of ranked structures:

$(\mu =) \; X \subseteq Y, \mu(Y) \cap X \neq \emptyset \Rightarrow \mu(Y) \cap X = \mu(X)$

or its logical form

$(\mathrel{|\!\sim}=) \; T \vdash T', Con(\overline{\overline{T'}}, T) \Rightarrow \overline{\overline{T}} = \overline{\overline{T'}} \cup T.$

$\mu(Y)$ or its analogue $\overline{\overline{T'}}$ (set $X := M(T)$, $Y := M(T')$) speak about the limit, the "ideal", and this, of course, is not what we have in the limit version. This limit version was intoduced precisely to avoid speaking about the ideal.

So, first, we have to translate $\mu(Y) \cap X \neq \emptyset$ to something else, and the natural candidate seems to be

$\forall B \in \Lambda(Y). B \cap X \neq \emptyset.$

In logical terms, we have replaced the set of consequences of Y by some $Th(B)$ where $T' \subseteq Th(B) \subseteq \overline{\overline{T'}}$. The conclusion can now be translated in a similar way to $\forall B \in \Lambda(Y). \exists A \in \Lambda(X). A \subseteq B \cap X$ and $\forall A \in \Lambda(X). \exists B \in \Lambda(Y). B \cap X \subseteq A$. The total translation reads now:

$(\Lambda =)$ Let $X \subseteq Y$. Then

$\Big(\forall B \in \Lambda(Y). B \cap X \neq \emptyset\Big) \Rightarrow \Big(\forall B \in \Lambda(Y). \exists A \in \Lambda(X). A \subseteq B \cap X$ and $\forall A \in \Lambda(X). \exists B \in \Lambda(Y). B \cap X \subseteq A\Big).$

By Fact 1.6.1 (5) and (7), we see that this holds in ranked structures. Thus, the limit reading seems to provide a correct algebraic limit.

Yet, Example 1.6.2 below shows the following:

1.6 The Limit Variant   121

Let $m' \neq m$ be arbitrary. For $T' := Th(\{m, m'\})$, $T := \emptyset$, we have $T' \vdash T$, $\overline{\overline{T'}} = Th(\{m'\})$, $\overline{\overline{T}} = Th(\{m\})$, $Con(\overline{\overline{T}}, T')$, but $Th(\{m\}) = \overline{\overline{T}} \cup T' \neq \overline{\overline{T'}}$.

Thus:

(1) The prerequisite holds, though usually for $A \in \Lambda(T)$, $A \cap M(T') = \emptyset$.

(2) (PR) fails, which is independent of the prerequisite $Con(\overline{\overline{T}}, T')$, so the problem is not just due to the prerequisite.

(3) Both inclusions of $(\mid\!\sim=)$ fail.

We will see below in Corollary 1.6.6 a sufficient condition to make $(\mid\!\sim=)$ hold in ranked structures. It has to do with definability or formulas, more precisely, the crucial property is to have sufficiently often $\widehat{A \cap M(T')} = \widehat{A} \cap M(T')$ for $A \in \Lambda(T)$ - see Section 1.7.2.1 for reference.

Example 1.6.2

(Taken from [Sch04], Example 3.10.1 (1) there.)

Take an infinite propositional language $p_i : i \in \omega$. We have ω_1 models (assume for simplicity CH).

Take the model m which makes all p_i true, and put it on top. Next, going down, take all models which make p_0 false, and then all models which make p_0 true, but p_1 false, etc. in a ranked construction. So, successively more p_i will become (and stay) true. Consequently, $\emptyset \models_\Lambda p_i$ for all i. But the structure has no minimum, and the "logical" limit m is not in the set wise limit. Let $T := \emptyset$ and $m' \neq m$, $T' := Th(\{m, m'\})$, then $\overline{\overline{T}} = Th(\{m\})$, $\overline{\overline{T'}} = Th(\{m'\})$, and $\overline{\overline{T'}} \cup T = \overline{\overline{T'}} = Th(\{m'\})$ and $\overline{\overline{T}} \cup T' = \overline{\overline{T}} = Th(\{m\})$.

□

This example shows that our translation is not perfect, but it is half the way. Note that the minimal variant faces the same problems (definability and others), so the problems are probably at least not totally due to our perhaps insufficient translation.

We turn to other rules.

$(\Lambda\cap)$ If $A, B \in \Lambda(X)$ then there is $C \subseteq A \cap B$, $C \in \Lambda(X)$

seems a minimal requirement for an appropriate limit. It holds in transitive structures by Fact 1.6.1 (2).

The central logical condition for minimal smooth structures is

(CUM) $T \subseteq T' \subseteq \overline{\overline{T}} \Rightarrow \overline{\overline{T}} = \overline{\overline{T'}}$

It would again be wrong - using the limit - to translate this only partly by: If $T \subseteq T' \subseteq \overline{\overline{T}}$, then for all $A \in \Lambda(M(T))$ there is $B \in \Lambda(M(T'))$ such that $A \subseteq B$ - and vice versa. Now, smoothness is in itself a wrong condition for limit structures, as it speaks about minimal elements, which we will not necessarily have. This cannot guide us. But when we consider a more modest version of cumulativity, we see what to do.

$(CUMfin)$ If $T \mathrel|\!\sim \phi$, then $\overline{\overline{T}} = \overline{\overline{T \cup \{\phi\}}}$.

This translates into algebraic limit conditions as follows - where $Y = M(T)$, and $X = M(T \cup \{\phi\})$:

$(\Lambda CUMfin)$ Let $X \subseteq Y$. If there is $B \in \Lambda(Y)$ such that $B \subseteq X$, then:

$\left(\forall A \in \Lambda(X) \exists B' \in \Lambda(Y).B' \subseteq A \text{ and } \forall B' \in \Lambda(Y) \exists A \in \Lambda(X).A \subseteq B' \right)$.

Note, that in this version, we do not have the "ideal" limit on the left of the implication, but one fixed approximation $B \in \Lambda(Y)$. We can now prove that $(\Lambda CUMfin)$ holds in transitive structures: The first part holds by Fact 1.6.1 (2), the second, as $B \cap B' \in \Lambda(Y)$ by Fact 1.6.1 (1). This is true without additional properties of the structure, which might at first sight seem surprising. But note that the initial segments play a similar role as the set of minimal elements: an initial segment has to minimize the other elements, just as the set of minimal elements in the smooth case does.

The central algebraic property of minimal preferential structures is

(μPR) $X \subseteq Y \Rightarrow \mu(Y) \cap X \subseteq \mu(X)$

This translates naturally and directly to

(ΛPR) $X \subseteq Y \Rightarrow \forall A \in \Lambda(X) \exists B \in \Lambda(Y).B \cap X \subseteq A$

(ΛPR) holds in transitive structures: $Y - X \in \Lambda(Y - X)$, so the result holds by Fact 1.6.1 (3).

The central algebraic condition of ranked minimal structures is

$(\mu =)$ $X \subseteq Y, \mu(Y) \cap X \neq \emptyset \Rightarrow \mu(Y) \cap X = \mu(X)$

We saw above how to translate this condition to $(\Lambda =)$, we also saw that $(\Lambda =)$ holds in ranked structures.

We will see in Corollary 1.6.6 that the following logical version holds in ranked structures:

$T \mathrel|\!\!\not\sim \neg \gamma$ implies $\overline{\overline{T}} = \overline{\overline{T \cup \{\gamma\}}}$

We generalize above translation results to a recipe:

Translate

(1) $\mu(X) \subseteq \mu(Y)$ to $\forall B \in \Lambda(Y) \exists A \in \Lambda(X).A \subseteq B$, and thus
(2) $\mu(Y) \cap X \subseteq \mu(X)$ to $\forall A \in \Lambda(X) \exists B \in \Lambda(Y).B \cap X \subseteq A$,
(3) $\mu(X) \subseteq Y$ to $\exists A \in \Lambda(X).A \subseteq Y$, and thus
(4) $\mu(Y) \cap X \neq \emptyset$ to $\forall B \in \Lambda(Y).B \cap X \neq \emptyset$
(5) $X \subseteq \mu(Y)$ to $\forall B \in \Lambda(Y).X \subseteq B$,

and quantify expressions separately, thus we repeat:

1.6 The Limit Variant

(6) (μCUM) $\mu(Y) \subseteq X \subseteq Y \Rightarrow \mu(X) = \mu(Y)$ translates to

(7) $(\Lambda CUM fin)$ Let $X \subseteq Y$. If there is $B \in \Lambda(Y)$ such that $B \subseteq X$, then:
$$\left(\forall A \in \Lambda(X) \exists B' \in \Lambda(Y).B' \subseteq A \text{ and } \forall B' \in \Lambda(Y) \exists A \in \Lambda(X).A \subseteq B' \right).$$

(8) $(\mu =)$ $X \subseteq Y, \mu(Y) \cap X \neq \emptyset \Rightarrow \mu(Y) \cap X = \mu(X)$ translates to

(9) $(\Lambda =)$ Let $X \subseteq Y$. If $\forall B \in \Lambda(Y).B \cap X \neq \emptyset$, then
$$\left(\forall A \in \Lambda(X) \exists B' \in \Lambda(Y).B' \cap X \subseteq A, \text{ and } \forall B' \in \Lambda(Y) \exists A \in \Lambda(X).A \subseteq B' \cap X \right).$$

We collect now for easier reference the definitions and some algebraic properties which we saw above to hold:

Definition 1.6.2

$(\Lambda \cap)$ If $A, B \in \Lambda(X)$ then there is $C \subseteq A \cap B$, $C \in \Lambda(X)$,

(ΛPR) $X \subseteq Y \Rightarrow \forall A \in \Lambda(X) \exists B \in \Lambda(Y).B \cap X \subseteq A$,

$(\Lambda CUM fin)$ Let $X \subseteq Y$. If there is $B \in \Lambda(Y)$ such that $B \subseteq X$, then:
$$\left(\forall A \in \Lambda(X) \exists B' \in \Lambda(Y).B' \subseteq A \text{ and } \forall B' \in \Lambda(Y) \exists A \in \Lambda(X).A \subseteq B' \right).$$

$(\Lambda =)$ Let $X \subseteq Y$. If $\forall B \in \Lambda(Y).B \cap X \neq \emptyset$, then
$$\left(\forall A \in \Lambda(X) \exists B' \in \Lambda(Y).B' \cap X \subseteq A, \text{ and } \forall B' \in \Lambda(Y) \exists A \in \Lambda(X).A \subseteq B' \cap X \right).$$

Fact 1.6.3

In transitive structures hold:

(1) $(\Lambda \cap)$

(2) (ΛPR)

(3) $(\Lambda CUM fin)$

In ranked structures holds:

(4) $(\Lambda =)$

Proof

Exercise, solution in the Appendix.

To summarize the discussion:

Just as in the minimal case, the algebraic laws may hold, but not the logical ones, due in both cases to definability problems. Thus, we cannot expect a clean proof of correspondence. But we can argue that we did a correct translation, which shows its limitation, too. The part with $\mu(X)$ and $\mu(Y)$ on both sides of $\subseteq$ is obvious, we will have a perfect correspondence. The part with $X \subseteq \mu(Y)$ is obvious, too. The problem is in the part with $\mu(X) \subseteq Y$. As we cannot use the limit, but only its approximation, we are limited here to one (or finitely many) consequences of T, if $X = M(T)$, so we obtain only $T \mathrel{|\!\sim} \phi$, if $Y \subseteq M(\phi)$, and if there is $A \in \Lambda(X).A \subseteq Y$.

We consider a limit only appropriate, if it is an algebraic limit which preserves algebraic properties of the minimal version in above translation.

The advantage of such limits is that they allow - with suitable caveats - to show that they preserve the logical properties of the minimal variant, and thus are equivalent to the minimal case (with, of course, perhaps a different relation). Thus, they allow a straightforward trivialization.

1.6.3.2 Logical Properties of the Limit Variant

We begin with some simple logical facts about the limit version.

We abbreviate $\Lambda(T) := \Lambda(M(T))$ etc., assume transitivity.

Fact 1.6.4

(1) $A \in \Lambda(T) \Rightarrow M(\overline{\overline{T}}) \subseteq \widehat{A}$

(2) $M(\overline{\overline{T}}) = \bigcap \{ \widehat{A} : A \in \Lambda(T) \}$

(2a) $M(\overline{\overline{T'}}) \models \sigma \Rightarrow \exists B \in \Lambda(T').\ \widehat{B} \models \sigma$

(3) $M(\overline{\overline{T'}}) \cap M(T) \models \sigma \Rightarrow \exists B \in \Lambda(T').\ \widehat{B} \cap M(T) \models \sigma.$

Proof

(1) Note that $A \models \phi \Rightarrow T \mathrel{|\!\sim} \phi$ by definition, see Definition 1.2.15.

Let $M(\overline{\overline{T}}) \not\subseteq \widehat{A}$, so there is ϕ, $\widehat{A} \models \phi$, so $A \models \phi$, but $M(\overline{\overline{T}}) \not\models \phi$, so $T \not\mathrel{|\!\sim} \phi$, $contradiction$.

(2) "⊆" by (1). "⊇": Let $x \in \bigcap \{ \widehat{A} : A \in \Lambda(T) \} \Rightarrow \forall A \in \Lambda(T).x \models Th(A)$
$\Rightarrow x \models \overline{\overline{T}}.$

(2a) $M(\overline{\overline{T'}}) \models \sigma \Rightarrow T' \mathrel{|\!\sim} \sigma \Rightarrow \exists B \in \Lambda(T').B \models \sigma$. But $B \models \sigma \Rightarrow \widehat{B} \models \sigma.$

1.6 The Limit Variant

(3) $M(\overline{\overline{T'}}) \cap M(T) \models \sigma \Rightarrow \overline{\overline{T'}} \cup T \vdash \sigma \Rightarrow \exists \tau_1 \ldots \tau_n \in \overline{\overline{T'}}$ such that $T \cup \{\tau_1, \ldots, \tau_n\} \vdash \sigma$, so $\exists B \in \Lambda(T').Th(B) \cup T \vdash \sigma$. So $M(Th(B)) \cap M(T) \models \sigma \Rightarrow \widehat{B} \cap M(T) \models \sigma$.

□

We saw in Example 1.6.2 and its discussion the problems which might arise in the limit version, even if the algebraic behaviour is correct.

This analysis leads us to consider the following facts:

Fact 1.6.5

(1) Let $\forall B \in \Lambda(T') \exists A \in \Lambda(T).A \subseteq B \cap M(T)$, then $\overline{\overline{T' \cup T}} \subseteq \overline{\overline{T}}$.

Let, in addition, $\{B \in \Lambda(T') : \widehat{B} \cap \widehat{M(T)} = \widehat{B \cap M(T)}\}$ be cofinal in $\Lambda(T')$. Then

(2) $Con(\overline{\overline{T'}}, T)$ implies $\forall A \in \Lambda(T').A \cap M(T) \neq \emptyset$.

(3) $\forall A \in \Lambda(T) \exists B \in \Lambda(T').B \cap M(T) \subseteq A$ implies $\overline{\overline{T}} \subseteq \overline{\overline{T' \cup T}}$.

Note that $M(T) = \widehat{M(T)}$, so we could also have written $\widehat{B} \cap M(T) = \widehat{B \cap M(T)}$, but above way of writing stresses more the essential condition $\widehat{X} \cap \widehat{Y} = \widehat{X \cap Y}$.

Proof

(1) Let $\overline{\overline{T'}} \cup T \vdash \sigma$, so $\exists B \in \Lambda(T').\widehat{B} \cap M(T) \models \sigma$ by Fact 1.6.4, (3) above (using compactness). Thus $\exists A \in \Lambda(T).A \subseteq B \cap M(T) \models \sigma$ by prerequisite, so $\sigma \in \overline{\overline{T}}$.

(2) Exercise, solution in the Appendix.

(3) Let $\sigma \in \overline{\overline{T}}$, so $T \hspace{1mm}|\hspace{-1mm}\sim \sigma$, so $\exists A \in \Lambda(T).A \models \sigma$, so $\exists B \in \Lambda(T').B \cap M(T) \subseteq A$ by prerequisite, so $\exists B \in \Lambda(T').(B \cap M(T) \subseteq A$ and $\widehat{B} \cap \widehat{M(T)} = \widehat{B \cap M(T)})$. So for such B $\widehat{B} \cap \widehat{M(T)} = \widehat{B \cap M(T)} \subseteq \widehat{A} \models \sigma$. By Fact 1.6.4 (1) $M(\overline{\overline{T'}}) \subseteq \widehat{B}$, so $M(\overline{\overline{T'}}) \cap M(T) \models \sigma$, so $\overline{\overline{T'}} \cup T \vdash \sigma$.

□

We obtain now as easy corollaries of a more general situation the following properties shown in [Sch04] and below by direct proofs. Thus, we have the trivialization results shown there.

Corollary 1.6.6

Let the structure be transitive.

(1) Let $\{B \in \Lambda(T') : \widehat{B} \cap \widetilde{M(T)} = \widetilde{B \cap M(T)}\}$ be cofinal in $\Lambda(T')$, then
$(PR)\ T \vdash T' \Rightarrow \overline{\overline{T}} \subseteq \overline{\overline{T'}} \cup T$ holds.

(2) $\overline{\overline{\phi \wedge \phi'}} \subseteq \overline{\overline{\phi}} \cup \{\phi'\}$ holds.

If the structure is ranked, then also:

(3) Let $\{B \in \Lambda(T') : \widehat{B} \cap \widetilde{M(T)} = \widetilde{B \cap M(T)}\}$ be cofinal in $\Lambda(T')$, then
$(\hspace{-1pt}\sim\hspace{-4pt}=)\ T \vdash T', Con(\overline{\overline{T'}}, T) \Rightarrow \overline{\overline{T}} = \overline{\overline{T'}} \cup T$ holds.

(4) $T \not\vdash \neg\gamma \Rightarrow \overline{\overline{T}} = \overline{T \cup \{\gamma\}}$ holds.

Proof

(1) $\forall A \in \Lambda(M(T)) \exists B \in \Lambda(M(T')).B \cap M(T) \subseteq A$ by Fact 1.6.3 (2). So the result follows from Fact 1.6.5 (3).

(2) Set $T' := \{\phi\}$, $T := \{\phi, \phi'\}$. Then for $B \in \Lambda(T')$ $\widehat{B} \cap M(T) = \widehat{B} \cap M(\phi') = \widetilde{B \cap M(\phi')}$ by Fact 1.2.1 $(Cl \cap +)$, so the result follows by (1).

(3) Let $Con(\overline{\overline{T'}}, T)$, then by Fact 1.6.5 (2) $\forall A \in \Lambda(T').A \cap M(T) \neq \emptyset$, so by Fact 1.6.3 (4) $\forall B \in \Lambda(T') \exists A \in \Lambda(T).A \subseteq B \cap M(T)$, so $\overline{\overline{T'}} \cup T \subseteq \overline{\overline{T}}$ by Fact 1.6.5 (1).

The other direction follows from (1).

(4) Set $T := T' \cup \{\gamma\}$. Then for $B \in \Lambda(T')$ $\widehat{B} \cap M(T) = \widehat{B} \cap M(\gamma) = \widetilde{B \cap M(\gamma)}$ again by Fact 1.2.1 $(Cl \cap +)$, so the result follows from (3).

$\square$

1.6.4 Simplifications of the General Transitive Limit Case

Our main result here is that the transitive limit version for formulas - but not for full theories - is essentially equivalent to the minimal version. At the same time, the limit version sometimes separates the finitary and infinitary versions, see Example 1.6.3, Fact 1.6.2, Fact 1.6.7, and Example 1.6.4.

1.6 The Limit Variant

More details can be found in [Sch04], section 3.4.1.

Example 1.6.3

$\overline{\overline{T \cup T'}} \subseteq \overline{\overline{T}} \cup T'$ can be wrong in the transitive limit version.

Any not definability preserving structure, where (PR) fails, serves as a counterexample, as minimal structures are special cases of the limit variant. Here is still another example.

Let $v(\mathcal{L}) := \{p_i : i < \omega\}$. Let $m \models p_i : i < \omega$, and $m' \models \neg p_0$, $m' \models p_i : 0 < i < \omega$, with $m \prec m'$ (this is the entire relation).

Let $T := \emptyset$, $T' := Th(\{m, m'\})$, then $T \cup T' = T'$, $\overline{\overline{T'}} = Th(\{m\})$, so $T \cup T' \hspace{1pt}\mid\hspace{-3pt}\sim\hspace{1pt} p_0$, $\overline{\overline{T}} = \overline{T} = \emptyset$, and $\overline{\overline{T}} \cup T' = T' = T'$, but $p_0 \notin T'$, contradiction.

□

Note

The structure is not definability preserving, and (PR) holds neither in the minimal nor in the limit variant.

Fact 1.6.7

Finite cumulativity holds in transitive limit structures: If $\phi \hspace{1pt}\mid\hspace{-3pt}\sim\hspace{1pt} \psi$, then $\overline{\overline{\phi}} = \overline{\overline{\phi \wedge \psi}}$.

Proof

Exercise, solution see [Sch04], Fact 3.4.5.

Example 1.6.4

Infinitary cumulativity may fail in transitive limit structures.

Consider the same language as in Example 1.6.3, set again $m < m'$, so $Th(\{m, m'\}) \hspace{1pt}\mid\hspace{-3pt}\sim\hspace{1pt} p$, but this time, we add more pairs to the relation: m and m' will now be the topmost models, and we put below all other models, making more and more p_i, $i \neq 0$, true, but alternating p_0 with $\neg p_0$, resulting in a total order (i.e. a ranked structure). Set $\phi := p_0 \vee \neg p_0$. Thus $\overline{\overline{\phi}} = Th(\{m, m'\})$, so $\hspace{1pt}\mid\hspace{-3pt}\sim\hspace{1pt}$ is not even idempotent, $\overline{\overline{\phi}} \neq \overline{\overline{(\overline{\overline{\phi}})}}$, as $\overline{\overline{Th(\{m, m'\})}} = Th(m)$.

□

See the comment after Example 3.4.2 in [Sch04] for a discussion.

We conclude with

Fact 1.6.8

Having cofinally many definable sets trivializes the problem (again in the transitive case).

Proof

Exercise, solution see [Sch04], Fact 3.4.6.

We summarize our main positive results on the limit variant of general preferential structures:

Proposition 1.6.9

Let the relation be transitive. Then

(1) Every instance of the the limit version, where the definable closed minimizing sets are cofinal in the closed minimizing sets, is equivalent to an instance of the minimal version.
(2) If we consider only formulas on the left of $\mid\!\sim$, the resulting logic of the limit version can also be generated by the minimal version of a (perhaps different) preferential structure. Moreover, the structure can be chosen smooth.

Proof

Exercise, solution see [Sch04], Proposition 3.4.7.

1.6.5 Ranked Structures Without Copies

1.6.5.1 Introduction

We consider here ranked relations, and see again that the limit version for formulas is equivalent to the minimal version, but not for general theories.

More introductory details can be found in [Sch04], section 3.10.3, and section 3.10.3.1.

1.6.5.2 Representation

We first note some elementary facts. Recall Definition 1.2.15.

Remark 1.6.10

In ranked structures, the following hold:

1.6 The Limit Variant

(1) If $\emptyset \neq A \subseteq X$, and $\forall a \in A \forall x \in X(x \prec a\ x \perp a \rightarrow x \in A)$, then A minimizes X.

(2) Thus, for $X \neq \emptyset$ $\Lambda(X) = \{\emptyset \neq A \subseteq X \colon \forall a \in A \forall x \in X(x \prec a\ x \perp a \rightarrow x \in A)\}$, $\Lambda(X)$ consists of all nonempty, and downward and horizontally closed subsets of X.

(3) If $\bigcap \Lambda(X) \neq \emptyset$, then $\bigcap \Lambda(X) = \mu(X)$ (where $\mu = \mu_\prec$, of course).

(4) If X is finite, $\bigcap \Lambda(X) = \mu(X)$.

(5) If $x, y \in \bigcap \Lambda(X)$, then $x \perp y$.

(6) As the order is fully determined by considering pairs, we can recover all information about Λ by considering $\Lambda(X)$, or, alternatively, $\mu(X)$ for pairs $X = \{a, b\}$ - whenever $\mathcal{Y}$ contains all pairs.

Proof

Exercise, solution see [Sch04], Remark 3.10.15.

We will use for representation:

Definition 1.6.3

We define the following conditions (Λi):

$(\Lambda 1)$ $\Lambda(X) \subseteq \mathcal{P}(X)$,

$(\Lambda 2)$ $X \in \Lambda(X)$,

$(\Lambda 3)$ $X \neq \emptyset \rightarrow \emptyset \notin \Lambda(X)$,

$(\Lambda 4)$ $A, B \in \Lambda(X) \rightarrow A \subseteq B$ or $B \subseteq A$,

$(\Lambda 5)$ $A \in \Lambda(X), Y \subseteq X, Y \cap A \neq \emptyset \rightarrow Y \cap A \in \Lambda(Y)$,

$(\Lambda 6)$ if there are X and A s.t. $A \in \Lambda(X)$, $a \in A$, $b \in X$-A, then: $a, b \in Y \rightarrow \exists B \in \Lambda(Y)(a \in B, b \notin B)$,

$(\Lambda 7)$ $\Lambda' \subseteq \Lambda(X), \bigcap \Lambda' \neq \emptyset \rightarrow \bigcap \Lambda' \in \Lambda(X)$,

$(\Lambda 8)$ $X \subseteq Y, A \in \Lambda(X) \Rightarrow \exists B \in \Lambda(Y).B \cap X = A$.

The conditions for the limit case (see Definition 1.6.3) can be separated into four groups, the first two are essentially independent of the particular case:

(a) Trivial conditions like $X \in \Lambda(X)$, conditions $(\Lambda 1) - (\Lambda 4)$ in the ranked preferential case.

(b) Conditions which express that the systems are sufficiently rich, conditions $(\Lambda 6) - (\Lambda 7)$ in the ranked preferential case.

(c) Conditions which reflect the limit case to the finite one, condition $(\Lambda 5)$ in the ranked preferential case.

(d) Conditions which express the specificities of the finite case - they can either be general ones, which hold for the infinite case, too, or conditions which directly treat the finite case, again condition $(\Lambda 5)$ in the preferential case.

Note that condition $(\Lambda 5)$ thus serves in the ranked preferential case a double purpose. On the one hand, Y can be chosen finite, which permits to go down, on the other hand, it expresses the basic coherence property of ranked preferential models.

We show that conditions $(\Lambda 1) - (\Lambda 7)$ are sound and complete for the limit variant of ranked structures without copies, where the domain is closed under finite unions and contains all finite sets.

Completeness means here the following: If $\prec$ is the relation constructed, Λ the original set of systems satisfying $(\Lambda 1) - (\Lambda 7)$, $\Lambda_\prec$ the set of $\prec$-initial segments, then for all $X \in \mathcal{Y}$ $\Lambda(X) \subseteq \Lambda_\prec(X)$, and for $A \in \Lambda_\prec(X)$ there is $A' \subseteq A$ $A' \in \Lambda(X)$. This is sufficient, as we are only interested in what finally holds.

More details can be found in [Sch04], section 3.10.3.

The Details

Proposition 1.6.11 is the main representation result for the general limit case.

Proposition 1.6.11

$(\Lambda 1)-(\Lambda 7)$ are sound and complete for the limit variant of ranked structures without copies, where the domain is closed under finite unions and contains all finite sets.

Proof

Exercise, solution see [Sch04], Proposition 3.10.16.

1.6.5.3 Partial Equivalence of Limit and Minimal Ranked Structures

Definition 1.6.4

$T \models_\Lambda \phi$ iff there is $A \in \Lambda(M(T))$ s.t. $A \models \phi$ We shall also write $T \mathrel{|\!\sim} \phi$ for $\models_\Lambda$, and $\overline{\overline{T}} := \{\phi : T \models_\Lambda \phi\}$.

The problem with the logical variant is that we do not "see" directly the closed sets. We see only the ϕ, but not the A - moreover, A need not be the model set of any theory.

1.6 The Limit Variant

Example 1.6.5

Take an infinite propositional language $p_i : i \in \omega$. We have ω_1 models (assume for simplicity CH).

(1) Take the model m_0 which makes all p_i true, and put it on top. Next, going down, take all models which make p_0 false, and then all models which make p_0 true, but p_1 false, etc. in a ranked construction. So, successively more p_i will become (and stay) true. Consequently, $\emptyset \models_\Lambda p_i$ for all i. But the structure has no minimum, and the logical limit m_0 is not in the set wise limit - but, of course, a model of the theory. (Recall compactness, so $\overline{\overline{T}} = T \cup \{\phi : T \models_\Lambda \phi\}$ is consistent by inclusion, so it has a model, which must be in the set of all T-models.)

(2) Take exactly the same set structure, but enumerate the models differently: each consistent formula is made unboundedly often true (this is possible, as each consistent formula has ω_1 many models), so $\emptyset \models_\Lambda \phi$ iff ϕ is a tautology.

The behavior is as different as possible (under consistency - from the empty theory to a consistent complete one).

The first example shows in particular that $M(T \cup \{\phi\})$ need not be closed in $M(T)$, if $T \models_\Lambda \phi$ - the topmost model satisfies ϕ.

□

Note that the situation is quite asymmetric: If $T \models_\Lambda \phi$, then we know that all $\neg \phi$ models are minimized, from some level onward, there will be no more $\neg \phi$ models, but we do not know whether any $\overline{\overline{T}}$-model is very low, as we saw, it might be in the worst position. The best guess we had for a minimal model was the worst one.

Fact 1.6.12

The following laws hold in ranked structures interpreted as in Definition 1.6.4:

(1) $\overline{\overline{T}}$ is consistent, if T is,
(2) $\overline{T} \subseteq \overline{\overline{T}}$,
(3) $\overline{\overline{T}}$ is classically closed,
(4) $T \mathrel{\mid\!\sim} \phi, T' \mathrel{\mid\!\sim} \phi \to T \vee T' \mathrel{\mid\!\sim} \phi$,
(5) If $T \mathrel{\mid\!\sim} \phi$, then $T \mathrel{\mid\!\sim} \phi' \leftrightarrow T \cup \{\phi\} \mathrel{\mid\!\sim} \phi'$.

Proof

Trivial.

Exercise, solution see [Sch04], Fact 3.10.17.

We have a first trivialization result:

Fact 1.6.13

Having cofinally many definable sets in the Λ's trivializes the problem - it becomes equivalent to the minimal variant.

Proof

Suppose each $\Lambda(X)$ contains cofinally many definable sets, let $\Lambda'(X)$ be this subset. Then $\bigcap \Lambda(X) = \bigcap \Lambda'(X)$. As $\Lambda(X)$ is totally ordered by $\subseteq$, by compactness of the standard topology, and $\emptyset \notin \Lambda$, $\bigcap \Lambda'(X) \neq \emptyset$, but then $\emptyset \neq \bigcap \Lambda'(X) = \mu(X)$, so we are in the simple μ-case.

$\square$

The following example shows the difference between considering full theories and considering just formulas (on the left of $\mid\!\sim$). If we consider full theories, we can "grab" single models, and thus determine the full order. As long as we restrict ourselves to formulas, we are much more shortsighted. In particular, we can make sequences of models to converge to some model, but put this model elsewhere. Suitable such manipulations will pass unobserved by formulas. The example also shows that there are structures whose limit version for theories is unequal to any minimal structure.

Example 1.6.6

Let $\mathcal{L}$ be given by the propositional variables p_i, $i < \omega$. Order the atomic formulas by $p_i \prec \neg p_i$, and then order all sequences $s = \langle +/-p_0, +/-p_1, \ldots \rangle$, $i < n \leq \omega$ lexicographically, identify models with such sequences of length ω. So, in this order, the biggest model is the one making all p_i false, the smallest the one making all p_i true. Any finite sequence (an initial segment) $s = \langle +/-p_0, +/-p_1, \ldots +/-p_n \rangle$ has a smallest model $\langle +/-p_0, +/-p_1, \ldots +/-p_n, p_{n+1}, p_{n+2}, \ldots \rangle$, which continues all positive, call it m_s. As there are only countably many such finite sequences, the number of m_s is countable, too (and $m_s = m_{s'}$ for different s, s' can happen). Take now any formula ϕ, it can be written as a finite disjunction of sequences s of fixed length n $\langle +/-p_0, +/-p_1, \ldots +/-p_n \rangle$, choose wlog. n minimal, and denote s_ϕ the smallest (in our order) of these s. E.g., if $\phi = (p_0 \wedge p_1) \vee (p_1 \wedge \neg p_2) = (p_0 \wedge p_1 \wedge p_2)$ $(p_0 \wedge p_1 \wedge \neg p_2)$ $(p_0 \wedge p_1 \wedge \neg p_2)$ $(\neg p_0 \wedge p_1 \wedge \neg p_2)$, and $s_\phi = \langle p_0, p_1, p_2 \rangle$.

(1) Consider now the initial segments defined by this order. In this order, the initial segments of the models of ϕ are fully determined by the smallest (in our order) s of ϕ, moreover, they are trivial, as they all contain the minimal model $m_s = s_\phi + \langle p_{n+1}, p_{n+2}, \ldots \rangle$ - where $+$ is concatenation. It is important to note that even when we take away m_s, the initial segments will still converge to m_s - but it is not there any more. Thus, in both cases, m_s there or not, $\phi \models_\Lambda s_\phi + \langle p_{n+1}, p_{n+2}, \ldots \rangle$ - written a little sloppily. (A more formal argument: If $\phi \models_\Lambda \psi$, with the m_s present, then ψ holds in m_s, but ψ has finite length, so beyond some p_k the values do not matter, and we can make them negative - but such sequences did not change their rank, they stay there.)

1.6 The Limit Variant
133

(2) Modify the order now. Put all m_s on top of the construction. As there are only countably many, all consistent ϕ will have most of their models in the part left untouched - the m_s are not important for formulas and their initial segments.

The reordered structure (in (2)) is not equivalent to any minimal structure when considering full theories: Suppose it were. We have $\emptyset \hspace{1pt}\vert\!\sim +p_i$ for all i, so the whole structure has to have exactly one minimal model, but this model is minimized by other models, a contradiction.

More comments can be found in [Sch04], example 3.10.2 there.

Proposition 1.6.14

When considering just formulas, in the ranked case without copies, Λ is equivalent to μ - so Λ is trivialized in this case. More precisely:

Let a logic $\phi \hspace{1pt}\vert\!\sim \psi$ be given by the limit variant without copies, i.e. by Definition 1.6.4. Then there is a ranked structure, which gives exactly the same logic, but interpreted in the minimal variant.

(As Example 1.6.6 has shown, this is NOT necessarily true if we consider full theories T and $T \hspace{1pt}\vert\!\sim \psi$.)

Proof

Assume $\hspace{1pt}\vert\!\sim$ is given by initial segments Λ, i.e. $\phi \hspace{1pt}\vert\!\sim \psi$ iff ψ finally holds in all initial segments of the ϕ-models.

We show that, if we define $f(M(\phi)) := M(\overline{\overline{\phi}})$, f has the properties:

$(\mu \subseteq)$ $f(X) \subseteq X$,
$(\mu \emptyset)$ $X \neq \emptyset \to f(X) \neq \emptyset$,
$(\mu =)$ $X \subseteq Y, f(Y) \cap X \neq \emptyset \to f(X) = f(Y) \cap X$.

Obviously, the set of $M(\phi)$'s is closed under finite unions.

The result is then a consequence of the representation result Proposition 1.4.8.

$(\mu \subseteq)$ and $(\mu \emptyset)$ are trivial.

$(\mu =)$ Assume $M(\psi) \subseteq M(\phi)$ and $M(\overline{\overline{\phi}}) \cap M(\psi) \neq \emptyset$, so $\vdash \psi \to \phi$ and $Con(\overline{\overline{\phi}}, \psi)$. We show $\overline{\overline{\psi}} = \overline{\overline{\phi}} \cup \{\psi\}$, thus $f(M(\psi)) = M(\overline{\overline{\psi}}) = M(\overline{\overline{\phi}} \cup \{\psi\}) = M(\overline{\overline{\phi}}) \cap M(\psi) = f(M(\phi)) \cap M(\psi)$.

$Con(\overline{\overline{\phi}}, \psi)$ implies $\neg\psi \notin \overline{\overline{\phi}}$, so any initial segment A of $M(\phi)$ contains a ψ-model. Thus, $M(\psi) \cap A \neq \emptyset$, and $M(\psi) \cap A$ is an initial segment of $M(\psi)$ by $(\Lambda 5)$. Thus, if $\phi' \in \overline{\overline{\phi}}$, ϕ' will finally hold in $M(\phi)$, so $\phi' \wedge \psi$ will finally hold in $M(\psi)$. Thus, if $\sigma \in \overline{\overline{\phi}} \cup \{\psi\}$, then $\overline{\overline{\phi}} \cup \{\psi\} \vdash \sigma$, so $\overline{\overline{\phi}} \vdash \psi \to \sigma$, so $\psi \to \sigma \in \overline{\overline{\phi}}$, so $\psi \wedge (\psi \to \sigma) \in \overline{\overline{\psi}}$, and $\sigma \in \overline{\overline{\psi}}$. Conversely, if ϕ' holds finally in $M(\psi)$, as any initial segment A' of $M(\psi)$ can be completed to an initial segment A of $M(\phi)$ (complete all levels of A') s.t. $A \cap M(\psi) = A'$, in ϕ, finally $\phi' \vee \neg \psi$ holds. (This is the only place where the fact that ψ is a formula is important.) So $\psi \mathrel{|\!\sim} \phi'$ implies $\phi \mathrel{|\!\sim} \phi' \vee \neg\psi$, so $\phi' \in \overline{\overline{\phi}} \cup \{\psi\}$.

(The important fact is here the closure of the domain under complements.)

□

See the end of Chapter 3 in [Sch04] for more discussion.

1.7 Preferential Structures Without Definability Preservation

1.7.1 Introduction

This section treats situations where we can approximate the result only. If the result is not definable (by a theory or formula), we can describe it only roughly, i.e. by approximation. We see similar phenomena also elsewhere, for instance $\mu(X \times Y)$ may not be some $X' \times Y'$ any more. Such cases were treated, e.g., in [BLS99], or [Sch95-3], see also [Sch04], Section 6.3 there. As a consequence, representation theorems are not so nice any more.

Example 1.7.1

This example was first given in [Sch92]. It shows that condition (PR) may fail in preferential structures which are not definability preserving.

Let $v(\mathcal{L}) := \{p_i : i \in \omega\}$, $n, n' \in M_{\mathcal{L}}$ be defined by $n \models \{p_i : i \in \omega\}$, $n' \models \{\neg p_0\} \cup \{p_i : 0 < i < \omega\}$.

Let $\mathcal{M} := \langle M_{\mathcal{L}}, \prec \rangle$ where only $n \prec n'$, i.e., just two models are comparable. Note that the structure is transitive and smooth. Thus, by Fact 1.3.10 $(\mu \subseteq)$, (μPR), (μCUM) hold.

Let $\mu := \mu_{\mathcal{M}}$, and $\mathrel{|\!\sim}$ be defined as usual by μ.

Set $T := \emptyset$, $T' := \{p_i : 0 < i < \omega\}$. We have $M_T = M_{\mathcal{L}}$, $f(M_T) = M_{\mathcal{L}} - \{n'\}$, $M_{T'} = \{n, n'\}$, $f(M_{T'}) = \{n\}$. So by the result of Example 1.2.1, f is not definability preserving, and, furthermore, $\overline{\overline{T}} = T$, $\overline{\overline{T'}} = \{p_i : i < \omega\}$, so $p_0 \in \overline{\overline{T \cup T'}}$, but $\overline{\overline{T}} \cup T' = \overline{\overline{T}} \cup T' = T'$, so $p_0 \notin \overline{\overline{T}} \cup T'$, contradicting (PR), which holds in all definability preserving preferential structures.

□

1.7 Preferential Structures Without Definability Preservation

Example 1.7.1 showed that in the general case without definability preservation, (PR) fails, and the following Example 1.7.2 shows that in the ranked case, $(\mathrel|\!\sim=)$ may fail. So failure is not just a consequence of the very liberal definition of general preferential structures.

Example 1.7.2

Take $\{p_i : i \in \omega\}$ and put $m := m_{\bigwedge p_i}$, the model which makes all p_i true, in the top layer, all the other in the bottom layer. Let $m' \neq m$, $T' := \emptyset$, $T := Th(m, m')$. Then Then $\overline{\overline{T'}} = T'$, so $Con(\overline{\overline{T'}}, T)$, $\overline{\overline{T}} = Th(m')$, $\overline{\overline{T'}} \cup T = T$.

□

We now give an example of a definability preserving non-compact preferential logic - in answer to a question by D. Makinson (personal communication):

Example 1.7.3

Take an infinite language, p_i, $i < \omega$. Fix one model, m, which makes p_0 true (and, say, for definiteness, all the others true, too), and m' which is just like m, but it makes p_0 false. Well-order all the other p_0-models, and all the other $\neg p_0$-models separately.

Construct now the following ranked structure:

On top, put m, directly below it m'. Further down put the bloc of the other $\neg p_0$-models, and at the bottom the bloc of the other p_0-models.

As the structure is well-ordered, it is definability preserving (singletons are definable).

Let T be the theory defined by m, m', then $T \mathrel|\!\sim \neg p_0$.

Let ϕ be such that $M(T) \subseteq M(\phi)$, then $M(\phi)$ contains a p_0-model other than m, so $\phi \mathrel|\!\sim p_0$.

□

1.7.1.1 The Problem

If a structure or a function is not definability preserving, we cannot describe the result of the function (or structure) precisely, we do not "see" the missing elements. The size of the invisible gaps depends on the size of the language, there is no uniform cardinality. Thus, "small sets" are not defined by cardinality, but by definability (in the language at hand). This results in the lack of fixed size characterizations.

Much more discussion is found in [Sch04], section 5.1.1.

Example 1.7.4

Let m be any $\mathcal{L}$-model of an infinite language $\mathcal{L}$. Then $M_\mathcal{L} - \{m\}$ is not definable, as shown in Example 1.2.1. Thus, if we define f by $f(M_\mathcal{L}) := M_\mathcal{L} - \{m\}$, and $f(X) := X$ for any other set $X \subset M_\mathcal{L}$, f is not definability preserving.

□

Comment 1.7.1

We discuss now the negative result, that there is no normal characterization of general preferential structures possible. Similar results hold for the ranked case, and for distance defined theory revision.

Let κ be any infinite cardinal. We show that there is no characterization Φ of general (i.e. not necessarily definability preserving) preferential structures which has size $\leq \kappa$. We suppose there were one such characterization Φ of size $\leq \kappa$, and construct a counterexample. The idea of the proof is very simple.

Take the language $\mathcal{L}$ defined by $p_i : i < \kappa$. We show that it suffices to consider for any given instantiation of $\Phi \leq \kappa$ many pairs $m \prec m^-$ in a case not representable by a preferential structure, and that $\leq \kappa$ many such pairs give the same result in a true preferential structure. Thus, every instantiation is true in an "illegal" and a "legal" example, so Φ cannot discern between legal and illegal examples. The main work is to show that $\leq \kappa$ many pairs suffice in the illegal example. This is, again, in principle, easy, we show that there is a "best" set of size $\leq \kappa$ which calculates $\overline{\overline{T}}$ for all T considered in the instantiation.

For any model m with $m \models p_0$, let m^- be exactly like m with the exception that $m^- \models \neg p_0$.

Define the logic $\mid\!\sim$ as follows in two steps:

(1) $\overline{\overline{Th(\{m, m^-\})}} := Th(\{m\})$ (Speaking preferentially, $m \prec m^-$, for all such m, m^-, this will be the entire relation. The relation is thus extremely simple, $\prec$-paths have at most length 1, so $\prec$ is automatically transitive.)

We now look at (in terms of preferential models only some!) consequences:

(2) $\overline{\overline{T}} := $ Th $(\bigcap\{M(Th(M(T)\text{-}A)): card(A) \leq \kappa, A \subseteq M(T), \forall n(n \in A \to n = m^-$ and $m, m^- \in M(T))\})$.

This, without the size condition, would be exactly the preferential consequence of part (1) of the definition, but this logic as it stands is not preferential.

Suppose there were a characterization of size $\leq \kappa$. It has to say "no" for at least one instance of the universally quantified condition Φ. We will show that we find a true preferential structure where this instance of Φ has the same truth value, a contradiction. To demonstrate it, we consider the preferential structure where we do not make all $m \prec m^-$, but only the κ many of them we have used in the instance of Φ. We will see that the expression Φ still fails with our instances.

1.7.2 Characterisations Without Definability Preservation

1.7.2.1 Introduction

General Remarks, Affected Conditions

We assume now - unless explicitly stated otherwise - $\mathcal{Y} \subseteq \mathcal{P}(Z)$ to be closed under arbitrary intersections (this is used for the definition of $\frown$) and finite unions, and $\emptyset, Z \in \mathcal{Y}$. This holds, of course, for $\mathcal{Y} = \mathbf{D}_\mathcal{L}$, $\mathcal{L}$ any propositional language.

The aim of the present Chapter is to present the results of [Sch04] connected to problems of definability preservation in a uniform way, stressing the crucial condition $\widehat{X \cap Y} = \widehat{X} \cap \widehat{Y}$. This presentation shall help and guide future research concerning similar problems.

For motivation, we first consider the problem with definability preservation for the rules

$(PR)\ \overline{\overline{T \cup T'}} \subseteq \overline{\overline{T}} \cup T'$, and

$(\mathrel{\mid\!\sim} =)\ T \vdash T',\ Con(\overline{\overline{T'}}, T) \Rightarrow \overline{\overline{T}} = \overline{\overline{T' \cup T}}$ holds.

which are consequences of

$(\mu PR)\ X \subseteq Y \Rightarrow \mu(Y) \cap X \subseteq \mu(X)$ or

$(\mu =)\ X \subseteq Y, \mu(Y) \cap X \neq \emptyset \Rightarrow \mu(Y) \cap X = \mu(X)$ respectively

and definability preservation.

We remind the reader of Definition 1.2.2 and Fact 1.2.1, partly taken from [Sch04]. We turn to the central condition.

The Central Condition

We analyze the problem of (PR), seen in Example 1.7.2 (1) above, working in the intended application.

(PR) is equivalent to $M(\overline{\overline{T}} \cup T') \subseteq M(\overline{\overline{T \cup T'}})$. To show (PR) from (μPR), we argue as follows, the crucial point is marked by "?":

$M(\overline{\overline{T \cup T'}}) = M(Th(\mu(M_{T \cup T'}))) = \widehat{\mu(M_{T \cup T'})} \supseteq \mu(M_{T \cup T'}) = \mu(M_T \cap M_{T'}) \supseteq$
(by (μPR)) $\mu(M_T) \cap M_{T'} ?\ \widehat{\mu(M_T)} \cap M_{T'} = M(Th(\mu(M_T))) \cap M_{T'} = M(\overline{\overline{T}}) \cap M_{T'} = M(\overline{\overline{T}} \cup T')$. If μ is definability preserving, then $\mu(M_T) = \widehat{\mu(M_T)}$, so "?"

above is equality, and everything is fine. In general, however, we have only $\mu(M_T) \subseteq \widehat{\mu(M_T)}$, and the argument collapses.

But it is not necessary to impose $\mu(M_T) = \widehat{\mu(M_T)}$, as we still have room to move: $\widehat{\mu(M_{T \cup T'})} \supseteq \mu(M_{T \cup T'})$. (We do not consider here $\mu(M_T \cap M_{T'}) \supseteq \mu(M_T) \cap M_{T'}$ as room to move, as we are now interested only in questions related to definability preservation.) If we had $\widehat{\mu(M_T) \cap M_{T'}} \subseteq \widehat{\mu(M_T)} \cap \widehat{M_{T'}}$, we could use $\mu(M_T) \cap M_{T'} \subseteq \mu(M_T \cap M_{T'}) = \mu(M_{T \cup T'})$ and monotony of $\frown$ to obtain $\widehat{\mu(M_T)} \cap M_{T'} \subseteq \widehat{\mu(M_T) \cap M_{T'}} \subseteq \widehat{\mu(M_T \cap M_{T'})} = \widehat{\mu(M_{T \cup T'})}$. If, for instance, $T' = \{\psi\}$, we have $\widehat{\mu(M_T)} \cap M_{T'} = \widehat{\mu(M_T) \cap M_{T'}}$ by Fact 1.2.1 $(Cl \cap +)$. Thus, definability preservation is not the only solution to the problem.

We have seen in Fact 1.2.1 that $\widehat{X \cup Y} = \widehat{X} \cup \widehat{Y}$, moreover $X - Y = X \cap CY$ (CY the set complement of Y), so, when considering boolean expressions of model sets (as we do in usual properties describing logics), the central question is whether

$(\sim \cap)$ $\widehat{X \cap Y} = \widehat{X} \cap \widehat{Y}$

holds.

We take a closer look at this question.

$\widehat{X \cap Y} \subseteq \widehat{X} \cap \widehat{Y}$ holds by Fact 1.2.1 (6). Using $(Cl \cup)$ and monotony of $\frown$, we have $\widehat{X} \cap \widehat{Y} = \widehat{((X \cap Y) \cup (X - Y))} \cap \widehat{((X \cap Y) \cup (Y - X))} = \widehat{((X \cap Y) \cup (X - Y))} \cap \widehat{((X \cap Y) \cup (Y - X))} = \widehat{X \cap Y} \cup (\widehat{X - Y} \cap \widehat{Y - X})$, thus $\widehat{X} \cap \widehat{Y} \subseteq \widehat{X \cap Y}$ iff

$(\sim \cap')$ $\widehat{Y - X} \cap \widehat{X - Y} \subseteq \widehat{X \cap Y}$ holds.

Intuitively speaking, the condition holds iff we cannot approximate any element both from $X - Y$ and X-Y, which cannot be approximated from $X \cap Y$, too.

Note that in above Example 1.7.2 (1) $X := \mu(M_T) = M_{\mathcal{L}} - \{n'\}$, $Y := M_{T'} = \{n, n'\}$, $\widehat{X - Y} = M_{\mathcal{L}}$, $\widehat{Y - X} = \{n'\}$, $\widehat{X \cap Y} = \{n\}$, and $\widehat{X} \cap \widehat{Y} = \{n, n'\}$.

We consider now particular cases:

(1) If $X \cap Y = \emptyset$, then by $\emptyset \in \mathcal{Y}$, $(\sim \cap)$ holds iff $\widehat{X} \cap \widehat{Y} = \emptyset$.

(2) If $X \in \mathcal{Y}$ and $Y \in \mathcal{Y}$, then $\widehat{X - Y} \subseteq X$ and $\widehat{Y - X} \subseteq Y$, so $\widehat{X - Y} \cap \widehat{Y - X} \subseteq X \cap Y \subseteq \widehat{X \cap Y}$ and $(\sim \cap)$ trivially holds.

1.7 Preferential Structures Without Definability Preservation

(3) $X \in \mathcal{Y}$ and $CX \in \mathcal{Y}$ together also suffice - in these cases $\widetilde{Y-X} \cap \widetilde{X-Y} = \emptyset : \widetilde{Y-X} = \widetilde{Y \cap CX} \subseteq CX$, and $\widetilde{X-Y} \subseteq X$, so $\widetilde{Y-X} \cap \widetilde{X-Y} \subseteq X \cap CX = \emptyset \subseteq \widetilde{X \cap Y}$. (The same holds, of course, for Y.) (In the intended application, such X will be $M(\phi)$ for some formula ϕ. But, a warning, $\mu(M(\phi))$ need not again be the $M(\psi)$ for some ψ.)

We turn to the properties of various structures and apply our results.

Application to Various Structures

We now take a look at other frequently used logical conditions. First, in the context on nonmonotonic logics, the following rules will always hold in smooth preferential structures, even if we consider full theories, and not necessarily definability preserving structures:

Fact 1.7.1

Also for full theories, and not necessarily definability preserving structures hold:

(1) $(LLE), (RW), (AND), (REF)$, by definition and $(\mu \subseteq)$,

(2) (OR),

(3) (CM) in smooth structures,

(4) the infinitary version of (CUM) in smooth structures.

In definability preserving structures, but also when considering only formulas hold:

(5) (PR),

(6) $(\mid\sim=)$ in ranked structures.

Proof

We use the corresponding algebraic properties. The result then follows from Proposition 1.2.18.

□

We turn to theory revision. The following definition and example, taken from [Sch04] shows, that the usual AGM axioms for theory revision fail in distance based structures in the general case, unless we require definability preservation. See Section 4.3 for discussion and motivation.

Definition 1.7.1

We summarize the AGM postulates $(K * 7)$ and $(K * 8)$ in $(*4)$:

$(*4)$ If $T * T'$ is consistent with T'', then $T * (T' \cup T'') = \overline{(T * T') \cup T''}$.

Example 1.7.5

Consider an infinite propositional language $\mathcal{L}$.

Let X be an infinite set of models, m, m_1, m_2 be models for $\mathcal{L}$. Arrange the models of $\mathcal{L}$ in the real plane such that all $x \in X$ have the same distance < 2 (in the real plane) from m, m_2 has distance 2 from m, and m_1 has distance 3 from m.

Let T, T_1, T_2 be complete (consistent) theories, T' a theory with infinitely many models, $M(T) = \{m\}$, $M(T_1) = \{m_1\}$, $M(T_2) = \{m_2\}$. $M(T') = X \cup \{m_1, m_2\}$, $M(T'') = \{m_1, m_2\}$.

Assume $Th(X) = T'$, so X will not be definable by a theory.

Then $\overline{M(T) \mid M(T')} = X$, but $T * T' = Th(X) = T'$. So $T * T'$ is consistent with T'', and $\overline{(T * T') \cup T''} = T''$. But $T' \cup T'' = T''$, and $T * (T' \cup T'') = T_2 \neq T''$, contradicting $(*4)$.

$\square$

We show now that the version with formulas only holds here, too, just as does above (PR), when we consider formulas only - this is needed below for T'' only. This was already shown in [Sch04], we give now a proof based on our new principles.

Fact 1.7.2

$(*4)$ holds when considering only formulas.

Proof

Exercise, solution in the Appendix.

1.7.2.2 General and Smooth Structures Without Definability Preservation

Introduction

Note that in Section 3.2 and Section 3.3 of [Sch04], as well as in Proposition 4.2.2 of [Sch04] we have characterized $\mu : \mathcal{Y} \to \mathcal{Y}$ or $\mid : \mathcal{Y} \times \mathcal{Y} \to \mathcal{Y}$, but a closer inspection of the proofs shows that the destination can as well be assumed $\mathcal{P}(Z)$, consequently we can simply re-use above algebraic representation results also for the not definability preserving case. (Note that the easy direction of all these results work for destination $\mathcal{P}(Z)$, too.) In particular, also the proof for the not definability preserving case of revision in [Sch04] can be simplified - but we will not go into details here.

$(\cup)$ and $(\cap)$ are again assumed to hold now - we need $(\cap)$ for $\frown$.

The central functions and conditions to consider are summarized in the following definition.

1.7 Preferential Structures Without Definability Preservation

Definition 1.7.2

Let $\mu : \mathcal{Y} \to \mathcal{Y}$, we define $\mu_i : \mathcal{Y} \to \mathcal{P}(Z)$:

$\mu_0(U) := \{x \in U : \neg \exists Y \in \mathcal{Y}(Y \subseteq U \text{ and } x \in Y - \mu(Y))\}$,

$\mu_1(U) := \{x \in U : \neg \exists Y \in \mathcal{Y}(\mu(Y) \subseteq U \text{ and } x \in Y - \mu(Y))\}$,

$\mu_2(U) := \{x \in U : \neg \exists Y \in \mathcal{Y}(\mu(U \cup Y) \subseteq U \text{ and } x \in Y - \mu(Y))\}$

(note that we use $(\cup)$ here),

$\mu_3(U) := \{x \in U : \forall y \in U. x \in \mu(\{x, y\})\}$

(we use here $(\cup)$ and that singletons are in $\mathcal{Y}$).

"Small" is now in the sense of Definition 1.2.2.

$(\mu PR0)\ \mu(U) - \mu_0(U)$ is small,

$(\mu PR1)\ \mu(U) - \mu_1(U)$ is small,

$(\mu PR2)\ \mu(U) - \mu_2(U)$ is small,

$(\mu PR3)\ \mu(U) - \mu_3(U)$ is small.

$(\mu PR0)$ with its function will be the one to consider for general preferential structures, $(\mu PR2)$ the one for smooth structures.

We compare the present notation to that in Condition 5.2.2 in [Sch04]:

$\mu_0(U)$ above is the first $\mu'(U)$ there,

$\mu_2(U)$ above is the second $\mu'(U)$ there,

$(\mu PR0)$ is $(\mu 2)$ there,

$(\mu PR2)$ is $(\mu 2s)$ there.

A Non-trivial Problem

Unfortunately, we cannot use $(\mu PR0)$ in the smooth case, too, as Example 1.7.7 below will show. This sheds some doubt on the possibility to find an easy common approach to all cases of not definability preserving preferential, and perhaps other, structures. The next best guess, $(\mu PR1)$ will not work either, as the same example shows - or by Fact 1.7.3 (10), if μ satisfies (μCum), then $\mu_0(U) = \mu_1(U)$. $(\mu PR3)$ and μ_3 are used for ranked structures.

We will now see that this first impression of a difficult situation is indeed well founded.

First, note that in our context, μ will not necessarily respect (μPR). Thus, if e.g., $x \in Y - \mu(Y)$, and $\mu(Y) \subseteq U$, we cannot necessarily conclude that $x \notin \mu(U \cup Y)$ - the fact that x is minimized in $U \cup Y$ might be hidden by the bigger $\mu(U \cup Y)$.

Consequently, we may have to work with small sets (Y in the case of μ_2 above) to see the problematic elements - recall that the smaller the set $\mu(X)$ is, the less it can "hide" missing elements - but will need bigger sets ($U \cup Y$ in above example) to recognize the contradiction.

Second, "problematic" elements are those involved in a contradiction, i.e., contradicting the representation conditions. Now, a negation of a conjunction is a disjunction of negations, so, generally, we will have to look at various possibilities of violated conditions. But the general situation is much worse, still.

Example 1.7.6

Look at the ranked case, and assume no closure properties of the domain. Recall that we might be unable to see $\mu(X)$, but see only $\widehat{\mu(X)}$. Suppose we have $\widehat{\mu(X_1)} \cap (X_2 - \widehat{\mu(X_2)}) \neq \emptyset$, $\widehat{\mu(X_2)} \cap (X_3 - \widehat{\mu(X_3)}) \neq \emptyset$, $\widehat{\mu(X_{n-1})} \cap (X_n - \widehat{\mu(X_n)}) \neq \emptyset$, $\widehat{\mu(X_n)} \cap (X_1 - \widehat{\mu(X_1)}) \neq \emptyset$, which seems to be a contradiction. (It only is a real contradiction if it still holds without the closures.) But, we do not know where the contradiction is situated. It might well be that for all but one i really $\mu(X_i) \cap (X_{i+1} - \mu(X_{i+1})) \neq \emptyset$, and not only that for the closure $\widehat{\mu(X_i)}$ of $\mu(X_i)$ $\widehat{\mu(X_i)} \cap (X_{i+1} - \widehat{\mu(X_{i+1})}) \neq \emptyset$, but we might be unable to find this out. So we have to branch into all possibilities, i.e., for one, or several i $\widehat{\mu(X_i)} \cap (X_{i+1} - \widehat{\mu(X_{i+1})}) \neq \emptyset$, but $\mu(X_i) \cap (X_{i+1} - \mu(X_{i+1})) = \emptyset$.

□

The situation might even be worse, when those $\widehat{\mu(X_i)} \cap (X_{i+1} - \widehat{\mu(X_{i+1})}) \neq \emptyset$ are involved in several cycles, etc. Consequently, it seems very difficult to describe all possible violations in one concise condition, and thus we will examine here only some specific cases, and do not pretend that they are the only ones, that other cases are similar, or that our solutions (which depend on closure conditions) are the best ones.

Outline of Our Solutions in Some Particular Cases

The strategy of representation without definability preservation will in all cases be very simple: Under sufficient conditions, among them smallness (μPRi) as described above, the corresponding function μ_i has all the properties to guarantee representation by a corresponding structures, and we can just take our representation theorems for the dp case, to show this. Using smallness again, we can show that we have obtained a sufficient approximation - see Proposition 1.7.5, Proposition 1.7.6, Proposition 1.7.9.

1.7 Preferential Structures Without Definability Preservation

We first show some properties for the μ_i, $i = 0, 1, 2$. A corresponding result for μ_3 is given in Fact 1.7.7 below. (The conditions and results are sufficiently different for μ_3 to make a separation more natural.)

Property (9) of the following Fact 1.7.3 fails for μ_0 and μ_1, as Example 1.7.7 below will show. We will therefore work in the smooth case with μ_2.

Results

Fact 1.7.3

(This is partly Fact 5.2.6 in [Sch04].)

Recall that $\mathcal{Y}$ is closed under $(\cup)$, and $\mu : \mathcal{Y} \to \mathcal{Y}$. Let A, B, U, U', X, Y be elements of $\mathcal{Y}$ and the μ_i be defined from μ as in Definition 1.7.2. i will here be 0, 1, or 2, but not 3.

(1) Let μ satisfy $(\mu \subseteq)$, then $\mu_1(X) \subseteq \mu_0(X)$ and $\mu_2(X) \subseteq \mu_0(X)$,

(2) Let μ satisfy $(\mu \subseteq)$ and (μCum), then $\mu(U \cup U') \subseteq U \Leftrightarrow \mu(U \cup U') = \mu(U)$,

(3) Let μ satisfy $(\mu \subseteq)$, then $\mu_i(U) \subseteq \mu(U)$, and $\mu_i(U) \subseteq U$,

(4) Let μ satisfy $(\mu \subseteq)$ and one of the (μPRi), then $\mu(A \cup B) \subseteq \mu(A) \cup \mu(B)$,

(5) Let μ satisfy $(\mu \subseteq)$ and one of the (μPRi), then $\mu_2(X) \subseteq \mu_1(X)$,

(6) Let μ satisfy $(\mu \subseteq)$, (μPRi), then $\mu_i(U) \subseteq U' \Leftrightarrow \mu(U) \subseteq U'$,

(7) Let μ satisfy $(\mu \subseteq)$ and one of the (μPRi), then $X \subseteq Y, \mu(X \cup U) \subseteq X \Rightarrow \mu(Y \cup U) \subseteq Y$,

(8) Let μ satisfy $(\mu \subseteq)$ and one of the (μPRi), then $X \subseteq Y \Rightarrow X \cap \mu_i(Y) \subseteq \mu_i(X)$ - so (μPR) holds for μ_i, (more precisely, only for μ_2 we need the prerequisites, in the other cases the definition suffices)

(9) Let μ satisfy $(\mu \subseteq)$, $(\mu PR2)$, (μCum), then $\mu_2(X) \subseteq Y \subseteq X \Rightarrow \mu_2(X) = \mu_2(Y)$ - so (μCum) holds for μ_2.

(10) $(\mu \subseteq)$ and (μCum) for μ entail $\mu_0(U) = \mu_1(U)$.

Proof

(1) $\mu_1(X) \subseteq \mu_0(X)$ follows from $(\mu \subseteq)$ for μ. For μ_2: By $Y \subseteq U$, $U \cup Y = U$, so $\mu(U) \subseteq U$ by $(\mu \subseteq)$.

(2) $\mu(U \cup U') \subseteq U \subseteq U \cup U' \Rightarrow_{(\mu CUM)} \mu(U \cup U') = \mu(U)$.

(3) $\mu_i(U) \subseteq U$ by definition. To show $\mu_i(U) \subseteq \mu(U)$, take in all three cases $Y := U$, and use for $i = 1, 2$ $(\mu \subseteq)$.

(4) Exercise, solution see [Sch04], Fact 5.2.6 (3).

(5) Let $Y \in \mathcal{Y}$, $\mu(Y) \subseteq U$, $x \in Y - \mu(Y)$, then (by (4)) $\mu(U \cup Y) \subseteq \mu(U) \cup \mu(Y) \subseteq U$.

(6) Exercise, solution see [Sch04], Fact 5.2.6 (6).

(7) $\mu(Y \cup U) = \mu(Y \cup X \cup U) \subseteq_{(4)} \mu(Y) \cup \mu(X \cup U) \subseteq Y \cup X = Y$.

(8) For $i = 0, 1$: Let $x \in X - \mu_0(X)$, then there is A such that $A \subseteq X$, $x \in A - \mu(A)$, so $A \subseteq Y$. The case $i = 1$ is similar. We need here only the definitions. For $i = 2$: Let $x \in X - \mu_2(X)$, A such that $x \in A - \mu(A)$, $\mu(X \cup A) \subseteq X$, then by (7) $\mu(Y \cup A) \subseteq Y$.

(9) Exercise, solution see [Sch04], Fact 5.2.6 (9).

(10) $\mu_1(U) \subseteq \mu_0(U)$ by (1). Let Y such that $\mu(Y) \subseteq U$, $x \in Y - \mu(Y)$, $x \in U$. Consider $Y \cap U$, $x \in Y \cap U$, $\mu(Y) \subseteq Y \cap U \subseteq Y$, so $\mu(Y) = \mu(Y \cap U)$ by (μCum), and $x \notin \mu(Y \cap U)$. Thus, $\mu_0(U) \subseteq \mu_1(U)$.

☐

Fact 1.7.4

In the presence of $(\mu \subseteq)$, (μCum) for μ, we have:

$(\mu PR0) \Leftrightarrow (\mu PR1)$,

and $(\mu PR2) \Rightarrow (\mu PR1)$.

If (μPR) also holds for μ, then so will $(\mu PR1) \Rightarrow (\mu PR2)$.

(Recall that $(\cup)$ and $(\cap)$ are assumed to hold.)

Proof

$(\mu PR0) \Leftrightarrow (\mu PR1)$: By Fact 1.7.3, (10), $\mu_0(U) = \mu_1(U)$ if (μCum) holds for μ.

$(\mu PR2) \Rightarrow (\mu PR1)$: Suppose $(\mu PR2)$ holds. By $(\mu PR2)$ and (5), $\mu_2(U) \subseteq \mu_1(U)$, so $\mu(U) - \mu_1(U) \subseteq \mu(U) - \mu_2(U)$. By $(\mu PR2)$, $\mu(U) - \mu_2(U)$ is small, then so is $\mu(U) - \mu_1(U)$, so $(\mu PR1)$ holds.

$(\mu PR1) \Rightarrow (\mu PR2)$: Suppose $(\mu PR1)$ holds, and $(\mu PR2)$ fails. By failure of $(\mu PR2)$, there is $X \in \mathcal{Y}$ such that $\mu_2(U) \subseteq X \subset \mu(U)$. Let $x \in \mu(U)$-X, as $x \notin \mu_2(U)$, there is Y such that $\mu(U \cup Y) \subseteq U$, $x \in Y - \mu(Y)$. Let $Z := U \cup Y \cup X$. By (μPR), $x \notin \mu(U \cup Y)$, and $x \notin \mu(U \cup Y \cup X)$. Moreover, $\mu(U \cup X \cup Y) \subseteq \mu(U \cup Y) \cup \mu(X)$ by Fact 1.7.3 (4), $\mu(U \cup Y) \subseteq U$, $\mu(X) \subseteq X \subseteq \mu(U) \subseteq U$ by prerequisite, so $\mu(U \cup X \cup Y) \subseteq U \subseteq U \cup Y \subseteq U \cup X \cup Y$, so $\mu(U \cup X \cup Y) = \mu(U \cup Y) \subseteq U$. Thus, $x \notin \mu_1(U)$, and $\mu_1(U) \subseteq X$, too, a contradiction.

☐

Here is an example which shows that Fact 1.7.3, (9) may fail for μ_0 and μ_1.

1.7 Preferential Structures Without Definability Preservation

Example 1.7.7

Consider $\mathcal{L}$ with $v(\mathcal{L}) := \{p_i : i \in \omega\}$. Let $m \not\models p_0$, let $m' \in M(p_0)$ arbitrary. Make for each $n \in M(p_0) - \{m'\}$ one copy of m, likewise of m', set $\langle m, n \rangle \prec \langle m', n \rangle$ for all n, and $n \prec \langle m, n \rangle$, $n \prec \langle m', n \rangle$ for all n. The resulting structure $\mathcal{Z}$ is smooth and transitive. Let $\mathcal{Y} := \boldsymbol{D}_{\mathcal{L}}$, define $\mu(X) := \widehat{\mu_{\mathcal{Z}}(X)}$ for $X \in \mathcal{Y}$.

Let $m' \in X - \mu_{\mathcal{Z}}(X)$. Then $m \in X$, or $M(p_0) \subseteq X$. In the latter case, as all m'' such that $m'' \neq m'$, $m'' \models p_0$ are minimal, $M(p_0) - \{m'\} \subseteq \mu_{\mathcal{Z}}(X)$, so $m' \in \widehat{\mu_{\mathcal{Z}}(X)} = \mu(X)$. Thus, as $\mu_{\mathcal{Z}}(X) \subseteq \mu(X)$, if $m' \in X - \mu(X)$, then $m \in X$.

Define now $X := M(p_0) \cup \{m\}$, $Y := M(p_0)$.

We first show that μ_0 does not satisfy (μCum). $\mu_0(X) := \{x \in X : \neg \exists A \in \mathcal{Y}(A \subseteq X : x \in A - \mu(A))\}$. $m \not\in \mu_0(X)$, as $m \not\in \mu(X) = \widehat{\mu_{\mathcal{Z}}(X)}$. Moreover, $m' \not\in \mu_0(X)$, as $\{m, m'\} \in \mathcal{Y}$, $\{m, m'\} \subseteq X$, and $\mu(\{m, m'\}) = \mu_{\mathcal{Z}}(\{m, m'\}) = \{m\}$. So $\mu_0(X) \subseteq Y \subseteq X$. Consider now $\mu_0(Y)$. As $m \not\in Y$, for any $A \in \mathcal{Y}$, $A \subseteq Y$, if $m' \in A$, then $m' \in \mu(A)$, too, by above argument, so $m' \in \mu_0(Y)$, and μ_0 does not satisfy (μCum).

We turn to μ_1.

By Fact 1.7.3 (1), $\mu_1(X) \subseteq \mu_0(X)$, so $m, m' \not\in \mu_1(X)$, and again $\mu_1(X) \subseteq Y \subseteq X$. Consider again $\mu_1(Y)$. As $m \not\in Y$, for any $A \in \mathcal{Y}$, $\mu(A) \subseteq Y$, if $m' \in A$, then $m' \in \mu(A)$, too: if $M(p_0) - \{m'\} \subseteq A$, then $m' \in \widehat{\mu_{\mathcal{Z}}(A)}$, if $M(p_0) - \{m'\} \not\subseteq A$, but $m' \in A$, then either $m' \in \mu_{\mathcal{Z}}(A)$, or $m \in \mu_{\mathcal{Z}}(A) \subseteq \mu(A)$, but $m \not\in Y$. Thus, (μCum) fails for μ_1, too.

It remains to show that μ satisfies $(\mu \subseteq)$, (μCum), $(\mu PR0)$, $(\mu PR1)$. Note that by Fact 1.5.14 (3) and Proposition 1.5.19 $\mu_{\mathcal{Z}}$ satisfies (μCum), as $\mathcal{Z}$ is smooth. $(\mu \subseteq)$ is trivial. We show (μPRi) for $i = 0, 1$. As $\mu_{\mathcal{Z}}(A) \subseteq \mu(A)$, by (μPR) and (μCum) for $\mu_{\mathcal{Z}}$, $\mu_{\mathcal{Z}}(X) \subseteq \mu_0(X)$ and $\mu_{\mathcal{Z}}(X) \subseteq \mu_1(X)$: To see this, we note $\mu_{\mathcal{Z}}(X) \subseteq \mu_0(X)$: Let $x \in X - \mu_0(X)$, then there is Y such that $x \in Y - \mu(Y).Y \subseteq X$, but $\mu_{\mathcal{Z}}(Y) \subseteq \mu(Y)$, so by $Y \subseteq X$ and (μPR) for $\mu_{\mathcal{Z}}$ $x \not\in \mu_{\mathcal{Z}}(X)$. $\mu_{\mathcal{Z}}(X) \subseteq \mu_1(X)$: Let $x \in X - \mu_1(X)$, then there is Y such that $x \in Y - \mu(Y)$, $\mu(Y) \subseteq X$, so $x \in Y - \mu_{\mathcal{Z}}(Y)$ and $\mu_{\mathcal{Z}}(Y) \subseteq X$. $\mu_{\mathcal{Z}}(X \cup Y) \subseteq \mu_{\mathcal{Z}}(X) \cup \mu_{\mathcal{Z}}(Y) \subseteq X \subseteq X \cup Y$, so $\mu_{\mathcal{Z}}(X \cup Y) = \mu_{\mathcal{Z}}(X)$ by (μCum) for $\mu_{\mathcal{Z}}$. $x \in Y - \mu_{\mathcal{Z}}(Y) \Rightarrow x \not\in \mu_{\mathcal{Z}}(X \cup Y)$ by (μPR) for $\mu_{\mathcal{Z}}$, so $x \not\in \mu_{\mathcal{Z}}(X)$.

But by Fact 1.7.3, (3) $\mu_i(X) \subseteq \mu(X)$. As by definition, $\mu(X) - \mu_{\mathcal{Z}}(X)$ is small, (μPRi) hold for $i = 0, 1$. It remains to show (μCum) for μ. Let $\mu(X) \subseteq Y \subseteq X$, then $\mu_{\mathcal{Z}}(X) \subseteq \mu(X) \subseteq Y \subseteq X$, so by (μCum) for $\mu_{\mathcal{Z}}$ $\mu_{\mathcal{Z}}(X) = \mu_{\mathcal{Z}}(Y)$, so by definition of μ, $\mu(X) = \mu(Y)$.

(Note that by Fact 1.7.3 (10), $\mu_0 = \mu_1$ follows from (μCum) for μ, so we could have demonstrated part of the properties also differently.)

$\square$

By Fact 1.7.3 (3) and (8) and Proposition 1.3.5, μ_0 has a representation by a (transitive) preferential structure, if $\mu : \mathcal{Y} \to \mathcal{Y}$ satisfies $(\mu \subseteq)$ and $(\mu PR0)$, and μ_0 is defined as in Definition 1.7.2.

We thus have (taken from [Sch04], Proposition 5.2.5 there):

Proposition 1.7.5

Let Z be an arbitrary set, $\mathcal{Y} \subseteq \mathcal{P}(Z)$, $\mu : \mathcal{Y} \to \mathcal{Y}$, $\mathcal{Y}$ closed under arbitrary intersections and finite unions, and $\emptyset, Z \in \mathcal{Y}$, and let $\frown$ be defined with respect to $\mathcal{Y}$.

(a) If μ satisfies $(\mu \subseteq), (\mu PR0)$, then there is a transitive preferential structure $\mathcal{Z}$ over Z such that for all $U \in \mathcal{Y}$ $\mu(U) = \widehat{\mu_{\mathcal{Z}}(U)}$.

(b) If $\mathcal{Z}$ is a preferential structure over Z and $\mu : \mathcal{Y} \to \mathcal{Y}$ such that for all $U \in \mathcal{Y}$ $\mu(U) = \widehat{\mu_{\mathcal{Z}}(U)}$, then μ satisfies $(\mu \subseteq), (\mu PR0)$.

Proof

(a) Let μ satisfy $(\mu \subseteq), (\mu PR0)$. μ_0 as defined in Definition 1.7.2 satisfies properties $(\mu \subseteq), (\mu PR)$ by Fact 1.7.3, (3) and (8). Thus, by Proposition 1.3.5, there is a transitive structure $\mathcal{Z}$ over Z such that $\mu_0 = \mu_{\mathcal{Z}}$, but by $(\mu PR0)$ $\mu(U) = \widehat{\mu_0(U)} = \widehat{\mu_{\mathcal{Z}}(U)}$ for $U \in \mathcal{Y}$.

(b) $(\mu \subseteq) : \mu_{\mathcal{Z}}(U) \subseteq U$, so by $U \in \mathcal{Y}$ $\mu(U) = \widehat{\mu_{\mathcal{Z}}(U)} \subseteq U$.

$(\mu PR0)$: If $(\mu PR0)$ is false, there is $U \in \mathcal{Y}$ such that for $U' := \bigcup \{Y' - \mu(Y') : Y' \in \mathcal{Y}, Y' \subseteq U\}$ $\widehat{\mu(U) - U'} \subset \mu(U)$. By $\mu_{\mathcal{Z}}(Y') \subseteq \mu(Y')$, $Y' - \mu(Y') \subseteq Y' - \mu_{\mathcal{Z}}(Y')$. No copy of any $x \in Y' - \mu_{\mathcal{Z}}(Y')$ with $Y' \subseteq U$, $Y' \in \mathcal{Y}$ can be minimal in $\mathcal{Z} \upharpoonright U$. Thus, by $\mu_{\mathcal{Z}}(U) \subseteq \mu(U)$, $\mu_{\mathcal{Z}}(U) \subseteq \mu(U) - U'$, so $\widehat{\mu_{\mathcal{Z}}(U)} \subseteq \widehat{\mu(U) - U'} \subset \mu(U)$, contradiction.

□

We turn to the smooth case.

If $\mu : \mathcal{Y} \to \mathcal{Y}$ satisfies $(\mu \subseteq), (\mu PR2), (\mu CUM)$ and μ_2 is defined from μ as in Definition 1.7.2, then μ_2 satisfies $(\mu \subseteq), (\mu PR), (\mu Cum)$ by Fact 1.7.3 (3), (8), and (9), and can thus be represented by a (transitive) smooth structure, by Proposition 1.3.18, and we finally have (taken from [Sch04], Proposition 5.2.9 there):

Proposition 1.7.6

Let Z be an arbitrary set, $\mathcal{Y} \subseteq \mathcal{P}(Z)$, $\mu : \mathcal{Y} \to \mathcal{Y}$, $\mathcal{Y}$ closed under arbitrary intersections and finite unions, and $\emptyset, Z \in \mathcal{Y}$, and let $\frown$ be defined with respect to $\mathcal{Y}$.

1.7 Preferential Structures Without Definability Preservation

(a) If μ satisfies $(\mu \subseteq), (\mu PR2), (\mu CUM)$, then there is a transitive smooth preferential structure $\mathcal{Z}$ over Z such that for all $U \in \mathcal{Y}$ $\mu(U) = \widetilde{\mu_{\mathcal{Z}}(U)}$.

(b) If $\mathcal{Z}$ is a smooth preferential structure over Z and $\mu : \mathcal{Y} \to \mathcal{Y}$ such that for all $U \in \mathcal{Y}$ $\mu(U) = \widetilde{\mu_{\mathcal{Z}}(U)}$, then μ satisfies $(\mu \subseteq), (\mu PR2), (\mu CUM)$.

Proof

Exercise, solution see [Sch04], Proposition 5.2.9.

1.7.2.3 Ranked Structures

We recall from Section 1.2.4 the basic properties of ranked structures.

We give now an easy version of representation results for ranked structures without definability preservation.

Notation 1.7.1

We abbreviate $\mu(\{x, y\})$ by $\mu(x, y)$ etc.

Fact 1.7.7

Let the domain contain singletons and be closed under $(\cup)$.

Let for $\mu : \mathcal{Y} \to \mathcal{Y}$ hold:

$(\mu =)$ for finite sets, $(\mu \in), (\mu PR3), (\mu \emptyset fin)$.

Then the following properties hold for μ_3 as defined in Definition 1.7.2:

(1) $\mu_3(X) \subseteq \mu(X)$,
(2) for finite X, $\mu(X) = \mu_3(X)$,
(3) $(\mu \subseteq)$,
(4) (μPR),
(5) $(\mu \emptyset fin)$,
(6) $(\mu =)$,
(7) $(\mu \in)$,
(8) $\mu(X) = \widetilde{\mu_3(X)}$.

Proof

(1) Suppose not, so $x \in \mu_3(X)$, $x \in X - \mu(X)$, so by $(\mu \in)$ for μ, there is $y \in X$, $x \notin \mu(x, y)$, contradiction.

(2) By $(\mu PR3)$ for μ and (1), for finite U $\mu(U) = \mu_3(U)$.

(3) $(\mu \subseteq)$ is trivial for μ_3.

(4) Let $X \subseteq Y$, $x \in \mu_3(Y) \cap X$, suppose $x \in X - \mu_3(X)$, so there is $y \in X \subseteq Y$, $x \notin \mu(x,y)$, so $x \notin \mu_3(Y)$.

(5) $(\mu \emptyset fin)$ for μ_3 follows from $(\mu \emptyset fin)$ for μ and (2).

(6) Let $X \subseteq Y$, $y \in \mu_3(Y) \cap X$, $x \in \mu_3(X)$, we have to show $x \in \mu_3(Y)$. By (4), $y \in \mu_3(X)$. Suppose $x \notin \mu_3(Y)$. So there is $z \in Y.x \notin \mu(x,z)$. As $y \in \mu_3(Y)$, $y \in \mu(y,z)$. As $x \in \mu_3(X)$, $x \in \mu(x,y)$, as $y \in \mu_3(X)$, $y \in \mu(x,y)$. Consider $\{x,y,z\}$. Suppose $y \notin \mu(x,y,z)$, then by $(\mu \in)$ for μ, $y \notin \mu(x,y)$ or $y \notin \mu(y,z)$, $contradiction$. Thus $y \in \mu(x,y,z) \cap \mu(x,y)$. As $x \in \mu(x,y)$, and $(\mu =)$ for μ and finite sets, $x \in \mu(x,y,z)$. Recall that $x \notin \mu(x,z)$. But for finite sets $\mu = \mu_3$, and by (4) (μPR) holds for μ_3, so it holds for μ and finite sets. $contradiction$

(7) Let $x \in X - \mu_3(X)$, so there is $y \in X.x \notin \mu(x,y) = \mu_3(x,y)$.

(8) As $\mu(X) \in \mathcal{Y}$, and $\mu_3(X) \subseteq \mu(X)$, $\widetilde{\mu_3(X)} \subseteq \mu(X)$, so by $(\mu PR3)$ $\widetilde{\mu_3(X)} = \mu(X)$.

$\square$

Fact 1.7.8

If $\mathcal{Z}$ is ranked, and we define $\mu(X) := \widetilde{\mu_{\mathcal{Z}}(X)}$, and $\mathcal{Z}$ has no copies, then the following hold:

(1) $\mu_{\mathcal{Z}}(X) = \{x \in X : \forall y \in X.x \in \mu(x,y)\}$, so $\mu_{\mathcal{Z}}(X) = \mu_3(X)$ for $X \in \mathcal{Y}$,

(2) $\mu(X) = \mu_{\mathcal{Z}}(X)$ for finite X,

(3) $(\mu =)$ for finite sets for μ,

(4) $(\mu \in)$ for μ,

(5) $(\mu \emptyset fin)$ for μ,

(6) $(\mu PR3)$ for μ.

Proof

(1) holds for ranked structures.

(2) and (6) are trivial. (3) and (5) hold for $\mu_{\mathcal{Z}}$, so by (2) for μ.

(4) If $x \notin \mu(X)$, then $x \notin \mu_{\mathcal{Z}}(X)$, $(\mu \in)$ holds for $\mu_{\mathcal{Z}}$, so there is $y \in X$ such that $x \notin \mu_{\mathcal{Z}}(x,y) = \mu(x,y)$ by (2).

$\square$

We summarize:

1.7 Preferential Structures Without Definability Preservation

Proposition 1.7.9

Let Z be an arbitrary set, $\mathcal{Y} \subseteq \mathcal{P}(Z)$, $\mu : \mathcal{Y} \to \mathcal{Y}$, $\mathcal{Y}$ closed under arbitrary intersections and finite unions, contain singletons, and $\emptyset, Z \in \mathcal{Y}$, and let $\widehat{}$ be defined with respect to $\mathcal{Y}$.

(a) If μ satisfies $(\mu =)$ for finite sets, $(\mu \in)$, $(\mu PR3)$, $(\mu \emptyset fin)$, then there is a ranked preferential structure $\mathcal{Z}$ without copies over Z such that for all $U \in \mathcal{Y}$
$$\mu(U) = \widehat{\mu_{\mathcal{Z}}(U)}.$$

(b) If $\mathcal{Z}$ is a ranked preferential structure over Z without copies and $\mu : \mathcal{Y} \to \mathcal{Y}$ such that for all $U \in \mathcal{Y}$ $\mu(U) = \widehat{\mu_{\mathcal{Z}}(U)}$, then μ satisfies $(\mu =)$ for finite sets, $(\mu \in)$, $(\mu PR3)$, $(\mu \emptyset fin)$.

Proof

(a) Let μ satisfy $(\mu =)$ for finite sets, $(\mu \in)$, $(\mu PR3)$, $(\mu \emptyset fin)$, then μ_3 as defined in Definition 1.7.2 satisfies properties $(\mu \subseteq)$, $(\mu \emptyset fin)$, $(\mu =)$, $(\mu \in)$ by Fact 1.7.7. Thus, by Proposition 1.4.9, there is a transitive structure $\mathcal{Z}$ over Z such that $\mu_3 = \mu_{\mathcal{Z}}$, but by Fact 1.7.7 (8) $\mu(U) = \widehat{\mu_3(U)} = \widehat{\mu_{\mathcal{Z}}(U)}$ for $U \in \mathcal{Y}$.

(b) This was shown in Fact 1.7.8.

$\square$

1.7.2.4 The Logical Results

We turn to (propositional) logic.

The main result here is Proposition 1.7.10. Recall Fact 1.2.12. The conditions are formulated or recalled in Conditions 1.7.1, the auxiliary Lemma 1.7.11 is the main step in the proof of Proposition 1.7.10.

We work now in $\mathcal{Y} := \mathbf{D}_\mathcal{L}$, so $\widehat{U} = M(Th(U))$ for $U \subseteq M_\mathcal{L}$ and the prerequisites of Fact 1.2.1 will hold.

Condition 1.7.1

(CP) $Con(T) \to Con(\overline{\overline{T}})$,

(LLE) $\overline{T} = \overline{T'} \to \overline{\overline{T}} = \overline{\overline{T'}}$,

(CCL) $\overline{\overline{T}}$ is classically closed,

(SC) $T \subseteq \overline{\overline{T}}$,

($\hspace{0.2em}\sim\hspace{-0.9em}\mid\hspace{0.4em}$ 4) Let $T, T_i, i \in I$ be theories s.t. $\forall i \; T_i \vdash T$, then there is no ϕ s.t. $\phi \notin \overline{\overline{T}}$ and $M(\overline{\overline{T}} \cup \{\neg\phi\}) \subseteq \bigcup\{M(T_i) - M(\overline{\overline{T_i}}) : i \in I\}$,

$\hspace{0.2em}\sim\hspace{-0.9em}\mid\hspace{0.4em}$ 4s) Let $T, T_i, i \in I$ be theories s.t. $\forall i \; T \subseteq \overline{\overline{T_i \vee T}}$, then there is no ϕ s.t. $\phi \notin \overline{\overline{T}}$ and $M(\overline{\overline{T}} \cup \{\neg\phi\}) \subseteq \bigcup\{M(T_i) - M(\overline{\overline{T_i}}) : i \in I\}$,

($\hspace{0.2em}\sim\hspace{-0.9em}\mid\hspace{0.4em}$ 5) $\overline{\overline{T}} \vee \overline{\overline{T'}} \subseteq \overline{\overline{T \vee T'}}$,

(CUM) $T \subseteq \overline{\overline{T'}} \subseteq \overline{\overline{T}} \rightarrow \overline{\overline{T}} = \overline{\overline{T'}}$

for all T, T', T_i.

Condition (CP) is auxiliary and corresponds to the nonemptiness condition $(\mu\emptyset)$ of μ in the smooth case: $U \neq \emptyset \rightarrow \mu(U) \neq \emptyset$ - or to the fact that all models occur in the structure.

We formulate now the logical representation theorem for not necessarily definability preserving preferential structures.

Proposition 1.7.10

Let $\hspace{0.2em}\sim\hspace{-0.9em}\mid\hspace{0.4em}$ be a logic for $\mathcal{L}$. Then:

(a.1) If $\mathcal{M}$ is a classical preferential model over $M_\mathcal{L}$ and $\overline{\overline{T}} = Th(\mu_\mathcal{M}(M(T)))$, then (LLE), (CCL), (SC), ($\hspace{0.2em}\sim\hspace{-0.9em}\mid\hspace{0.4em}$ 4) hold for the logic so defined.

(a.2) If (LLE), (CCL), (SC), ($\hspace{0.2em}\sim\hspace{-0.9em}\mid\hspace{0.4em}$ 4) hold for a logic, then there is a transitive classical preferential model over $M_\mathcal{L}$ $\mathcal{M}$ s.t. $\overline{\overline{T}} = Th(\mu_\mathcal{M}(M(T)))$.

(b.1) If $\mathcal{M}$ is a smooth classical preferential model over $M_\mathcal{L}$ and $\overline{\overline{T}} = Th(\mu_\mathcal{M}(M(T)))$, then (CP), (LLE), (CCL), (SC), ($\hspace{0.2em}\sim\hspace{-0.9em}\mid\hspace{0.4em}$ 4s), ($\hspace{0.2em}\sim\hspace{-0.9em}\mid\hspace{0.4em}$ 5), (CUM) hold for the logic so defined.

(b.2) If (CP), (LLE), (CCL), (SC), ($\hspace{0.2em}\sim\hspace{-0.9em}\mid\hspace{0.4em}$ 4s), ($\hspace{0.2em}\sim\hspace{-0.9em}\mid\hspace{0.4em}$ 5), (CUM) hold for a logic, then there is a smooth transitive classical preferential model $\mathcal{M}$ over $M_\mathcal{L}$ s.t. $\overline{\overline{T}} = Th(\mu_\mathcal{M}(M(T)))$.

The proof is an easy consequence of Proposition 1.7.5, Proposition 1.7.6, and Lemma 1.7.11, and will be shown after the proof of the latter.

Lemma 1.7.11

(a) If $\mu : \boldsymbol{D}_\mathcal{L} \rightarrow \boldsymbol{D}_\mathcal{L}$ satisfies $(\mu \subseteq)$, $(\mu PR0)$ (for $\mathcal{Y} = \boldsymbol{D}_\mathcal{L}$), then $\hspace{0.2em}\sim\hspace{-0.9em}\mid\hspace{0.4em}$ defined by $\overline{\overline{T}} := Th(\mu(M(T)))$ satisfies (LLE), (CCL), (SC), ($\hspace{0.2em}\sim\hspace{-0.9em}\mid\hspace{0.4em}$ 4).

(b) If $\mu : \boldsymbol{D}_\mathcal{L} \rightarrow \boldsymbol{D}$ satisfies $(\mu\emptyset)$, $(\mu \subseteq)$, $(\mu PR2)$, (μCUM) (for $\mathcal{Y} = \boldsymbol{D}_\mathcal{L}$), then $\hspace{0.2em}\sim\hspace{-0.9em}\mid\hspace{0.4em} \mathcal{L}$ defined by $\overline{\overline{T}} := Th(\mu(M(T)))$ satisfies (CP), (LLE), (CCL), (SC), ($\hspace{0.2em}\sim\hspace{-0.9em}\mid\hspace{0.4em}$ 4s), ($\hspace{0.2em}\sim\hspace{-0.9em}\mid\hspace{0.4em}$ 5), (CUM).

1.7 Preferential Structures Without Definability Preservation

(c) If $\mathrel|\!\sim$ satisfies (LLE), (CCL), (SC), ($\mathrel|\!\sim$ 4), then there is $\mu : \mathbf{D}_{\mathcal{L}} \to \mathbf{D}_{\mathcal{L}}$ such that $\overline{\overline{T}} = Th(\mu(M(T)))$ for all T and μ satisfies $(\mu \subseteq)$, $(\mu PR0)$ (for $\mathcal{Y} = \mathbf{D}_{\mathcal{L}}$).

(d) If $\mathrel|\!\sim$ satisfies (CP), (LLE), (CCL), (SC), ($\mathrel|\!\sim$ 4s), ($\mathrel|\!\sim$ 5), (CUM), then there is $\mu : \mathbf{D}_{\mathcal{L}} \to \mathbf{D}_{\mathcal{L}}$ such that $\overline{\overline{T}} = Th(\mu(M(T)))$ for all T and μ satisfies $(\mu\emptyset)$, $(\mu \subseteq)$, $(\mu PR2)$, (μCUM) (for $\mathcal{Y} = \mathbf{D}_{\mathcal{L}}$).

Proof

Exercise, solution see [Sch04], Lemma 5.2.12.

Proof of Proposition 1.7.10:

Exercise, solution see [Sch04], Proposition 5.2.11.

1.7.3 The General Case and the Limit Version Cannot Be Characterized

1.7.3.1 Introduction

We show more than what the headline announces:

- general, not necessarily definability preserving preferential structures,
- the general limit version of preferential structures,
- not necessarily definability preserving ranked preferential structures,
- the limit version of ranked preferential structures,
- general, not necessarily definability preserving distance based revision,
- the general limit version of distance based revision

all have no "normal" characterization by logical means of any size.

This negative result for the limit version, together with the reductory results of Section 1.6.4 and Section 1.6.5, casts a heavy doubt on the utility of the limit version as a reasoning tool. It seems either hopelessly, or unnecessarily, complicated.

A similar result is shown in Proposition 1.7.15 for not necessarily definability preserving ranked structures, and in Proposition 4.3.13 for not necessarily definability preserving distance based revision.

More discussion can be found in [Sch04], section 5.2.3.

1.7.3.2 The Details

We will work in a propositional language $\mathcal{L}$ with κ many (κ an infinite cardinal) propositional variables $p_i : i < \kappa$. As p_0 will have a special role, we will set $p := p_0$. In the revision case, we will use another special variable, which we will call q.

In all cases, we will show that there is no normal characterization of size $\leq \kappa$. As κ was arbitrary, we will have shown the results.

Given any model set $X \subseteq M_{\mathcal{L}}$, we define again $\widehat{X} := M(Th(X))$ - the closure of X in the standard topology.

Fact 1.7.12

(1) $X \subseteq \widehat{X}$.

(2) Let T be any $\mathcal{L}$-theory, and $A \subseteq M_{\mathcal{L}}$, then $\widehat{M(T) - A} = M(T \cup T_A)$ for some T_A. Of course, T_A may be empty or a subset of $\overline{T}$, if $\widehat{M(T) - A} = M(T)$. Thus, for $\mathcal{X} \subseteq \mathcal{P}(M_{\mathcal{L}}) \bigcap \{\widehat{M(T) - A} : A \in \mathcal{X}\} = \bigcap \{M(T \cup T_A) : A \in \mathcal{X}\}$
$= M(\bigcup \{T \cup T_A : A \in \mathcal{X}\})$ for suitable T_A.

(3) If $\widehat{M(T) - A} \neq M(T)$, then $Th(M(T) - A) \supset \overline{T}$, so $\widehat{M(T) - A} = M(T \cup T_A)$ for some T_A s.t. $T \nvdash T_A$.

(Trivial).

□

We now state and prove our main technical lemma.

Lemma 1.7.13

Let $\mathcal{L}$ be a language of κ many (κ an infinite cardinal) propositional variables. Let a theory T be given, $\mathcal{E}_T \subseteq \{X \subseteq M_{\mathcal{L}} : card(X) \leq \kappa\}$ be closed under unions of size $\leq \kappa$ and subsets, and $\overline{\overline{T}}$ be defined by $\overline{\overline{T}} := Th(\bigcap \{\widehat{M(T) - A} : A \in \mathcal{E}_T\})$. Then there is an (usually not unique) "optimal" $A_T \in \mathcal{E}_T$ s.t.

(1) $\overline{\overline{T}} = Th(M(T) - A_T)$,

(2) for all $A \in \mathcal{E}_T$ $\widehat{M(T) - A_T} \subseteq \widehat{M(T) - A}$.

1.7 Preferential Structures Without Definability Preservation

Proof

Before we give the details, we describe the (simple) idea. The proof shows essentially how to do the right counting.

We cannot work directly with the $A \in \mathcal{E}_T$, and take the union, there might be too many of them, and the resulting set might be too big. But the $A \in \mathcal{E}_T$ give mostly the same results $\widehat{M(T) - A}$, and there are not very many interesting ones of them, or of their corresponding theories: To each A corresponds a theory T_A with $\widehat{M(T) - A} = M(T \cup T_A)$. As we are only interested in those A or T_A which change $M(T)$, we will successively add formulas to some initial T_A, until we have found a maximal $T_{A'}$ s.t. $\overline{\overline{T}} = Th(M(T \cup T_{A'}))$, and A' will be the A_T. Thus, we work neither directly with all A, nor with all T_A, but count formulas, and there are only $\leq \kappa$ many of them. We will then take the union A_T of the corresponding A (i.e. which add new formulas), this will have size $\leq \kappa$ again.

Now the details.

By Fact 1.7.12, $\bigcap \{\widehat{M(T) - A} : A \in \mathcal{E}_T\} = M(\bigcup \{T \cup T_A : A \in \mathcal{E}_T\})$, so $\overline{\overline{T}} = Th(M(\bigcup \{T \cup T_A : A \in \mathcal{E}_T\})) = \overline{\bigcup \{T \cup T_A : A \in \mathcal{E}_T\}}$ for suitable T_A.

We have to show that we can obtain $\overline{\overline{T}}$ with one single $A_T \in \mathcal{E}_T$, i.e. $\overline{\overline{T}} = Th(M(T) - A_T)$.

Let $\mathcal{E} := \mathcal{E}_T$, and let Ψ_i be an (arbitrary) enumeration of $\{T_A : A \in \mathcal{E}\}$.

We define an increasing chain $\Gamma_i : i \leq \mu$ ($\mu \leq \kappa$) of sets of formulas by induction, and show that for each Γ_i there is $A_i \in \mathcal{E}$ s.t. $\widehat{M(T) - A_i} = M(T \cup \Gamma_i)$, and $\overline{\overline{T}} = \overline{T \cup \bigcup \{\Gamma_i : i \leq \mu\}}$.

$\Gamma_0 := \Psi_0$.

$\Gamma_{i+1} := \Gamma_i \cup \Psi_j$, where Ψ_j is the first $\Psi_l \not\subseteq \Gamma_i$ - if this does not exist, as Γ_i contains already all Ψ_l, we stop the construction.

$\Gamma_\lambda := \bigcup \{\Gamma_i : i < \lambda\}$ for limits λ.

Note that the chain of Γ's has length $\leq \kappa$, as we always add at least one of the κ many formulas of $\mathcal{L}$ in the successor step (the construction will stop at a successor step).

We now show that there is $A_i \in \mathcal{E}$ s.t. $\widehat{M(T) - A_i} = M(T \cup \Gamma_i)$ by induction.

By construction,

$\widehat{M(T) - A_0} = M(T \cup \Gamma_0)$ – where $A_0 \in \mathcal{E}$ is one of the $A \in \mathcal{E}$ which correspond to Ψ_0 (usually, there are many of them).

Suppose $\overline{M(T) - A_i} = M(T \cup \Gamma_i)$ by induction, and $\overline{M(T) - A_j} = M(T \cup \Psi_j)$. Then $M(T) - (A_i \cup A_j) \models T \cup \Gamma_i \cup \Psi_j$, so there is a subset A_{i+1} of $A_i \cup A_j$, thus of size $\leq \kappa$, and $A_{i+1} \in \mathcal{E}$, s.t. $\overline{M(T) - A_{i+1}} = M(T \cup \Gamma_i \cup \Psi_j) = M(T \cup \Gamma_{i+1})$, as $\Gamma_{i+1} = \Gamma_i \cup \Psi_j$.

Suppose $\overline{M(T) - A_i} = M(T \cup \Gamma_i)$ for $i < \lambda \leq \kappa$ by induction. Then $M(T) - \bigcup \{A_i : i < \lambda\} \models T \cup \bigcup \{\Gamma_i : i < \lambda\}$, so there is a subset A_λ of $\bigcup \{A_i : i < \lambda\}$, i.e. of size $\leq \kappa$, and $A_\lambda \in \mathcal{E}$, s.t. $\overline{M(T) - A_\lambda} = M(T \cup \bigcup \{\Gamma_i : i < \lambda\}) = M(T \cup \Gamma_\lambda)$.

This is also true for the last element Γ_μ, as the entire chain has length $\leq \kappa$.

Consequently, there is $A_T := A_\mu \in \mathcal{E}$ s.t. $\overline{M(T) - A_T} = M(T \cup \Gamma_\mu) = M(\overline{\overline{T}})$, as $\overline{\overline{T}} = \overline{\bigcup \{T \cup T_A : A \in \mathcal{E}\}}$ and $\Gamma_\mu = \bigcup \{T_A : A \in \mathcal{E}\}$, and for each $A \in \mathcal{E}$ $\overline{M(T) - A} \supseteq \overline{M(T) - A_T}$, and $\overline{\overline{T}} = Th(\overline{M(T) - A_T}) = Th(M(T) - A_T)$, as Γ_μ contains all Ψ corresponding to some $A \in \mathcal{E}$. Thus, (1) and (2) hold.

(Loosely speaking, $A_T := A_\mu$ is a maximal element of $\mathcal{E}_T$, more precisely, its Ψ is maximal. The important fact is that such A_T exists, and still has size $\leq \kappa$.)

□

We are now ready to state and prove the negative result for general, not necessarily definability preserving preferential structures and the general limit variant.

Proposition 1.7.14

(1) There is no "normal" characterization of any fixed size of not necessarily definability preserving preferential structures.

(2) There is no "normal" characterization of any fixed size of the general limit variant of preferential structures.

Proof

Recall that the "small sets of exceptions" can be arbitrarily big unions of exceptions.

Proof of (2):

It is easy to see that (2) is a consequence of (1): Any minimal variant of suitable preferential structures can also be read as a degenerate case of the limit variant: There is a smallest closed minimizing set, so both variants coincide. This is in particular true for the structurally extremely simple cases we consider here - the relation will be trivial, as the paths in the relation have length at most 1, we work with quantity. On the other hand, it is easily seen that the logic we define first is not preferential, neither in the minimal, nor in the limit reading.

1.7 Preferential Structures Without Definability Preservation

Proof of (1):

Let then κ be any infinite cardinal. We show that there is no characterization of general (i.e. not necessarily definability preserving) preferential structures which has size $\leq \kappa$. We suppose there were one such characterization Φ of size $\leq \kappa$, and construct a counterexample.

> The idea of the proof is very simple. We show that it suffices to consider for any given instantiation of $\Phi \leq \kappa$ many pairs $m \prec m^-$ in a case not representable by a preferential structure, and that $\leq \kappa$ many such pairs give the same result in a true preferential structure for this instantiation. Thus, every instantiation is true in an "illegal" and a "legal" example, so Φ cannot discern between legal and illegal examples. The main work is to show that $\leq \kappa$ many pairs suffice in the illegal example, this was done in Lemma 1.7.13.

We first note some auxiliary facts and definitions, and then define the logic, which, as we show, is not representable by a preferential structure. We then use the union of all the "optimal" sets A_T guaranteed by Lemma 1.7.13 to define the preferential structure, and show that in this structure $\overline{\overline{T}}$ for $T \in \mathcal{T}$ is the same as in the old logic, so the truth value of the instantiated expression is the same in the old logic and the new structure.

Writing down all details properly is a little complicated.

As any formula ϕ in the language has finite size, ϕ uses only a finite number of variables, so ϕ has 0 or 2^κ different models.

For any model m with $m \models p$, let m^- be exactly like m with the exception that $m^- \models \neg p$. (If $m \not\models p$, m^- is not defined.)

Let $\mathcal{A} := \{X \subseteq M(\neg p) : card(X) \leq \kappa\}$. For given T, let $\mathcal{A}_T := \{X \in \mathcal{A} : X \subseteq M(T) \wedge \forall m^- \in X.m \in M(T)\}$. Note that $\mathcal{A}_T$ is closed under subsets and under unions of size $\leq \kappa$. For T, let $\mathcal{B}_T := \{X \in \mathcal{A}_T : \widetilde{M(T) - X} \neq M(T)\}$, the (in the logical sense) "big" elements of $\mathcal{A}_T$. For $X \subseteq M_\mathcal{L}$, let $X \lceil M(T) := \{m^- \in X : m^- \in M(T) \wedge m \in M(T)\}$. Thus, $\mathcal{A}_T = \{X \lceil M(T) : X \in \mathcal{A}\}$.

Define now the logic $\mathrel|\!\sim$ as follows in two steps:

(1) $\overline{\overline{Th(\{m, m^-\})}} := Th(\{m\})$

> (Speaking preferentially, $m \prec m^-$, for all pairs m, m^-, this will be the entire relation. The relation is thus extremely simple, $\prec$-paths have length at most 1, so $\prec$ is automatically transitive.)

> We now look at (in terms of preferential models only some!) consequences:

(2) $\overline{\overline{T}} := Th(\bigcap \{\widetilde{M(T) - A} : A \in \mathcal{B}_T\}) = Th(\bigcap \{\widetilde{M(T) - A} : A \in \mathcal{A}_T\})$.

We note:

(a) This - with exception of the size condition - would be exactly the preferential consequence of part (1) of the definition.

(b) (1) is a special case of (2), we have seperated them for didactic reasons.

(c) The prerequisites of Lemma 1.7.13 are satisfied for $\overline{\overline{T}}$ and $\mathcal{A}_T$.

(d) It is crucial that we close before intersecting.

(Remark: We discuss a similar idea - better "protection" of single models by bigger model sets - in Section 1.8.3 where we give a counterexample to the KLM characterization.)

This logic is not preferential. We give the argument for the minimal case, the argument for the limit case is the same.

Take $T := \emptyset$. Take any $A \in \mathcal{A}_T$. Then $Th(M_\mathcal{L}) = Th(M_\mathcal{L} - A)$, as any ϕ, which holds in A, will have 2^κ models, so there must be a model of ϕ in $M_\mathcal{L} - A$, so we cannot separate A or any of its subsets. Thus, $\widetilde{M(\emptyset) - A} = M(\emptyset)$ for all A of size $\leq \kappa$, so $\overline{\overline{\emptyset}} = \overline{\emptyset}$, which cannot be if $\hspace{-2pt}\sim$ is preferential, for then $\overline{\overline{\emptyset}} = \overline{p}$.

Suppose there were a characterization Φ of size $\leq \kappa$. It has to say "no" for at least one instance $\mathcal{T}$ (i.e. a set of size $\leq \kappa$ of theories) of the universally quantified condition Φ. We will show that we find a true preferential structure where this instance $\mathcal{T}$ of Φ has the same truth value, more precisely, where all $T \in \mathcal{T}$ have the same $\overline{\overline{T}}$ in the old logic and in the preferential structure, a contradiction, as this instance evaluates now to "false" in the preferential structure, too.

Suppose $T \in \mathcal{T}$.

If $\overline{\overline{T}} = \overline{T}$, we do nothing (or set $A_T := \emptyset$). When $\overline{\overline{T}}$ is different from $\overline{T}$, this is because $\mathcal{B}_T \neq \emptyset$.

By Lemma 1.7.13, for each of the $\leq \kappa$ $T \in \mathcal{T}$, it suffices to consider a set A_T of size $\leq \kappa$ of suitable models of $\neg p$ to calculate $\overline{\overline{T}}$, i.e. $\overline{\overline{T}} = Th(M(T) - A_T)$, so, all in all, we work just with at most κ many such models. More precisely, set

$B := \bigcup \{A_T : \overline{\overline{T}} = Th(M(T) - A_T) \neq \overline{T}, T \in \mathcal{T}\}$.

Note that for each T with $\overline{\overline{T}} \neq \overline{T}$, $B \lceil M(T) \in \mathcal{B}_T$, as B has size $\leq \kappa$, and B contains A_T, so $\widetilde{M(T) - B \lceil M(T)} \neq M(T)$. But we also have $\overline{\overline{T}} = Th(M(T) - A_T) = Th(M(T) - B \lceil M(T))$, as A_T was optimal in $\mathcal{B}_T$.

Consider now the preferential structure where we do not make all $m \prec m^-$, but only the κ many of them featuring in B, i.e. those we have used in the instance T of Φ. We have to show that the instance T of Φ still fails in the new structure. But this is now trivial. Things like $\overline{T}$, etc. do not change, the only problem might be $\overline{\overline{T}}$. As we work in a true preferential structure, we now have to consider not subsets of size at most κ, but all of $B \lceil\lceil M(T)$ at once - which also has size $\leq \kappa$. But, by definition of the new structure, $\overline{\overline{T}} = Th(M(T) - B\lceil\lceil M(T)) = Th(M(T) - A_T)$. On the other hand, if $\overline{\overline{T}} = \overline{T}$ in the old structure, the same will hold in the new structure, as $B\lceil\lceil M(T)$ is one of the sets considered, and they did not change $\overline{\overline{T}}$.

Thus, the $\overline{\overline{T}}$ in the new and in the old structure are the same. So the instance T of Φ fails also in a suitable preferential structure, contradicting its supposed discriminatory power.

The limit reading of this simple structure gives the same result.

□

For discussion and proof of the ranked case, we refer the reader to [Sch04], Proposition 5.2.16, and the comment preceding this proposition.

Proposition 1.7.15

(1) There is no "normal" characterization of any fixed size of not necessarily definability preserving ranked preferential structures.

(2) There is no "normal" characterization of any fixed size of the general limit version of ranked preferential structures.

Proof

The proof follows closely the proof of Proposition 1.7.14.

Exercise, solution see [Sch04], Proposition 5.2.16.

1.8 Various Results and Approaches

1.8.1 Introduction

- In Section 1.8.2, we discuss the role of copies (or non-injective labelling functions in KLM terminology) in preferential structures. ("KLM" stands for [KLM90] or its authors.)
- In Section 1.8.3, we show that the KLM characterization cannot be extended to the infinite case.

- In Section 1.8.4, we show how to obtain cumulativity by a topological construction, and not through smoothness of the structure.
- In Section 1.8.5, we replace the partial orders between models of preferential structures by unions of total orders, and discuss the consequences.
- In Section 1.8.6, we discuss a joint article with S. Berger and D. Lehmann on preferred update histories.
- In Section 1.8.7, we reconstruct completeness proofs for preferential structures as done by KLM, to facilitate comparison with our own constructions.
- In Section 1.8.8, we discuss preferential choice for branching time structures, and extend results by Katsuno and Mendelzon.

[Sch04] also contains a short anaysis of X-Logics, introduced by P. Siegel et al., the interested reader is referred to section 3.2.4 there.

1.8.2 The Role of Copies in Preferential Structures

We discuss the importance of copies in preferential structures in more detail. The material in this Section was published in [Sch96-1] and [Sch04].

Most representation results for preferential structures use in their constructions several copies of logically identical models (see, e.g. [KLM90], or Section 1.3.2 and Section 1.3.3. Thus, we may have in those constructions m and m' with the same logical properties, but with different "neighborhoods" in the preferential structure, for example, there may be some m'' with $m'' \prec m$, but $m'' \not\prec m'$. David Makinson and Hans Kamp had asked the author whether such repetitions of models are sometimes necessary to represent a logic, we now give a (positive) answer. For the connection of the question to ranked structures see Section 1.4.1, in particular Lemma 1.4.3.

We have already given a simple example (see Example 1.3.1 above) illustrating the importance in the finite case. We discuss now more subtle situations.

1.8.2.1 The Infinite Case

Let κ, λ be infinite cardinals. Let $\mathcal{L}$ have κ propositional variables, p_i, $i < \kappa$. Consider any $\vdash$-consistent $\mathcal{L}$-theory T, a model m s.t. $m \not\models T$, and the following structure $\mathcal{M}$: $\mathcal{X} := \{\langle m, n\rangle : n \models T\} \cup \{\langle n, 0\rangle : n \models T\}$, with $\langle n, 0\rangle \prec \langle m, n\rangle$. Let $\phi \in T$ be s.t. $m \not\models \phi$. Obviously, $T \vee Th(m) \models_{\mathcal{M}} \phi$, as all copies of m are destroyed by the full set of models of T, but no T' truly stronger than T will do, as some copy of m will not be destroyed.

1.8 Various Results and Approaches

In general, however, the same logic as defined by $\mathcal{M}$ can be represented by structures with considerably less copies. It suffices to find a set of models $M \subset M_T$, where exactly the formulas of the classical closure of T hold - i.e. $M \models \phi$ iff $T \vdash \phi$, we shall then call M dense in M_T - and to take as $\mathcal{M}'$ the structure $\mathcal{X}' := \{\langle m, n \rangle : n \in M\} \cup \{\langle n, 0 \rangle : n \in M\}$, again with $\langle n, 0 \rangle \prec \langle m, n \rangle$. So we can rephrase the question to: What is the minimal size of M dense in M_T?

A Nice Case

Take for m the model that makes all p_i true, and $T := \{\neg p_0\}$, so $card(M_T) = 2^\kappa$, and the first construction of $\mathcal{M}$ as above will need 2^κ copies of m. As $\mathcal{L}$ has only κ formulas, and any subset of M_T makes all formulas of T true, we see that there is a dense subset $M \subseteq M_T$ of size κ: For any ϕ s.t. $T \not\vdash \phi$ take some $m_\phi \in M_T$ s.t. $m_\phi \not\models \phi$. But, in our nice case, considerably less than κ models might do: Assume there is $\lambda < \kappa$ s.t. $2^\lambda \geq \kappa$, so there is an injection $h : \{p_i : 0 < i < \kappa\} \to \mathcal{P}(\lambda)$. Let now $0 < i \neq j < \kappa$. For $\alpha < \lambda$, define the model m_α by $m_\alpha \models \neg p_0$ and $m_\alpha \models p_i :\leftrightarrow \alpha \in h(p_i)$. By $h(p_i) \neq h(p_j)$, there is $\alpha < \lambda$ s.t. $\alpha \in h(p_i) - h(p_j)$ or $\alpha \in h(p_j) - h(p_i)$, so $m_\alpha \models p_i \wedge \neg p_j$ or $m_\alpha \models \neg p_i \wedge p_j$, i.e. there is some m_α which discerns p_i, p_j. This is essentially enough: Let M be the closure of $\{m_\alpha : \alpha < \lambda\}$ under the finite operations $-, +, *$ defined by

$(-m) \models p_i :\leftrightarrow m \models \neg p_i$

$(m + m') \models p_i :\leftrightarrow m \models p_i$ or $m' \models p_i$

$(m * m') \models p_i :\leftrightarrow m \models p_i$ and $m' \models p_i$.

M still has cardinality λ, and $M \subseteq M_T$.

Let ϕ be s.t. $\neg p_0 \not\vdash \phi$, we have to find $m \in M$ s.t. $m \models \neg \phi$. Let $\neg \phi \equiv \phi_0 \vee \ldots \vee \phi_n$, where each $\phi_k = \pm p_{i_0} \wedge \ldots \wedge \pm p_{i_r}$ for some $i_0 \ldots i_r$. By $\neg p_0 \not\vdash \phi$, $Con(\neg p_0, \neg \phi)$ ($\vdash$ —consistency), so $Con(\neg p_0, \phi_k)$ for some $0 \leq k \leq n$. Fix such $\phi_k = \pm p_{i_0} \wedge \ldots \wedge \pm p_{i_r}$, say $\phi_k = p_{j_0} \wedge \ldots \wedge p_{j_s} \wedge \neg p_{g_0} \wedge \ldots \wedge \neg p_{g_t}$. By $Con(\neg p_0, \phi_k)$, p_0 is none of the p_{j_x}. (If one of the $\neg p_{g_y}$ is $\neg p_0$, it can be neglected, it will come out true anyway.) Fix $0 \leq x \leq s$, let $0 \leq y \leq t$. Then there is m_α s.t. $m_\alpha \models p_{j_x} \wedge \neg p_{g_y}$ or $-m_\alpha \models p_{j_x} \wedge \neg p_{g_y}$. Let $m_{x,y}$ be the m_α or $-m_\alpha$, and set $m_x := m_{x,0} * \ldots * m_{x,t}$. Then $m_x \models p_{j_x} \wedge \neg p_{g_0} \wedge \ldots \wedge \neg p_{g_t}$. For $m := m_0 + \ldots + m_s$, $m \models \phi_k$, so $m \models \neg \phi$, and $m \in M$.

On the other hand, in our example, λ many models with $2^\lambda < \kappa$ will not do: Assume that for each $0 < i \neq j < \kappa$ there is $\alpha < \lambda$ and $m_\alpha \in M_T$ with $m_\alpha \models p_i \wedge \neg p_j$. Then there is a function $f : 2^\lambda \to \kappa - \{0\}$ onto: For $A \subseteq \lambda$, let $f(A) := \bigcup\{j : 0 < j < \kappa \wedge \forall \alpha \in A. m_\alpha \models p_j\}$. But, for $0 < i < \kappa$, and $A_i := \{\alpha < \lambda : m_\alpha \models p_i\}$ $f(A_i) = i$: Obviously, for $\alpha \in A_i$, $m_\alpha \models p_i$. But, if $i \neq j$, then there is $\alpha \in A_i$ with $m_\alpha \models p_i \wedge \neg p_j$.

There Are, However, Examples Where We Need the Full Size κ

Let $\mathcal{L}$ be as above, consider $m^- \models \{\neg p_j : j < \kappa\}$, $T := \{p_i \vee p_j : i \neq j < \kappa\}$, and let $m^+ \models \{p_j : j < \kappa\}$ and $m_i^- \models \{\neg p_i\} \cup \{p_j : i \neq j < \kappa\}$ for $i < \kappa$. Let the structure $\mathcal{M}$ be defined by $\mathcal{X} := \{\langle m^-, m^+ \rangle\} \cup \{\langle m^-, m_i^- \rangle : i < \kappa\} \cup \{\langle m^+, 0 \rangle\} \cup \{\langle m_i^-, 0 \rangle : i < \kappa\}$ and $\langle n, 0 \rangle \prec \langle m^-, n \rangle$ for $n = m^+$ or $n = m_i^-$, some $i < \kappa$. Then $Th(m^-) \vee T \models_{\mathcal{M}} T$. But there is no $M \subseteq M_T$ dense with $card(M) < \kappa$. Obviously, $M_T = \{m^+\} \cup \{m_i^- : i < \kappa\}$, and $\{m_i^- : i < \kappa\} \subseteq M_T$ is dense (see [Sch92]), but taking away any m_i^- will change T: p_i becomes true.

We later turn to a different approach to copies in Section 1.8.5.

1.8.2.2 One Copy Version

The following material is very simple, and does not require further comments.

The essential property of preferential structures with at most one copy each is that we never need two or more elements to kill one other element. This is expressed by the following property, which we give in a finitary and an infinitary version:

Definition 1.8.1

(1-fin) Let $X = A \cup B_1 \cup B_2$ and $A \cap \mu(X) = \emptyset$. Then $A \subseteq (A \cup B_1 - \mu(A \cup B_1)) \cup (A \cup B_2 - \mu(A \cup B_2))$.

(1-infin) Let $X = A \cup \bigcup\{B_i : i \in I\}$ and $A \cap \mu(X) = \emptyset$. Then $A \subseteq \bigcup\{A \cup B_i - \mu(A \cup B_i)\}$.

It is obvious that both hold in 1-copy structures, it is equally obvious that the second guarantees the 1-copy property (consider $X = \{\{x\} : x \in X\}$, if $x \notin \mu(X)$, we find at least one $x' \in X$ s.t. $x \notin \mu(\{x, x'\})$, and this gives the construction for representation, too. It is almost as obvious that the finitary version does not suffice:

Example 1.8.1

Take an infinitary language $\{p, q_i : i \in \omega\}$, and let every p-model be killed by any infinite set of $\neg p$-models, and nothing else. Now, if A is minimized by $B_1 \cup B_2$, $B_1 \cup B_2$ contains an infinite number of $\neg p$-models, so either B_1 or B_2 does, so (1-fin) holds, but, obviously, the structure is not equivalent to any structure with one copy at most.

□

We turn to transitivity in the 1-copy case. Consider $a \prec b \prec c$, but $a \not\prec c$. So $\mu(\{a, c\}) = \{a, c\}$. By $\mu(\{a\}) = \{a\}, \mu(\{b\}) = \{b\}, \mu(\{c\}) = \{c\}$, we see that all three elements are present, so each has to be there as one copy. By $\mu(\{a, b\}) = \{a\}$

1.8 Various Results and Approaches

and $\mu(\{b,c\}) = \{b\}$, we see that $a \prec b \prec c$ has to hold. But then transitivity requires $a \prec c$, thus $\mu(\{a,c\}) = \{a\}$ has to hold, so the present structure is not equivalent to any transitive structure with the 1-copy property. Thus, to have transitivity, we need a supplementary condition:

Definition 1.8.2

(T) $\mu(A \cup B) \subseteq A$, $\mu(B \cup C) \subseteq B \rightarrow \mu(A \cup C) \subseteq A$.

Taking $a \prec b \prec c$ and $A := \{a\}$, $B := \{b\}$, $C := \{c\}$, we see that (T) imposes transitivity on 1-copy structures.

Note, however, that (T) does not necessarily hold in transitive structures with more that one copy - see above Example.

1.8.3 A Counterexample to the KLM-System

1.8.3.1 Introduction

In [KLM90], S. Kraus, D. Lehmann, M. Magidor have shown that the finitary restrictions of all supraclassical, cumulative, and distributive inference operations are representable by preferential structures. In [Sch92], we have shown that this does not generalize to the arbitrary infinite case.

Definition 1.8.3

$\hspace{1em}\mid\!\sim$ satisfies Distributivity iff $\overline{\overline{A}} \cap \overline{\overline{B}} \subseteq \overline{\overline{A \cap B}}$ for all theories A, B of $\mathcal{L}$.

Leaving aside questions of definability preservation, it translates into the following model set condition, where μ is the model choice function:

$(\mu D)\ \mu(X \cup Y) \subseteq \mu(X) \cup \mu(Y)$

Fact 1.8.1

We have shown that condition

$(\mu PR)\ X \subseteq Y \rightarrow \mu(Y) \cap X \subseteq \mu(X)$

essentially characterizes preferential structures.

In these terms, the problem is whether $(\mu \subseteq) + (\mu CUM) + (\mu D)$ entail (μPR) in the general case. Now, we see immediately:

$(\mu PR) + (\mu \subseteq)$ entail (μD) : $\mu(X \cup Y) = (\mu(X \cup Y) \cap X) \cup (\mu(X \cup Y) \cap Y)$
$\subseteq \mu(X) \cup \mu(Y)$.

Second, if the domain is closed under set difference, then $(\mu D) + (\mu \subseteq)$ entail (μPR): Let $U \subseteq V, V = U \cup (V\text{-}U)$. Then $\mu(V) \cap U \subseteq (\mu(U) \cup \mu(V-U)) \cap U = \mu(U)$.

The condition of closure under set difference is satisfied for formula defined model sets, but not in the general case of theory defined model sets.

See [Sch04], Section 3.5, for more discussion.

1.8.3.2 The Formal Results

S. Kraus, D. Lehmann, and M. Magidor have shown that for any logic $\vdash\!\sim$ for $\mathcal{L}$, which is supraclassical, cumulative, and distributive, there is a D−smooth preferential model $\mathcal{M}$, s.t. for all finite $T \subseteq \mathcal{L}$ $T^{\mathcal{M}} = \overline{\overline{T}}$ - where $T^{\mathcal{M}}$ is the logic defined by the structure. ([KLM90], see also [Mak94], Observation 3.4.7.) We show that the restriction to finite T is necessary, by providing a counterexample for the infinite case. We start by quoting a Lemma by D. Makinson.

Both Lemma 1.8.2 and our counterexample Example 1.8.2 have appeared in [Mak94], (Lemma 3.4.9, Observation 3.4.10). The reader less familiar with transfinite ordinals can find there a more algebraic proof that our counterexample satisfies the logical properties claimed. Our technique of constructing a logic inductively by a mixed iteration of suitable length has, however, proved useful in other situations as well (see [Sch91-2]), moreover, it is very fast and straightforward: once you have the necessary ingredients, the machinery will run almost by itself.

Lemma 1.8.2

(Lemma and proof: David Makinson, personal communication): Let a logic $\vdash\!\sim$ on $\mathcal{L}$ be representable by a classical preferential model structure. Then, for all $A \subseteq \mathcal{L}$, $x \in \mathcal{L}$, $x \notin \overline{\overline{A}}$ there is a maximal consistent (under $\vdash$) $\Delta \subseteq \mathcal{L}$ s.t. $\overline{\overline{A}} \subseteq \Delta$, $x \notin \Delta$, and $\overline{\overline{\Delta}} \neq \mathcal{L}$.

Proof

Let $\mathcal{M} = (\mathcal{X}, \prec)$ be a representation of $\vdash\!\sim$, i.e. $\overline{\overline{A}} = A^{\mathcal{M}}$ for all $A \subseteq \mathcal{L}$. Let $A \subseteq \mathcal{L}$, $x \in \mathcal{L}$, and $x \notin \overline{\overline{A}}$. Then there is $\langle m, i \rangle$ minimal in $\mathcal{X} \upharpoonright M_A$, with $m \not\models x$. Note that by minimality, $m \models \overline{\overline{A}}$. $\Delta := \{y \in \mathcal{L}: m \models y\}$ is maximal consistent, $x \notin \Delta$, $\overline{\overline{A}} \subseteq \Delta$, and $\langle m, i \rangle$ is also minimal in $\mathcal{X} \upharpoonright M_\Delta$, by $M_\Delta \subseteq M_A$. Thus, $m \models \overline{\overline{\Delta}}$, and by classicality of the models, $\overline{\overline{\Delta}} \neq \mathcal{L}$.

$\square$

We now construct a supraclassical, cumulative, distributive logic, and show that the logic so defined fails to satisfy the condition of Lemma 1.8.2, and is thus not representable by a preferential structure.

1.8 Various Results and Approaches

Example 1.8.2

Let $v(\mathcal{L})$ contain the propositional variables $p_i : i \in \omega, r$. (Note that we do not require $\mathcal{L}$ to be countable, we leave plenty of room for modifications of the construction!) We shall violate compactness badly "in both directions" by adding the rules (infinitely many p_i) $\hspace{1pt}\sim r$ and (infinitely many $\neg p_i$) $\hspace{1pt}\sim r$. To account for distributivity, we shall add for all $\phi \in \mathcal{L}$ (infinitely many $p_i \vee \phi$) $\hspace{1pt}\sim r \vee \phi$ and (infinitely many $\neg p_i \vee \phi$) $\hspace{1pt}\sim r \vee \phi$. Closing under $\hspace{1pt}\sim$ and classical logic ω_1 many times to take care of the countably infinite rules will give the result.

The Details

We define the logic $\hspace{1pt}\sim$ by a mixed iteration: For $B \subseteq \mathcal{L}$ define $I^+_{B,\phi} := \{i < \omega: p_i \vee \phi \in B\}$, $I^-_{B,\phi} := \{i < \omega: \neg p_i \vee \phi \in B\}$. Define now inductively

$A_0 := A$

for successor ordinals (α a limit or 0, $i \in \omega$):

$A_{\alpha+2i+1} := \overline{A_{\alpha+2i}}$

$A_{\alpha+2i+2} := A_{\alpha+2i+1} \cup \{r \vee \phi: I^+_{A_{\alpha+2i+1},\phi}$ is infinite or $I^-_{A_{\alpha+2i+1},\phi}$ is $infinite\}$

for limit λ:

$A_\lambda := \bigcup\{A_i : i < \lambda\}$

$\overline{\overline{A}} := A_{\omega_1}$.

We show $\hspace{1pt}\sim$ is as desired. Note that the defined logic is monotone.

(1) $\overline{A} \subseteq \overline{\overline{A}}$ is trivial.
(2) $A \subseteq B \subseteq \overline{\overline{A}} \to \overline{\overline{A}} = \overline{\overline{B}}$:

   (2.1) $\overline{\overline{A}} \subseteq \overline{\overline{B}}$ by monotony
   (2.2) $\overline{\overline{B}} \subseteq \overline{\overline{A}}$: Let $\phi \in \overline{\overline{B}}$. In deriving ϕ in $\overline{\overline{B}}$, we have used only countably many elements from B. This is seen as follows. Let β be minimal such that $\phi \in B_\beta$. ϕ can be derived from at most countably many $\phi_i \in B_{\beta-1}$ (β has to be a successor ordinal). Arguing backwards, and using $\omega.\omega = \omega$ (cardinal multiplication), we see what we wanted. (This is, of course, the outline for an inductive proof.) As $B \subseteq \overline{\overline{A}}$, using regularity of ω_1, we see that there is some $\alpha < \omega_1$ s.t. all ϕ_j used in the derivation of ϕ from B are in A_α. But then $\phi \in A_{\alpha+\beta}$.

(3) Distributivity: We show by induction on the derivation of a, b that $a \in \overline{\overline{A}}, b \in \overline{\overline{B}} \to a \vee b \in \overline{\overline{A \cap B}}$. To get started, use $A_0 \subseteq A_1 = \overline{A}$, and $a \in \overline{A}, b \in \overline{B} \to a \vee b \in \overline{A \cap B}$. By symmetry, it suffices to consider the cases for a. Let $a_1, \ldots, a_n \vdash a$ by classical inference. By induction hypothesis, $a_1 \vee b, \ldots, a_n \vee b \in \overline{\overline{A \cap B}}$, but then $a \vee b \in \overline{\overline{A \cap B}}$, as the latter is closed under $\vdash$. Assume now $a = r \vee \phi \in A_\alpha$ has been derived from infinitely many $p_i \vee \phi$ ($i \in I$) in $A_{\alpha-1}$. By induction hypothesis, $p_i \vee \phi \vee b \in \overline{\overline{A \cap B}}$. So $p_i \vee \phi \vee b \in (\overline{A \cap B})_{\beta_i}$ for $\beta_i < \omega_1$. Again by regularity of ω_1, all $p_i \vee \phi \vee b \in (\overline{A \cap B})_\beta$ ($i \in I$) for some $\beta < \omega_1$. But then $r \vee \phi \vee b = a \vee b \in (\overline{A \cap B})_{\beta+2}$. The case $\neg p_i \vee \phi$ is similar.

□

We use the lemma to obtain the negative result, as the logic constructed above does not satisfy the lemma's condition:

Consider now $A := \emptyset$. Assume there is ϕ s.t. infinitely many $p_i \vee \phi \in \overline{A}$, thus there is ϕ s.t. infinitely many $p_i \vee \phi$ are tautologies. But then ϕ has to be a tautology (consider $(p_i \vee \phi) \leftrightarrow (\neg \phi \to p_i)$ and finiteness of ϕ!), thus ϕ and $\phi \vee r \in \overline{A}$. Likewise for $\neg p_i \vee \phi$. So, the rules (infinitely many $p_i \vee \phi$) $\mathrel{|\!\sim} r \vee \phi$, etc. give nothing new, and $\overline{\overline{A}} = \overline{A}$. In particular, $r \notin \overline{A}$.

Assume now $\Delta \subseteq \mathcal{L}$ to be maximal consistent. So Δ decides all $p_i : i \in \omega$. Thus either infinitely many p_i, or $\neg p_i$ in Δ. Thus, $r \in \overline{\Delta}$. Hence $\mathrel{|\!\sim}$ is not representable by a preferential structure.

□

Remark 1.8.3

In the last step, finiteness of ϕ seems to play a decisive role. But languages with infinite formulas will run into similar problems. If all formulas have size $< \beta$, a similar construction with β^+ p_i's and induction to β^{++} will give the same result. (β a cardinal, β^+ etc. cardinal successors.)

1.8.4 A Nonsmooth Model of Cumulativity

1.8.4.1 Introduction

If (CM) is violated, there are ϕ, ψ, τ such that $\phi \mathrel{|\!\sim} \psi$, $\phi \mathrel{|\!\sim} \tau$, but $\phi \wedge \psi \mathrel{|\!\not\sim} \tau$. As all minimal models of ϕ are then minimal models of $\phi \wedge \psi$, there must be a new minimal model of $\phi \wedge \psi$, which is not a minimal model of ϕ, weakening the set of consequences of $\phi \wedge \psi$, compared to the set of consequences of ϕ. Smoothness

1.8 Various Results and Approaches

assures that this cannot happen, as any model of $\phi \wedge \psi$ must be above some minimal model of ϕ. If smoothness cannot hold, we must prevent the existence of "dangerous" (i.e. consequence changing) new minimal $\phi \wedge \psi$-models by other means, which (by transitivity) can only be infinite descending chains of $\phi \wedge \psi$-models.

But smoothness cannot hold in an injective structure showing joint consistency of the system P, (WD), and $\neg(NR)$, and we have such "dangerous" $\phi \wedge \psi-$models, as we will see now. By failure of (NR), there are α, β, γ such that $\alpha \mid\!\sim \beta$, $\alpha \wedge \gamma \not\mid\!\sim \beta$, $\alpha \wedge \neg \gamma \not\mid\!\sim \beta$. Thus we have a minimal model m_1 of $\alpha \wedge \gamma \wedge \neg \beta$, and a minimal model m_2 of $\alpha \wedge \neg \gamma \wedge \neg \beta$. By (WD), there will be at most one $\alpha \wedge \neg \beta$-model, so they cannot both be minimal models of $\alpha \wedge \neg \beta$. Suppose m_1 is not, the other case is analogous. A simple analysis (in Fact 2.2 below) shows that there cannot be a minimal model of $\alpha \wedge \neg \beta$ below m_1, so Smoothness is indeed violated, and we must have an infinite descending chain X of $\alpha \wedge \neg \beta$-models below m_1. Let now $\phi := \alpha \wedge \neg \beta$, and m be the unique (if it exists - if not, a similar argument applies) minimal $\alpha \wedge \neg \beta$-model, and suppose $m \models \psi$, so $\phi \mid\!\sim \psi$. If there were now a minimal model m' of $\phi \wedge \psi$ in X, Cumulativity would be violated: By injectivity of the structure, m' is logically different from m, and the theory determined by $\{m\}$ is stronger than the one determined by $\{m, m'\}$ (finiteness of $\{m\}$ is crucial here). Thus, in X either ψ will be infinitely often true, or not at all. We will make it infinitely often true, so "X approximates m logically".

To summarize: The [BMP97] framework forces us to consider nonsmooth structures. It is natural to have cumulativity without smoothness through a topological construction. The topological view demonstrates thus again its utility and naturalness. It is a subtle bridge between the semantics and the logics.

(This is taken from [Sch99].)

1.8.4.2 The Formal Results

We recall some further rules, see [BMP97]:

Definition 1.8.4

(NR) (Negation Rationality): $\alpha \mid\!\sim \beta \Rightarrow \alpha \wedge \gamma \mid\!\sim \beta$ or $\alpha \wedge \neg \gamma \mid\!\sim \beta$ (for any γ),

(WD) (Weak Determinacy): $true \mid\!\sim \neg \alpha \Rightarrow \alpha \mid\!\sim \beta$ or $\alpha \mid\!\sim \neg \beta$ (for any β) (we say that such α decide),

(DR) (Disjunctive Rationality): $\alpha \vee \beta \mid\!\sim \gamma \Rightarrow \alpha \mid\!\sim \gamma$ or $\beta \mid\!\sim \gamma$.

We also recall the rule (which is part of the system P) :

(CM) (Cautious Monotony): $\alpha \mid\!\sim \beta$ and $\alpha \mid\!\sim \gamma \Rightarrow \alpha \wedge \gamma \mid\!\sim \beta$.

We use $\neg(NR)$ as shorthand for the existence of α, β, γ such that $\alpha \mid\!\sim \beta$, but neither $\alpha \wedge \gamma \mid\!\sim \beta$, nor $\alpha \wedge \neg \gamma \mid\!\sim \beta$.

Fact 1.8.4

There is no smooth injective preferential structure validating (WD) and $\neg(NR)$.

Proof

Suppose (NR) is false, so there are α, β, γ with $\alpha \mid\sim \beta$, $\alpha \wedge \gamma \not\mid\sim \beta$, $\alpha \wedge \neg\gamma \not\mid\sim \beta$. Let $\alpha \mid\sim \beta$, so $true \mid\sim \alpha \to \beta$. (If m is a minimal model of true, and if $m \models \alpha$, then m is a minimal model of α, so $m \models \beta$.) So $true \mid\sim \neg(\alpha \wedge \neg\beta)$. If $\vdash \phi \to \alpha \wedge \neg\beta$, then $true \mid\sim \neg\phi$. Thus, by (WD), if $\vdash \phi \to \alpha \wedge \neg\beta$, ϕ decides, thus, by injectivity, ϕ has at most one minimal model in the structure.

Let now $\alpha \wedge \gamma \not\mid\sim \beta$, $\alpha \wedge \neg\gamma \not\mid\sim \beta$, thus there is a minimal model m_1 of $\alpha \wedge \gamma$, where $\neg\beta$ holds, and a minimal model m_2 of $\alpha \wedge \neg\gamma$, where $\neg\beta$ holds. Thus, m_1 is a minimal model of $\alpha \wedge \gamma \wedge \neg\beta$, m_2 a minimal model of $\alpha \wedge \neg\gamma \wedge \neg\beta$.

(a) Suppose m_1 is not a minimal model of $\alpha \wedge \neg\beta$, then by smoothness, there is $m \prec m_1$, m a minimal model of $\alpha \wedge \neg\beta$. $\neg\gamma$ has to hold in m, so m is a minimal model of $\alpha \wedge \neg\beta \wedge \neg\gamma$. By uniqueness, $m = m_2$, so $m_2 \prec m_1$, and m_2 is a minimal model of $\alpha \wedge \neg\beta$.

(b) If m_2 is not a minimal model of $\alpha \wedge \neg\beta$, then, analogously, m_1 is, and $m_1 \prec m_2$.

(c) m_1 and m_2 are minimal models of $\alpha \wedge \neg\beta$: Impossible, as $\alpha \wedge \neg\beta$ decides.

Suppose now, e.g. m_2 is the minimal model of $\alpha \wedge \neg\beta$, and $m_2 \prec m_1$. As $\alpha \mid\sim \beta$, m_2 cannot be a minimal model of α, so there must be $m' \models \alpha$ below m_2. $m' \models \alpha \wedge \neg\beta$ is impossible (by minimality of m_2), so $m' \models \alpha \wedge \beta$. (Note that we did not need smoothness for this argument.) But $m' \models \gamma$, or $m' \models \neg\gamma$, contradicting minimality of m_1 or of m_2. The other case is analogous.

□

Thus, we have such "dangerous" $\phi \wedge \psi$-models, as we will see now. By failure of (NR), there are α, β, γ such that $\alpha \mid\sim \beta$, $\alpha \wedge \gamma \not\mid\sim \beta$, $\alpha \wedge \neg\gamma \not\mid\sim \beta$. Thus we have a minimal model m_1 of $\alpha \wedge \gamma \wedge \neg\beta$, and a minimal model m_2 of $\alpha \wedge \neg\gamma \wedge \neg\beta$. By (WD), there will be at most one minimal $\alpha \wedge \neg\beta$-model, so they cannot both be minimal models of $\alpha \wedge \neg\beta$. Suppose m_1 is not, the other case is analogous. A simple analysis shows that there cannot be a minimal model of $\alpha \wedge \neg\beta$ below m_1, so smoothness is indeed violated, and we must have an infinite descending chain X of $\alpha \wedge \neg\beta$-models below m_1. Let now $\phi := \alpha \wedge \neg\beta$, and m be the unique (if it exists - if not, a similar argument applies) minimal $\alpha \wedge \neg\beta$-model, and suppose $m \models \psi$, so $\phi \mid\sim \psi$. If there were now a minimal model m' of $\phi \wedge \psi$ in X, Cumulativity would be violated: By injectivity of the structure, m' is logically different from m, and the theory determined by $\{m\}$ is stronger than the one determined by $\{m, m'\}$ (finiteness of $\{m\}$ is crucial here). Thus, in X either ψ will be infinitely often true, or not at all. We will make it infinitely often true, so "X approximates m logically".

1.8 Various Results and Approaches

A Nonsmooth Injective Structure Validating P, (WD), $\neg(NR)$

Definition 1.8.5

A sequence f of models converges to a set of models M, $f \to M$, iff $\forall \phi (M \models \phi \to \exists i \forall j \geq i . f_j \models \phi)$. If $M = \{m\}$, we will also write $f \to m$.

Fact 1.8.5

Let f be a sequence composed of n subsequences $f^1, \ldots, f^n$, e.g. $f_{n*j+0} = f_j^1$, etc., and $f^i \to M_i$. Let ϕ be a formula unboundedly often true in f. Then there is $1 \leq i \leq n$ and $m \in M_i$ s.t. $m \models \phi$.

Proof

Exercise, solution in the Appendix.

Example 1.8.3

(A nonsmooth transitive injective structure validating system P, (WD), $\neg(NR)$)

As any transitive acyclic relation over a finite structure is necessarily smooth, and an injective structure over a finite language is finite, Fact 1.8.4 shows that we need an infinite language.

Take the language defined by the propositional variables $r, s, t, p_i : i < \omega$.

Take four models m_i, $i = 1, \ldots, 4$, where for all i, j $m_i \models p_j$ (to be definite), and let $m_0 \models r, \neg s, t$, $m_1 \models r, \neg s, \neg t$, $m_2 \models r, s, t$, $m_3 \models r, s, \neg t$. It is important to make m_2 and m_3 identical except for t, the other values for the p_j are unimportant.

Let $m_2 < m_1$. (The other m_i are incomparable.)

Define two sequences of models $f^1 \to m_1$, $f^3 \to m_3$ s.t. for all i, j $f_j^i \models r, \neg t$. This is possible, as $m_1 \models r, \neg t$, $m_3 \models r, \neg t$.

All models in these sequences can be chosen different, and different from the m_i - this is no problem, as we have for all consistent ϕ uncountably many models where ϕ holds.

Let f be the mixture of f^i, e.g. $f_{2n+0} := f_n^1$, etc.

Put m_0 above f, with f in descending order. Arrange the rest of the 2^ω models above m_0 ordered as the ordinals, i.e. every subset has a minimum. Thus, there is one long chain C (i.e. C is totally ordered) of models, at its lower end a descending countable chain f, directly above f m_0, above m_0 all other models except $m_1 - m_3$, arranged in a well-order. The models $m_1 - m_3$ form a separate group. See Figure 1.1.

Note that m_0 is a minimal model of t.

Obviously, (NR) is false, as $r \mathrel{\vert\!\sim} s$, but neither $r \wedge t \mathrel{\vert\!\sim} s$, nor $r \wedge \neg t \mathrel{\vert\!\sim} s$.

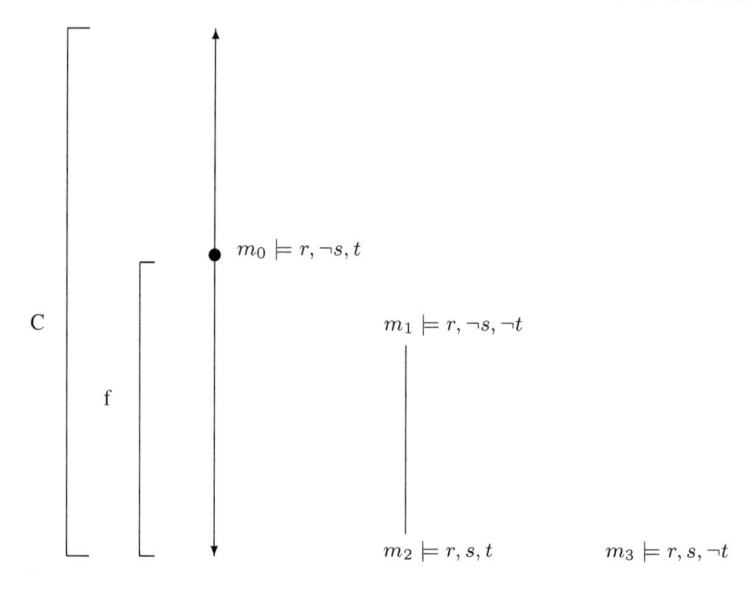

Fig. 1.1 The Structure of Example 1.8.3

The usual rules of P hold, as this is a preferential structure, except perhaps for (CM), which holds in smooth structures, and our construction is not smooth. (This is the real problem.)

Note that (CM) says $\phi \mathrel{|\!\sim} \psi \rightarrow \overline{\overline{\phi}} = \overline{\overline{\phi \wedge \psi}}$, so it suffices to show for all ϕ $\phi \mathrel{|\!\sim} \psi \rightarrow \mu(\phi) = \mu(\phi \wedge \psi)$. This is the point of the construction. The infinite descending chains converge to some minimal model, so if α holds in this minimal model, then α holds infinitely often in the chain, too. Thus there are no new minimal models of α, which might weaken the consequences.

For (WD), we have to show by Fact 1.8.4 that, if $M(\phi) \cap \mu(true) = \emptyset$, then $\mu(\phi)$ contains at most one model (where $\mu(true) = \{m_2, m_3\}$).

We examine the possible cases of $\mu(\phi)$ ($\emptyset$, $\{m_1\}$, $\{m_2\}$, $\{m_3\}$, $\{m_1, m_3\}$, $\{m_2, m_3\}$, and $\mu(\phi) \cap C \neq \emptyset$).

For (CM):

Case 1: $\mu(\phi) = \{m_2, m_3\}$: Then $\phi \mathrel{|\!\sim} \psi$ iff $\{m_2, m_3\} \models \psi$. So if $\phi \mathrel{|\!\sim} \psi$, then $\phi \wedge \psi$ holds in m_3, so by f^3, $\phi \wedge \psi$ is (downward) unboundedly often true in f, so $\mu(\phi \wedge \psi) = \{m_2, m_3\}$.

Case 2: $\mu(\phi) = \{m_1\}$ and Case 3: $\mu(\phi) = \{m_3\}$: as above, by f^1 and f^3.

Case 4: $\mu(\phi) = \{m_2\}$: As $m_2 \models \phi$, and $m_3 \not\models \phi$, ϕ is of the form $\phi' \wedge t$, so none of the f_i is a model of ϕ, so ϕ has a minimal model in the chain C, so this is impossible.

1.8 Various Results and Approaches 169

Case 5: $\mu(\phi) = \{m_1, m_3\}$: Then $\phi \hspace{-0.5mm}\mid\hspace{-2mm}\sim \psi$ iff $\{m_1, m_3\} \models \psi$. So as in Case 1, if $\phi \hspace{-0.5mm}\mid\hspace{-2mm}\sim \psi$, $\phi \wedge \psi$ is unboundedly often true in f^1 (and in f^3), and $\mu(\phi \wedge \psi) = \{m_1, m_3\}$.

Case 6: $\mu(\phi) = \emptyset$: This is impossible by Fact 1.8.5: If ϕ is unboundedly often true in C, then it must be true in one of m_1, m_3.

Case 7: $\mu(\phi) \cap C \neq \emptyset$: Then below each $m \models \phi$, there is $m' \in \mu(\phi)$. Thus, the usual argument which shows Cumulativity in smooth structures applies.

For (WD):

We only have to consider the cases where $m_2, m_3 \notin M(\phi)$, so the only possible cases are: Case 2, Case 7.

In Case 2, there is nothing to show, $\mu(\phi)$ is a singleton.

In Case 7, (WD) is trivial, we have a unique minimum: $m_2, m_3 \notin M(\phi)$ by prerequisite. But if $m_1 \models \phi$, then ϕ would be true unboundedly often in f, so it would not have a minimal model in C. Thus, $\mu(\phi)$ is a singleton.

□

1.8.5 A New Approach to Preferential Structures

1.8.5.1 Introduction

This section deals with some fundamental concepts and questions of preferential structures. A model for preferential reasoning will, in this section, be a total order on the models of the underlying classical language. Instead of working in completeness proofs with a canonical preferential structure as done traditionally, we work with sets of such total orders. We thus stay close to the way completeness proofs are done in classical logic. Our new approach will also justify multiple copies (or noninjective labelling functions) present in most work on preferential structures. A representation result for the finite case is given.

(This is taken from [SGMRT00].)

Main Concepts and Results

We address in this Section 1.8.5 some fundamental questions of preferential structures. Our guiding principle will be classical propositional (or first order) logic.

First, we reconsider the concept of a model for preferential reasoning. Traditionally, such a model is a strict partial order on the set of classical models of the underlying language. Instead, we will work here with strict _total_ orders on the set of classical models of the underlying language. Such structures have maximal preferential information, just as classical propositional models have maximal propositional information.

Second, we will work in completeness proofs with sets of such total orders and thus closely follow the strategy for classical logic, whereas the traditional approach for preferential models works with one canonical structure. More precisely, in classical logic, one shows $T \vdash \phi$ iff $T \models \phi$, by proving soundness and that for every ϕ s.t. $T \nvdash \phi$ there is a T-model $m_{T,\neg\phi}$, where ϕ fails. In traditional preferential logic, one constructs a canonical structure $\mathcal{M}$, which satisfies exactly the consequences of T, i.e. $T \models_\mathcal{M} \phi$ iff $T \mid\!\sim \phi$, simultaneously for all T and ϕ (where $T \models_\mathcal{M} \phi$ iff $\mu(T) \subseteq M(\phi)$, i.e. iff in all minimal models of T in $\mathcal{M}$ ϕ holds).

Third, our approach will also shed new light on the somewhat obscure question of multiple copies (equivalent to noninjective labelling functions) present in most constructions (see, e.g. the work of the author, or [KLM90], [LM92]). In our approach, it is natural to consider disjoint unions of sets of total orders over the classical models. They have (almost) the same properties as these sets have. As disjoint unions are structures with multiple copies, we have justified multiple copies of models or noninjective labelling functions in a natural way.

Strict Total Orders Are the Models of Preferential Reasoning

A classical propositional or first order model has maximal propositional or first order information: every formula is decided, either the formula or its negation holds. A set of models (corresponding to an incomplete formula, i.e. to a formula ϕ s.t. there is a formula ψ with neither $\phi \vdash \psi$, nor $\phi \vdash \neg\psi$) has less information. Preferential reasoning reasons about preferences between the classical models of a given language $\mathcal{L}$. Maximal preferential information is given by a strict total order between these classical models. A strict partial order can also be considered as the set of total orders which extend it (as a set of pairs). Thus, strict total orders on the set of classical models _are_, in this sense, the models of preferential reasoning, just as classical propositional models are the models of propositional reasoning.

Basic Definitions and Facts

Recall from Definition 1.6.1 that by a child (or successor) of an element x in a tree t we mean a direct child in t. A child of a child, etc. will be called an indirect child. Trees will be supposed to grow downwards, so the root is the top element.

1.8 Various Results and Approaches

Definition 1.8.6

For a given language $\mathcal{L}$, TO, etc. will stand for a strict total order on $M_\mathcal{L}$. Considering TO as a preferential model, we will slightly abuse notation here: as there will only be one copy per model, we will omit the indices i.

$\mathcal{O}$, etc. will stand for sets of such structures.

If $\mathcal{O}$ is such a set, we set $\mu_\mathcal{O}(X) := \bigcup\{\mu_\mathcal{M}(X) : \mathcal{M} \in \mathcal{O}\}$, and define $T \models_\mathcal{O} \phi$ iff $T \models_\mathcal{M} \phi$ for all $\mathcal{M} \in \mathcal{O}$.

Note that for all T and all strictly totally ordered structures TO, $\mu_{TO}(T)$ is either a singleton, or empty, so TO is definability preserving.

Definition 1.8.7

Let $\mathcal{Z} := \langle \mathcal{X}, \prec \rangle$ be a preferential structure. For $\langle x, i \rangle \in \mathcal{X}$, let

$\langle x, i \rangle_\mathcal{Z}^- := \{\langle y, j \rangle \in \mathcal{X} : \langle y, j \rangle \prec \langle x, i \rangle\}$, and
$\langle x, i \rangle_\mathcal{Z}^* := \{y : \exists \langle y, j \rangle \in \mathcal{X}.\langle y, j \rangle \prec \langle x, i \rangle\}$.

When the context is clear, we omit the index $\mathcal{Z}$.

Fact 1.8.6

Let $\mathcal{Z} := \langle \mathcal{X}, \prec \rangle$, $\mathcal{Z}' := \langle \mathcal{X}', \prec' \rangle$ be two preferential structures.

(1) Let $x \in X$. Then $x \in \mu_\mathcal{Z}(X)$ iff $\exists \langle x, i \rangle \in \mathcal{X}.X \cap \langle x, i \rangle_\mathcal{Z}^* = \emptyset$.
(2) If $\forall \langle x, i \rangle \in \mathcal{X} \exists \langle x, i' \rangle \in \mathcal{X}'.\langle x, i' \rangle_{\mathcal{Z}'}^* \subseteq \langle x, i \rangle_\mathcal{Z}^*$ and $\forall \langle x, i' \rangle \in \mathcal{X}' \exists \langle x, i \rangle \in \mathcal{X}.\langle x, i \rangle_\mathcal{Z}^* \subseteq \langle x, i' \rangle_{\mathcal{Z}'}^*$, then $\mu_\mathcal{Z} = \mu_{\mathcal{Z}'}$.

Proof

(1) $x \in \mu_\mathcal{Z} \Leftrightarrow \exists \langle x, i \rangle \in \mathcal{X}.\neg \exists \langle y, j \rangle \in \mathcal{X}.\langle y, j \rangle \prec \langle x, i \rangle \wedge y \in X \Leftrightarrow \exists \langle x, i \rangle \in \mathcal{X}.\langle x, i \rangle_\mathcal{Z}^* \cap X = \emptyset$.
(2) Let $x \in \mu_\mathcal{Z}(X)$, then by (1) $\exists \langle x, i \rangle \in \mathcal{X}.X \cap \langle x, i \rangle_\mathcal{Z}^* = \emptyset$. By prerequisite, $\exists \langle x, i' \rangle \in \mathcal{X}'.\langle x, i' \rangle_{\mathcal{Z}'}^* \subseteq \langle x, i \rangle_\mathcal{Z}^*$, so $x \in \mu_{\mathcal{Z}'}(X)$ by (1). The other direction is symmetrical.

□

Fact 1.8.7

If $\mathcal{O}$ is a set of preferential structures, then $T \models_\mathcal{O} \phi$ iff $\mu_\mathcal{O}(M_T) \models \phi$.

Proof

Exercise, solution in the Appendix.

Outline of Our Representation Results and Technique

We describe here the kind of representation result we will show in Section 1.8.5.4.

We have characterized in Section 1.3.3 and Section 1.3.4 usual smooth preferential structures first algebraically by conditions on their choice functions, and only then logically by corresponding conditions. More precisely, given a function μ satisfying certain conditions, we have shown that there is a preferential structure $\mathcal{Z}$, whose choice function $\mu_{\mathcal{Z}}$ is exactly μ. The choice functions correspond to the logics by the equation $\mu(M(T)) = M(\overline{T})$.

We will take a similar approach here, but will first analyze the form a representation theorem will have in our context.

Our starting point was that classical completeness proofs have the following form: For each ϕ s.t. $T \not\vdash \phi$, find $m_{T,\neg\phi}$ s.t. $m_{T,\neg\phi} \models T, \neg\phi$, or, equivalently, find a set of models $\mathbf{M}_T$ s.t. for each such ϕ there is suitable $m_{T,\neg\phi}$ in $\mathbf{M}_T$. Then, by soundness, $Th(\mathbf{M}_T) = \overline{T}$. Our construction will have a similar form.

First, given any strict total order TO (or any set $\mathcal{O}$ of strict total orders) over $M_\mathcal{L}$, the logic defined by $T \mathrel{\vert\!\sim} \phi :\Leftrightarrow T \models_{TO} \phi$ (or $:\Leftrightarrow T \models_{\mathcal{O}} \phi$) satisfies our conditions (LLE), (CCL), (SC), (PR), (CUM) (see Proposition 1.8.11). Second, given a logic $\mathrel{\vert\!\sim}$ satisfying (LLE), (CCL), (SC), (PR), (CUM), there is a set $\mathcal{O}$ of strict total orders over $M_\mathcal{L}$ s.t. $T \mathrel{\vert\!\sim} \phi \Leftrightarrow T \models_{\mathcal{O}} \phi$. Thus, the <u>set</u> $\mathcal{O}$ represents exactly $\mathrel{\vert\!\sim}$, contrary to usual preferential structures, where a single structure represents exactly $\mathrel{\vert\!\sim}$.

We work again first via an algebraic characterization, and show the following: Given any strict total order TO (or any set $\mathcal{O}$ of strict total orders), the choice function μ_{TO} (the choice function $\mu_{\mathcal{O}}$) satisfies our algebraic conditions $(\mu \subseteq)$, (μPR), (μCUM) (see Proposition 1.8.12). Conversely, given a choice function μ satisfying $(\mu \subseteq)$, (μPR), (μCUM), there is a set $\mathcal{O}$ of strict total orders s.t. $\mu = \mu_{\mathcal{O}}$.

The logical part will then follow easily via a standard argument.

The main open problem seems to be a characterization of the infinite case, or at least the infinite smooth case.

1.8.5.2 Validity in Traditional and in Our Preferential Structures

We distinguish here validity of type 1 and type 2, where type 1 validity is validity of entailment like $T \mathrel{\vert\!\sim} \phi$, and type 2 validity is validity of rules like $\phi \mathrel{\vert\!\sim} \psi \wedge \sigma \Rightarrow \phi \mathrel{\vert\!\sim} \psi$.

1.8 Various Results and Approaches

(The set $\mathcal{O}$ used in this section is motivated by Example 1.8.4, where we do not consider all totally ordered sets, but only those satisfying a certain property.)

Definition 1.8.8

(1) Validity of type 1:

This is validity of expressions like $\phi \mathrel{|\!\sim} \psi$ (or $T \mathrel{|\!\sim} \psi$), and is defined for a given preferential structure $\mathcal{M}$ in the usual sense by $\phi \models_\mathcal{M} \psi$ (or $T \models_\mathcal{M} \psi$). In our new interpretation we read this as: $\phi \models_\mathcal{O} \psi$ (or $T \models_\mathcal{O} \psi$).

(2) Validity of type 2:

This is validity of rules of, e.g. the type

(2.1) $\phi \mathrel{|\!\sim} \psi, \phi \mathrel{|\!\sim} \psi' \Rightarrow \phi \mathrel{|\!\sim} \psi \wedge \psi'$,

(2.2) $\phi \mathrel{|\!\sim} \psi \Rightarrow (\phi \mathrel{|\!\sim} \neg \phi'$ or $\phi \wedge \phi' \mathrel{|\!\sim} \psi)$,

(2.3) $\overline{\overline{T \cup T'}} \subseteq \overline{\overline{T}} \cup T'$.

As strict total orders are definability preserving, we can argue semantically when dealing with them. More precisely, there is a 1-1 correspondence between theories (and formulas) and sets of models: If $\mathcal{M}$ is a definability preserving preferential model, and T a theory, then $M(\{\phi : T \models_\mathcal{M} \phi\}) = \mu_\mathcal{M}(M(T))$, so setting $\overline{\overline{T}} := \{\phi : T \models_\mathcal{M} \phi\}$, we have for instance $T' \vdash \overline{\overline{T}}$ iff $M(T') \subseteq M(\overline{\overline{T}})$.

We now discuss the properties in Definition 1.8.8.

Discussion of (2.1)

In usual preferential structures, we read (2.1) as: If in a fixed structure $\mathcal{M}$ $\phi \models_\mathcal{M} \psi$ and $\phi \models_\mathcal{M} \psi'$ hold, then so will $\phi \models_\mathcal{M} \psi \wedge \psi'$.

In our new approach, we read (2.1) now as: If $\phi \models_\mathcal{O} \psi$ and $\phi \models_\mathcal{O} \psi'$ hold, then $\phi \models_\mathcal{O} \psi \wedge \psi'$ will also hold. In semantical terms: If $\mu_\mathcal{O}(\phi) \subseteq M(\psi)$ and $\mu_\mathcal{O}(\phi) \subseteq M(\psi')$, then $\mu_\mathcal{O}(\phi) \subseteq M(\psi) \cap M(\psi')$.

This is the exact analogue of the classical definition: $\alpha \models \beta$ iff in all classical models where α (and perhaps some other property, too) holds, β will also hold. Our α is here of the form $\phi \mathrel{|\!\sim} \psi$ (or $\phi \models_{TO} \psi$), etc.

Discussion of (2.2)

The usual approach is similar to the one for rule (2.1).

For the new approach, we have to be careful with distributivity. A comparison with classical logic helps. In all classical models it is true that if $\alpha \vee \beta$ holds, then α holds, or β holds (by definition of validity of "or"). But we do not say that $\alpha \vee \beta \models \alpha$ or $\alpha \vee \beta \models \beta$ holds, as this would imply either that in all models where $\alpha \vee \beta$ holds, α holds, or that in all models where $\alpha \vee \beta$ holds, β holds, which is usually false.

So a rule of type (2.2) holds iff $\phi \models_{\mathcal{O}} \psi$ implies $\phi \models_{\mathcal{O}} \neg \phi'$ or $\phi \wedge \phi' \models_{\mathcal{O}} \psi$. In semantical terms: A rule of type (2.2) holds iff $\mu_{\mathcal{O}}(\phi) \subseteq M(\psi)$ implies $\mu_{\mathcal{O}}(\phi) \subseteq M(\neg \phi')$ or $\mu_{\mathcal{O}}(\phi \wedge \phi') \subseteq M(\psi)$.

Note that (2.2) holds in all strict total orders on $M_{\mathcal{L}}$, as such structures are ranked. But in a set of such structures, it is usually wrong, as it is usually not true that either in all these structures $\phi \mathrel|\!\!\sim \neg \phi'$ holds, or that in all these structures $\phi \wedge \phi' \mathrel|\!\!\sim \psi$ holds.

Discussion of (2.3)

(2.3) stands for: If $T \cup T' \mathrel|\!\!\sim \phi$, then there are $\phi_1, \ldots, \phi_n$ and $\phi'_1, \ldots, \phi'_m$ s.t. $T \mathrel|\!\!\sim \phi_i$ and $\phi'_i \in T'$, and $\{\phi_1, \ldots, \phi_n, \phi'_1, \ldots, \phi'_m\} \vdash \phi$.

So, in usual preferential structures, (2.3) holds in structure $\mathcal{M}$, iff: If $T \cup T' \models_{\mathcal{M}} \phi$, then there are $\phi_1, \ldots, \phi_n$ and $\phi'_1, \ldots, \phi'_m$ s.t. $T \models_{\mathcal{M}} \phi_i$ and $\phi'_i \in T'$, and $\{\phi_1, \ldots, \phi_n, \phi'_1, \ldots, \phi'_m\} \vdash \phi$.

In our new approach, (2.3) holds iff in all strict total orders $TO \in \mathcal{O}$ $T \cup T' \models_{TO} \phi$, there are $\phi_1, \ldots, \phi_n$ and $\phi'_1, \ldots, \phi'_m$ s.t. $T \models_{TO} \phi_i$ and $\phi'_i \in T'$, and $\{\phi_1, \ldots, \phi_n, \phi'_1, \ldots, \phi'_m\} \vdash \phi$.

The discussion in semantical terms clarifies the role of the existential quantifiers (which are "ors" - see the discussion of (2.2)): Condition (2.3) reads now: in all $TO \in \mathcal{O}$ $\mu_{TO}(T) \cap M(T') \subseteq \mu_{TO}(T \cup T')$ holds (and thus also $\mu_{\mathcal{O}}(T) \cap M(T') \subseteq \mu_{\mathcal{O}}(T \cup T')$).

1.8.5.3 The Disjoint Union of Models and the Problem of Multiple Copies

Disjoint Unions and Preservation of Validity in Disjoint Unions

We introduce the disjoint union of preferential structures and examine the question whether a property Φ which holds in all $\mathcal{M}_i$, $i \in I$, will also hold in their disjoint union $\biguplus \{\mathcal{M}_i : i \in I\}$. This is true for type 1 validity, but not for type 2 validity in the general infinite case.

1.8 Various Results and Approaches

Preservation of Type 1 Validity

Definition 1.8.9

Let $\mathcal{M}_i := \langle M_i, \prec_i \rangle$ be a family of preferential structures.

Let then $\biguplus \{\mathcal{M}_i : i \in I\} := \langle M, \prec \rangle$, where $M := \{\langle x, \langle k, i \rangle \rangle : i \in I, \langle x, k \rangle \in M_i\}$, and $\langle x, \langle k, i \rangle \rangle \prec \langle x', \langle k', i' \rangle \rangle$ iff $i = i'$ and $\langle x, k \rangle \prec_i < x', k' >$.

Thus, $\biguplus \{\mathcal{M}_i : i \in I\}$ is the disjoint union of the sets and the relations, and we will call it so.

Fact 1.8.8

Let μ_i be the choice functions of the $\mathcal{M}_i$. Then $\mu_{\biguplus\{\mathcal{M}_i : i \in I\}}(X) = \bigcup\{\mu_i(X) : i \in I\}$, so $\mu_{\biguplus\{\mathcal{M}_i : i \in I\}} = \mu_{\{\mathcal{M}_i : i \in I\}}$.

Proof

Exercise, solution in the Appendix.

Fact 1.8.9

$T \models_{\biguplus\{\mathcal{M}_i : i \in I\}} \phi$ iff for all $i \in I$ $T \models_{\mathcal{M}_i} \phi$. Thus $T \models_{\biguplus\{\mathcal{M}_i : i \in I\}} \phi$ iff $T \models_{\{\mathcal{M}_i : i \in I\}} \phi$, and disjoint unions preserve type 1 validity.

Proof

(Trivial.) Let again $\mu := \mu_{\biguplus\{\mathcal{M}_i : i \in I\}}$. $T \models_{\biguplus\{\mathcal{M}_i : i \in I\}} \phi$ iff in all $m \in \mu(T)$ ϕ holds. If for all $i \in I$ in all $m \in \mu_i(T)$ ϕ holds, then ϕ holds in all $m \in \mu(T)$ by Fact 1.8.8. But if there is some $i \in I$ and $m \in \mu(T)$ i s.t. ϕ fails in m, then ϕ will fail in some $m \in \mu(T)$, too, again by Fact 1.8.8.

□

Preservation of Type 2 Validity

Rules of type (2.1) are preserved: This is a direct consequence of Fact 1.8.9, the argument is similar to the following one for type (2.2) rules.

Rules of type (2.2) are preserved: We show that if in all strict total orders TO where $\phi \mid\!\sim \psi$ (and perhaps some other property) holds, $\phi \mid\!\sim \neg \phi'$ holds, then $\phi \mid\!\sim \neg \phi'$ holds in the disjoint union $\mathcal{M}$ of these structures, and, if in all strict total orders TO where $\phi \mid\!\sim \psi$ (and perhaps some other property) holds, $\phi \wedge \phi' \mid\!\sim \psi$ holds, then $\phi \wedge \phi' \mid\!\sim \psi$ holds in the disjoint union $\mathcal{M}$ of these structures. But, it is a direct consequence of Fact 1.8.9 that in the first case $\phi \models_\mathcal{M} \neg \phi'$, and in the second case $\phi \wedge \phi' \models_\mathcal{M} \psi$.

Rules of type (2.3) are not necessarily preserved - at least not in the general infinite case:

Example 1.8.4

(This is the - slightly adapted - Example 1.7.1, which shows failure of infinite conditionalization in a case where definability preservation fails.)

Consider the language $\mathcal{L}$ defined by the propositional variables $p_i, i \in \omega$.
Let $T_0^+ := \{p_0\} \cup \{p_i : 0 < i < \omega\}$, $T_0^- := \{\neg p_0\} \cup \{p_i : 0 < i < \omega\}$,
set $T' := T_0^+ \vee T_0^-$, and $T := \emptyset$.
Let the classical model m_0^+ (m_0^-) be the unique model satisfying T_0^+ (T_0^-), so $M(T') = \{m_0^+, m_0^-\}$. Consider the set $\mathcal{O}$ of all strict total orders TO on $M_\mathcal{L}$ satisfying $T' \models_{TO} T_0^-$. Obviously $T' \models_{TO} T_0^-$ iff $m_0^- \prec_{TO} m_0^+$. If $TO \in \mathcal{O}$ has no (global) minimum, then $T \models_{TO} \bot$, so $\neg p_0 \in \overline{\overline{T}} \cup T'$ - where $\overline{\overline{T}} := \{\phi : T \models_{TO} \phi\}$. If TO has a minimum, which is neither m_0^+ nor m_0^-, then $\overline{\overline{T}} \cup T'$ is inconsistent, and again $\neg p_0 \in \overline{\overline{T}} \cup T'$. The minimum cannot be m_0^+, so in all cases $\neg p_0 \in \overline{\overline{T}} \cup T'$. But now every model except m_0^+ can be minimal, so in the disjoint union $\mathcal{M} := \biguplus \mathcal{O}$ of these structures, $\mu_\mathcal{M}(T) = M_\mathcal{L} - \{m_0^+\}$. Thus $\overline{\overline{T}} = \overline{T}$ (in $\mathcal{M}$), and $\overline{\overline{T}} \cup T' = \overline{T'}$, but $\neg p_0 \notin \overline{T'}$. In particular, the example shows that rule (2.3) of Section 1.8.5.2 might hold in all components of a disjoint union, but fail in the union: As any total order TO is definability preserving, (2.3) holds in TO, by the results of Section 1.3.4. On the other hand, $\neg p_0 \in \overline{T \cup T'}$ (in $\mathcal{M}$), so (2.3) fails in $\mathcal{M}$.

□

Remark 1.8.10

(1) Failure of definability preservation in $\mathcal{M}$ is crucial for our example. More generally, definability preserving disjoint unions preserve rule (2.3). We know this already from Section 1.3.4, but give a direct argument to illustrate which kinds of rules of type 2 will be preserved in definability preserving disjoint unions. Let $\mathcal{X}$ be some set of strict total orders and $\mathcal{M} = \biguplus \mathcal{X}$. We have to show $M(\overline{\overline{T}} \cup T') \subseteq M(\overline{T \cup T'})$ (in $\mathcal{M}$). If $\mathcal{M}$ is definability preserving, then $M(\overline{\overline{T}}) = \mu_\mathcal{M}(T)$, so $M(\overline{\overline{T}} \cup T') = M(\overline{\overline{T}}) \cap M(T') = \mu_\mathcal{M}(T) \cap M(T') = \bigcup\{\mu_{TO}(T) : TO \in \mathcal{X}\} \cap M(T') = \bigcup\{\mu_{TO}(T) \cap M(T') : TO \in \mathcal{X}\} \subseteq \bigcup\{\mu_{TO}(T \cup T') : TO \in \mathcal{X}\} = \mu_\mathcal{M}(T \cup T') = M(\overline{T \cup T'})$. (In the inclusion, we have used property ($\mu PR'$), which holds in all preferential structures.) Thus $\overline{T \cup T'} \subseteq \overline{\overline{T}} \cup T'$, and as $\neg p_0 \in \overline{T \cup T'}$ in our Example 1.8.4, the example would not work.

(2) The general argument showing preservation of a rule in a definability preserving structure will argue semantically as above, i.e. that the rule is preserved under union: $\Phi(\mu_i(X), \mu_i(Y), \ldots)$ implies $\Phi(\bigcup \mu_i(X), \bigcup \mu_i(Y), \ldots)$. The semantical argument is possible by $M(\overline{\overline{T}}) = \mu_\mathcal{M}(T)$.

1.8 Various Results and Approaches 177

Equivalence of General Preferential Structures with Sets of Total Orders

Ideally, one would like every preferential structure to be (or at least, to be equivalent for type 1 validity to) a disjoint union of strictly totally ordered structures. This is not the case.

Example 1.8.5

Consider the language defined by one variable, p. Let $m \models p$, $m' \models \neg p$, and consider the structure $\langle m, 0 \rangle \succ \langle m', 0 \rangle \succ \langle m', 1 \rangle \succ \langle m', 2 \rangle \succ \ldots$. Then $\mu(true) = \emptyset$, but $\mu(p) = \{m\}$. There are only two possible total orders: $m \prec m'$, $m' \prec m$. $m \prec m'$ gives $\mu(\emptyset) = \{m\}$, $m' \prec m$ gives $\mu(\emptyset) = \{m'\}$, $(m \prec m') \uplus (m' \prec m)$ gives $\mu(\emptyset) = \{m, m'\}$. (Omitting some models totally will not help, either.)

Thus, traditional preferential structures are more expressive than strict total orders (or their disjoint union).

In Section 1.8.5.4, we will construct an equivalent structure in the finite cumulative case.

Multiple Copies

The usual constructions with multiple copies (the author's notation) or noninjective labelling functions (notation, e.g. of Kraus, Lehmann, Magidor) have always intrigued the author for their intuitive justification, which seemed somewhat weak (e.g. different languages of description and reasoning, as discussed in [Imi87]). We give here a purely formal one. Recall that we have discussed in Section 1.8.2 the expressive strength of structures with multiple copies in more detail.

Fact 1.8.9 shows that we can construct a usual structure with multiple copies out of a set of strictly totally ordered sets of classical models (without multiple copies), preserving validity of type 1. Example 1.8.4 shows that validity of type 2 is usually not preserved. For its failure, we needed a not definability preserving structure, which exists only for infinite languages. We thus conjecture that validity of type 2 is also preserved in the case of finite languages.

Thus, considering sets of strict total orders of models leads us naturally to consider their disjoint unions - at least largely equivalent structures - which are constructions with multiple copies.

1.8.5.4 Representation in the Finite Case

We show in this Section 1.8.5.4 our main result, Proposition 1.8.11, a representation theorem for the finite cumulative case. The infinite case stays an open problem.

As done before, we first show an algebraic representation result, Proposition 1.8.12, whose proof is the main work, and translate this result by routine methods to the logical representation problem.

It is easily seen that the consequence relations of the structures examined will be cumulative: First, it is well known (see, e.g. [KLM90], or Section 1.3.4) that smooth structures define cumulative consequence relations. Second, transitive relations over finite sets are smooth, and, third, we will see that our structures will be finite (see the modifications in the proof of Proposition 1.8.12).

Let us explain why the result of Proposition 1.8.11 is precisely the result to be expected. Classical logic defines exactly one consequence relation, $\vdash$. The conditions for preferential structures (system P of [KLM90], or our conditions of Proposition 1.3.20) do not describe one consequence relation, but a whole class, which have to obey certain principles. The representation theorem of classical logic states $T \vdash \phi$ iff in all models, if T holds, then so will ϕ. This unrestricted universal quantifier fixes one consequence relation, $\vdash$. This cannot be expected in our case. In our case, each preferential consequence relation $\vdash\!\sim$, i.e. each relation $\vdash\!\sim$ satisfying our conditions, will have to correspond to one particular set $\mathcal{O}_{|\sim}$ of total orders, in the sense that $T \vdash\!\sim \phi$ iff in all $TO \in \mathcal{O}_{|\sim}$ $T \models_{TO} \phi$. The quantifier is restricted to $\mathcal{O}_{|\sim}$. This is the completeness part of Proposition 1.8.11. The soundness part shows that any set $\mathcal{O}$ of total orders satisfies the conditions, thus a fortiori any total order will do so. Looking back at traditional preferential structures, and, e.g. the classical paper [KLM90], we see the exact correspondence to our result. There, it was shown in the soundness part that every preferential structure satisfies the system P. The completeness part there shows that there is one preferential structure $\mathcal{M}$ s.t. $T \vdash\!\sim \phi$ iff $T \models_{\mathcal{M}} \phi$, if $\vdash\!\sim$ satisfies system P. As preferential structures in the usual sense correspond to sets of total orders, we see that our result is the exact analogue of, e.g. the KLM result. To summarize, we show the exact analogue to usual preferential structures, and the closest analogue possible to classical logic.

We state now our main result, logical characterization.

Proposition 1.8.11

Let $\mathcal{L}$ be a propositional language defined by a finite set of variables.

(A) (Soundness) Let $\mathcal{O}$ be a set of strict total orders over $M_{\mathcal{L}}$, defining a logic $\vdash\!\sim$ by $T \vdash\!\sim \phi :\Leftrightarrow T \models_{\mathcal{O}} \phi$. Then $\vdash\!\sim$ satisfies (LLE), (CCL), (SC), (PR), (CUM).

(B) (Completeness) If a logic $\vdash\!\sim$ for $\mathcal{L}$ satisfies (LLE), (CCL), (SC), (PR), (CUM), then there is a set $\mathcal{O}$ of strict total orders over $M_{\mathcal{L}}$ s.t. $T \vdash\!\sim \phi \Leftrightarrow T \models_{\mathcal{O}} \phi$.

For the algebraic representation result, we will consider some $\mathcal{Y} \subseteq \mathcal{P}(Z)$, closed under finite unions and finite intersections, and a function $\mu : \mathcal{Y} \to \mathcal{Y}$. $\mathcal{Y}$ is intended to be $D_{\mathcal{L}}$ for some propositional language $\mathcal{L}$.

The proof uses the following algebraic characterization.

1.8 Various Results and Approaches   179

Proposition 1.8.12

Let Z be a finite set, let $\mathcal{Y} \subseteq \mathcal{P}(Z)$ be closed under finite unions and finite intersections, and $\mu : \mathcal{Y} \to \mathcal{Y}$.

(A) (Soundness) Let $\mathcal{O}$ be a set of strict total orders over Z, then $\mu_\mathcal{O}$ satisfies $(\mu \subseteq)$, (μPR), (μCUM).

(B) (Completeness) Let μ satisfy $(\mu \subseteq)$, (μPR), (μCUM), then there is a set $\mathcal{O}$ of strict total orders over Z s.t. $\mu = \mu_\mathcal{O}$.

The proof of Proposition 1.8.12 is a modification of a proof for traditional preferential structures as shown in Section 1.3.3.

By Fact 1.8.8, $\mu_\mathcal{O} = \mu_{\biguplus \mathcal{O}}$, so we can work with the set or its disjoint union.

(A) Soundness:

Conditions $(\mu \subseteq)$ and (μPR) hold for arbitrary preferential structures, and (μCUM) holds for smooth preferential structures (see Section 1.3.2 and Section 1.3.3.) Strict total orders over finite sets are smooth, so is their disjoint union.

(B) Completeness:

We will modify the construction in the proof of Proposition 1.3.18. We have constructed there for a function μ satisfying $(\mu \subseteq)$, (μPR), (μCUM) a transitive smooth preferential structure $\mathcal{Z} = \langle \mathcal{X}, \prec \rangle$ representing μ. We first show in (a) that the construction is finite for finite languages. We then eliminate in (b) unnecessary copies, and construct in (c) for each remaining $\langle x, i \rangle$ a total order $TO_{\langle x,i \rangle}$ such that the set of all these $TO_{\langle x,i \rangle}$ represents μ.

(a) Finiteness of the construction:

First, if the language $\mathcal{L}$ is finite, the constructed structure is finite, too: As $v(\mathcal{L})$ is finite, $Z = M_\mathcal{L}$ is finite. For each nonminimal element $x \in Z$, there is one tree in T'_x, so this is easy. Now, for the set T_x. T_x consists of trees $t_{U,x}$ where the elements of $t_{U,x}$ are pairs $\langle U', x' \rangle$ with $U' \in \mathcal{Y} \subseteq \mathcal{P}(Z)$ and $x' \in Z$, so there are finitely many such pairs. Each element in the tree has at most $card(\mathcal{P}(Z))$ successors, and by Fact 1.3.19, (1), if $\langle U_m, x_m \rangle$ is a direct or indirect successor in the tree of $\langle U_n, x_n \rangle$, then $x_m \notin H(U_n)$, but $x_n \in U_n \subseteq H(U_n)$, so $x_n \neq x_m$, so branches have length at most $card(Z)$. So there is a uniform upper bound on the size of the trees, so there are only finitely many of such trees.

(b) Elimination of unnecessary copies:

Next, if, for each $x \in Z$, there is a finite number of copies, then "best" copies $\langle x, i \rangle$ in the sense that there is no $\langle x, i' \rangle \prec \langle x, i \rangle$ in $\mathcal{Z}$ exist, so we can eliminate the "not so good" copies $\langle x, i \rangle$ for which there is $\langle x, i' \rangle \prec \langle x, i \rangle$, without changing representation. (Note that, instead of arguing with finiteness, we can argue here with smoothness, too, as singletons are definable.)

Representation is not changed, as the following easy argument shows: Let $\mathcal{Z}' = \langle \mathcal{X}', \prec \rangle$ be the new structure, we have to show that $\mu_{\mathcal{Z}} = \mu_{\mathcal{Z}'}$. Suppose $X \in \mathcal{Y}$, and $x \in \mu_{\mathcal{Z}}(X)$. Then there is $\langle x, i \rangle$ minimal in $\mathcal{Z} \upharpoonright X$. But then $\langle x, i \rangle \in \mathcal{X}'$ too, and, as we have not introduced new smaller elements, $x \in \mu_{\mathcal{Z}'}(X)$. Suppose now $x \in \mu_{\mathcal{Z}'}$, then there is some $\langle x, i \rangle$ minimal in $\mathcal{Z}' \upharpoonright X$. If there were $\langle y, j \rangle$ smaller than $\langle x, i \rangle$ in $\mathcal{Z}$, $y \in X$, then $\langle y, j \rangle$ would have been eliminated, as there is minimal $\langle y, k \rangle$ below $\langle y, j \rangle$, but then, by transitivity, $\langle y, k \rangle$ is smaller than $\langle x, i \rangle$, too, but $\langle y, k \rangle$ is kept in $\mathcal{Z}'$, so $\langle x, i \rangle$ would not be minimal in $\mathcal{Z}'$, either. Thus, $\mu_{\mathcal{Z}} = \mu_{\mathcal{Z}'}$.

(c) Construction of the total orders:

We take now the modified construction $\mathcal{Z}'$ to construct a set of total orders. $\langle x, i \rangle^-$, etc. will now be relative to $\mathcal{Z}'$.

We construct for each $x \in Z$ a set $\mathcal{O}_x = \{TO_{\langle x, i \rangle} : \langle x, i \rangle \in \mathcal{X}'\}$ of total orders. $\biguplus \mathcal{O} := \biguplus \{TO : TO \in \mathcal{O}_x, x \in Z\}$ will be the final structure, equivalent to $\mathcal{Z}$. $TO_{\langle x, i \rangle}$ is constructed as follows: We first put all elements $y \in \langle x, i \rangle^*$ below x, and all $y \neq x$, $y \notin \langle x, i \rangle^*$ above x. We then order $\langle x, i \rangle^*$ totally, staying sufficiently close to the order of $\mathcal{Z}'$, and finally do the same with the remaining elements.

Fix now $\langle x, i \rangle$, and let $<:=<_{TO_{\langle x, i \rangle}}$ be the strict total order on Z to be constructed.

First, set $y < x$ iff $y \in \langle x, i \rangle^*$, and set $x < y$ iff $y \neq x$ and $y \notin \langle x, i \rangle^*$.

We construct in (α) the part of the total order below x, and then in (β) the part above x.

(α) Work now inside $\langle x, i \rangle^*$, and construct a total order $<$ on $\langle x, i \rangle^*$ in three steps.

(1) Extend the partial order $\prec$ on $\langle x, i \rangle^-$ to a total order $\triangleleft$.

(2) If $\langle y, j \rangle \triangleleft \langle y, j' \rangle$, eliminate $\langle y, j' \rangle$. By finiteness, one copy of y survives.

(3) For $y, z \in \langle x, i \rangle^*$, let $y < z$ iff there are $\langle y, j \rangle$, $\langle z, k \rangle$ with $\langle y, j \rangle \triangleleft \langle z, k \rangle$ left in step (2).

By step (2), $<$ in $\langle x, i \rangle^*$ is free from cycles, by elimination of unnecessary elements in the construction of $\mathcal{Z}'$ x does not occur in $\langle x, i \rangle^*$, so the entire relation constructed so far is free from cycles.

Note that for $y \in \langle x, i \rangle^*$, there is some $\langle y, j \rangle$ s.t. $\langle y, j \rangle^* \subseteq \{z : z < y\}$: Let $\langle y, j \rangle$ be the $\triangleleft$-least copy of y, i.e. the one which survives step (2). Then by (1), all $\langle z, k \rangle \in \langle y, j \rangle^-$ are $\triangleleft$-below $\langle y, j \rangle$. But if some such $\langle z, k \rangle$ is eliminated in (2), there is an even smaller $\langle z, k' \rangle \triangleleft \langle y, j \rangle$ which survives, so $z < y$ in step (3).

(β) Work now on $\mathcal{X}' - (\{\langle x, i\rangle\} \cup \langle x, i\rangle^-)$.

(1) Extend the order $\prec$ on $\mathcal{X}' - (\{\langle x, i\rangle\} \cup \langle x, i\rangle^-)$ to a total order $\lhd$.

(2) Eliminate again $\langle y, j'\rangle$, if $\langle y, j\rangle \lhd \langle y, j'\rangle$, but eliminate also all $\langle y, j'\rangle$ s.t. $y = x$ or $y \in \langle x, i\rangle^*$.

(3) Let $y < z$ iff there are $\langle y, j\rangle, \langle z, k\rangle$ with $\langle y, j\rangle \lhd \langle z, k\rangle$ left in step (2).

By the same argument as above, we see that for any $y <$-above x, there is some $\langle y, j\rangle$ s.t. $\langle y, j\rangle^* \subseteq \{z : z < y\}$.

Let finally $\mathcal{O} = \{TO_{\langle x,i\rangle} : x \in Z, \langle x, i\rangle \in \mathcal{X}'\}$ and consider $\biguplus \mathcal{O}$. Let $\langle x, i\rangle \in \mathcal{X}'$. By construction, $\langle x, TO_{\langle x,i\rangle}\rangle^*_{\mathcal{O}} = x^*_{TO_{\langle x,i\rangle}} = \langle x, i\rangle^*_{Z'}$. Consider now arbitrary $TO_{\langle x,i\rangle}$, and $y \in Z$. It was shown in the construction that there is $\langle y, j\rangle \in \mathcal{X}'$ s.t. $\langle y, j\rangle^*_{Z'} \subseteq y^*_{TO_{\langle x,i\rangle}} = \langle y, TO_{\langle x,i\rangle}\rangle^*_{\mathcal{O}}$. So by Fact 1.8.6, (2) $\mu_{Z'} = \mu_{\biguplus \mathcal{O}} = \mu_{\mathcal{O}}$.

$\square$ (Proposition 1.8.12)

Proposition 1.2.18 completes the proof of Proposition 1.8.11. (Note that by finiteness of the language, μ is definability preserving.)

Proof of Proposition 1.8.11:

Let $\mathcal{O}$ be any set of strict total orders over $M_{\mathcal{L}}$, then $\mu_{\mathcal{O}} = \mu_{\biguplus \mathcal{O}}$ satisfies $(\mu \subseteq)$, (μPR), (μCUM) by Proposition 1.8.12, so the logic defined by $T \hspace{1pt}\vert\hspace{-3pt}\sim\hspace{1pt} \phi :\Leftrightarrow \mu_{\biguplus \mathcal{O}}(M_T) \models \phi \Leftrightarrow T \models_{\biguplus \mathcal{O}} \phi \;(\Leftrightarrow T \models_{\mathcal{O}} \phi)$ satisfies (LLE), (CCL), (SC), (PR), (CUM) by Proposition 1.2.18. Conversely, given a logic $\hspace{1pt}\vert\hspace{-3pt}\sim\hspace{1pt}$ which satisfies (LLE), (CCL), (SC), (PR), (CUM), then the model choice function μ defined by $\mu(M_T) := M_{\overline{T}}$ satisfies $(\mu \subseteq)$, (μPR), (μCUM) and $T \hspace{1pt}\vert\hspace{-3pt}\sim\hspace{1pt} \phi \Leftrightarrow \mu(M_T) \models \phi$ by Proposition 1.2.18, so by Proposition 1.8.12, there is a set $\mathcal{O}$ of strict total orders over $M_{\mathcal{L}}$ s.t. $\mu = \mu_{\mathcal{O}} = \mu_{\biguplus \mathcal{O}}$, so $T \hspace{1pt}\vert\hspace{-3pt}\sim\hspace{1pt} \phi$ iff $\mu(M_T) = \mu_{\mathcal{O}}(M_T) = \mu_{\biguplus \mathcal{O}}(M_T) \models \phi$ iff $T \models_{\mathcal{O}} \phi$, the latter by Fact 1.8.7.

$\square$ (Proposition 1.8.11)

1.8.6 Preferred History Semantics for Iterated Updates

1.8.6.1 Introduction

Overview

This section continues work in [LMS01], see also Section 4.3, and [Sch04]. It contains joint work with Shai Berger and Daniel Lehmann, and was published as [BLS99]. We partly kept the original notation, to make comparisons easier.

We develop here an approach to update based on an abstract distance or ranking function. An agent has (incomplete, but reliable) information (observations) about a changing situation in the form of a sequence of formulas. At time 1, α_1 holds, at time 2, α_2 holds, ..., at time n, α_n holds. We are thus in a situation of iterated update. The agent tries to reason about the most likely outcome, i.e. to sharpen the information α_n by plausible reasoning. He knows that the real world has taken some trajectory, or history, that can be described by a sequence of models $\langle m_1, \ldots, m_n \rangle$, where $m_i \models \alpha_i$ (remember the observations were supposed to be reliable). We say that such a history explains the observations. For his reasoning, he makes two assumptions: First, an assumption of inertia: histories that stay constant are more likely than histories that change without necessity. For instance, if $n = 2$, and α_1 is consistent with α_2, $m_1 \models \alpha_1 \wedge \alpha_2$, $m_2 \models \alpha_2$, then the history $\langle m_1, m_1 \rangle$ is preferred to the history $\langle m_1, m_2 \rangle$. We do NOT assume that $\langle m_1, m_1 \rangle$ is more likely than some $\langle m_3, m_2 \rangle$, i.e. we do not compare the cardinality of changes, we only assume "sub-histories" to be more likely than longer ones. Second, the agent assumes that histories can be ranked by their likelihood, i.e. that there is an (abstract) scale, a total order, which describes this ranking. These assumptions are formalized below.

The agent then considers those models of α_n as most plausible, which are endpoints of preferred histories explaining the observations. Thus, his reasoning defines an operator [ ] from the set of sequences of observations to the set of formulas of the underlying language, s.t. $[\alpha_1, \ldots, \alpha_n] \models \alpha_n$.

The purpose of this section is to characterize the operators [ ] that correspond to such reasoning. Thus, we will give conditions for the operator [ ] which all operators based on history ranking satisfy, and, conversely, which allow to construct a ranking r from the operator [ ] such that the operator $[\ ]_r$ based on this ranking is exactly [ ]. The first part can be called the soundness, the second the completeness part.

Before giving a complete set of conditions in Section 1.8.6.3, we discuss in Section 1.8.6.2 some logical and intuitive properties of ranking based operators, in particular those properties which are related to the Alchourron, Gardenfors, Makinson postulates for theory revision, or to the update postulates of Katsuno, Mendelzon. In Section 1.8.6.3, we give a full characterization of these operators. We start from a result of Lehmann, Magidor, Schlechta [LMS01] on distance based revision, which has some formal similarity, and refine the techniques developed there.

In the rest of this section, we first compare briefly revision and update, and then emphasize the relevance of epistemic states for iterated update, i.e. that belief sets are in general insufficient to determine the outcome of our reasoning about iterated update. We then make our approach precise, give some basic definitions, and recall the AGM and KM postulates.

1.8 Various Results and Approaches 183

Revision and Update

Intuitively, belief revision (also called theory revision) deals with evolving knowledge about a static situation. Update, on the other hand, deals with knowledge about an evolving situation. It is not clear that this ontological distinction agrees with the semantical and proof-theoretic distinction between the AGM and the KM approaches. In this section, the distinction between revision and update must be understood as ontological, not as AGM vs. KM semantics or postulates.

In the case of belief revision, an agent receives successively different information about a situation, e.g. from different sources, and the union of this information may be inconsistent. The theory of belief revision describes "rational" ways to incorporate new information into a body of old information, especially when the new and old information together are inconsistent.

In the case of update, an agent is informed that at time t, a formula ϕ held, at time t' ϕ', etc. The agent tries, given this information and some background assumptions (e.g. of inertia: that things do not change unless forced to do so) to draw plausible conclusions about the probable development of the situation. This distinction goes back to Katsuno and Mendelzon [KM90].

Revision in this sense is formalized in Lehmann, Magidor and Schlechta [LMS01]. The authors have devised there a family of semantics for revision based on minimal change, where change is measured by distances between models of formulae of the background logic. More precisely, the operator defined by such functions revises a belief set T according to a new observation α by picking the models of α that are closest to models of T. Such revision operators have been shown to satisfy some of the common rationality postulates. Several weak forms of a distance function ("pseudo-distances") have been studied in [LMS01], and representation theorems for abstract revision operators by pseudo-distances have been proved. Since, in the weak forms, none of the notions usually connected with a distance (e.g. symmetry, the triangle inequality) are used, such pseudo-distances are actually no more than a preference function, or a ranked order, over pairs of models.

In the present section, we consider a setting that intuitively has an ontology of update. It sets a single belief change system for all sequences of observations. All pseudo-distances are between individual models, as in the semantics proposed in [KM90]. But, where Katsuno and Mendelzon's semantics takes a "local" approach and incorporates in the new belief set the best updating models for each model in the old belief set, we take a "global" approach and pick for the updated belief set only the ending models of the best overall histories. As a result, our system validates all the AGM postulates, and not all the KM ones. The introduction of an update system which follows the AGM postulates for revision may have some interesting ontological consequences, but these will not be dealt with in this work.

The formal definitions and assumptions are to be found in Definitions 1.8.10, 1.8.11, 1.8.12, and Condition 1.8.1.

We prove a representation theorem for update operators based on rankings of histories, similar to those in [LMS01]. Note that the approach taken here is more specific than that of [Sch95-3], which considers arbitrary (e.g. not necessarily ranked) preference relations between histories.

Epistemic States Are Not Belief Sets

The epistemic state of an agent, i.e., the state of its mind, is understood to include anything that influences its actions, its beliefs about what is true of the world, and the way it will update or revise those beliefs, depending on the information it gathers. The belief set of an agent, at any time, includes only the set of propositions it believes to be true about the world at this time. One of the components of epistemic states must therefore be the belief set of the agent. One of the basic assumptions of the AGM theory of belief revisions is that epistemic states are belief sets, i.e., they do not include any other information. At least, AGM do not formalize in their basic theory, as expressed by the AGM postulates, any incorporation of other information in the belief revision process.

In particular an agent that holds exactly the same beliefs about the state of the world, at two different instants in time is, at those times, in the same epistemic state and therefore, if faced with the same information, will revise its beliefs in the same way. Recent work on belief revision and update has shown this assumption has very powerful consequences, not always welcome [Leh95], [DP94], [FH96a]. Earlier work on belief base revision (see e.g. [Neb89]) expresses a similar concern about the fundamentals of belief revision.

We do not wish to take a stand on the question of whether this identification of epistemic states with belief sets is reasonable for the study of belief revision, but we want to point out that, in the study of belief update, with its natural sensitivity (by the principle of inertia) to the order in which the information is gathered, it is certainly unreasonable. This is illustrated by the following observation: Let ϕ and ψ be different atomic, i.e. logically independent, formulas. First scenario: update the trivial belief set (the set of all tautologies) by ϕ, then by ψ, then by $\neg\phi \vee \neg\psi$. Second scenario: update the trivial belief set by ψ, then by ϕ, then by $\neg\phi \vee \neg\psi$. We expect different belief sets: we shall most probably try to stick to the piece of information that is the most up-to-date, i.e., ψ in the first scenario and ϕ in the second scenario. But there is no reason for us to think that the belief sets obtained, in both scenarios, just before the last updates, should be different. We expect them to be identical: the consequences of $\phi \wedge \psi$. The same agent, in two different epistemic states, updates differently the same beliefs in the light of the same information.

Preferred History Semantics

We now make our approach more precise. The basic ontology is minimal: The agent makes a sequence of observations and interprets this sequence of observations in terms of possible histories of the world explaining the observations.

Assume a set $\mathcal{L}$ of formulas and a set $\mathcal{U}$ of models for $\mathcal{L}$. Formulas will be denoted by Greek letters from the beginning of the alphabet: α, β and so on, and models by m, n, and so on. We do not assume formulas are indexed by time. An observation is a *consistent* formula. (We assume observations to be consistent for two reasons: First, observations are assumed to be reliable; second, as we work with histories made of models, we need some model to explain every observation. Working with unreliable information would be the subject of another study.) Observing a formula means observing the formula holds.

A sequence of observations is here a *finite* sequence of observations. Sequences of observations will be denoted by Greek letters from the end of the alphabet: σ, τ and concatenation of such sequences by $\cdot$. Notice the empty sequence is a legal sequence of observations. We shall identify an observation with the sequence of observations of length one that contains it. What does a sequence of observations tell us about the present state of the world?

A history is a finite, non-empty sequence of models. Histories will be denoted by h, f, and so on.

Definition 1.8.10

A history $h = \langle m_0, \ldots, m_n \rangle$ *explains* a sequence of observations $\tau = \langle \alpha_0, \ldots, \alpha_k \rangle$ iff there are subscripts $0 \leq i_0 \leq i_1 \leq \ldots \leq i_k \leq n$ such that for any j, $0 \leq j \leq k$, $m_{i_j} \models \alpha_j$.

Thus, a history explains a sequence of observations if there is, in the history, a model that explains each of the observations in the correct order. Notice that n is in general different from k, that many consecutive i_j's may be equal, i.e., the same model may explain many consecutive observations, that some models of the history may not be used at all in the explanation, i.e., l, $0 \leq l \leq n$ may be equal to none of the i_j's, and that we do not require that j_k be equal to n, or that j_0 be equal to 0, i.e., there may be useless models even at the start or at the end of a history. Note also that if h explains a sequence σ of observations, it also explains any subsequence (not necessarily contiguous) of σ.

The set of histories that explain a sequence of observations give us information about the probable outcome. Monotonic logic is useless here: if we consider all histories explaining a sequence of observations, we cannot conclude anything. It is reasonable, therefore to assume the agent restricts the set of histories it considers to a subset of the explaining histories.

We shall assume the agent has some preferences among histories, some histories being more natural, simpler, more expected, than others. A sequence of observations defines thus a subset of the set of all histories that explain it: the set of all preferred histories that explain it. This set defines the set of beliefs that result from a sequence of observations: the set of formulas satisfied in all the models that may appear as last elements of a preferred history that explains the sequence. The beliefs held depend on the preferences, concerning histories, of the agent. The logical properties of update depend on the class of preferences we shall consider.

Formally, one assumes the agent's preferences are represented by a binary relation $<$ on histories. Intuitively, $h < f$ means that history h is strictly preferred, e.g. strictly more natural or strictly simpler, than history f. Note that our relation is on histories, not on models. We may now define preferred histories.

Definition 1.8.11

A history h is a preferred history for a sequence σ of observations iff

- h explains σ
- there is no history h' that explains σ such that $h' < h$.

In this work two assumptions are made concerning the preference relation $<$. First, we assume $<$ is a strict modular, well-ordering, i.e., $<$ is irreflexive, transitive, if $h < h'$, then for any f, either $h < f$ or $f < h'$ and there is no infinite descending chain. Secondly, we assume that partial histories (i.e., sub-histories) are preferred over (longer) histories:

Condition 1.8.1

If $h = \langle m_0, \ldots, m_n \rangle$ and $h' = \langle m_{j_0}, \ldots, m_{j_k} \rangle$, for $0 \leq j_0 < j_1 < \ldots < j_k \leq n$ and $0 \leq k < n$, then $h' < h$.

For instance, $h' = \langle m_2, m_4 \rangle$ is preferred to $h = \langle m_1, m_2, m_3, m_4 \rangle$.

This assumption is justified both by an epistemological concern: simpler explanations are better, and by an assumption of inertia concerning the way the universe evolves: things tend to stay as they are. This assumption, in a finite setting, trivializes the well-ordering assumption.

Finally, we formally define our operator $[\ ]$:

Definition 1.8.12

After a sequence σ of observations, the agent holds the beliefs $[\sigma]$, defined by $\alpha \in [\sigma]$ iff for every model m and every history h, if h is a preferred history explaining σ and m is the last element of h, then $m \models \alpha$.

Notice that, since histories are non-empty, this definition always makes sense.

The remainder of this section will show that the assumptions above have far reaching consequences: they are strong assumptions. Our purpose is indeed to look for a powerful logic, not for the minimal logic that agrees with any possible ontology.

Basic Definitions and Notation

We are dealing here only with finite and complete universes, so theories have logically equivalent formulas (and vice versa), and are isomorphic to sets of models, and we will use them in these senses interchangeably. We will see sequence concatenation also as an outer product (with respect to concatenation) of sets of histories. For technical reasons, most of the discussion will relate to sets of models. An exception, for easier readability and comparison with the AGM and KM conditions is @Section BLS-Prop@. We do not consistently differentiate singletons from the members that comprise them, because it is always clear from context which of them we are referring to.

A theory, or belief set, will be a deductively closed set of formulas.

For the convenience of the reader, we recall the AGM postulates for belief revision, (see e.g. [Gar92]), Table 4.1, and Proposition 4.2.2), and the Katsuno-Mendelzon postulates for update (see e.g. [KM90]).

$(K * 1)$ $K * \alpha$ is a deductively closed set of formulas.

$(K * 2)$ $\alpha \in K * \alpha$.

$(K * 3)$ $K * \alpha \subseteq Cn(K, \alpha)$.

$(K * 4)$ If $\neg \alpha \notin K$, then $Cn(K, \alpha) \subseteq K * \alpha$.

$(K * 5)$ If $K * \alpha$ is inconsistent then α is a logical contradiction.

$(K * 6)$ If $\models \alpha \leftrightarrow \beta$, then $K * \alpha = K * \beta$.

$(K * 7)$ $K * \alpha \wedge \beta \subseteq Cn(K * \alpha, \beta)$.

$(K * 8)$ If $\neg \beta \notin K * \alpha$, then $Cn(K * \alpha, \beta) \subseteq K * \alpha \wedge \beta$.

(U1) $\models (\psi \cdot \mu) \to \mu$.

(U2) If $\models \psi \to \mu$, then $\models (\psi \cdot \mu) \leftrightarrow \psi$.

(U3) If both ψ and μ are satisfiable, then so is $\psi \cdot \mu$.

(U4) If $\models \psi_1 \leftrightarrow \psi_2$ and $\models \mu_1 \leftrightarrow \mu_2$, then $\models (\psi_1 \cdot \mu_1) \leftrightarrow (\psi_2 \cdot \mu_2)$.

(U5) $\models ((\psi \cdot \mu) \wedge \phi) \to (\psi \cdot (\mu \wedge \phi))$.

(U6) If $\models (\psi \cdot \mu_1) \to \mu_2$ and $\models (\psi \cdot \mu_2) \to \mu_1$, then $\models (\psi \cdot \mu_1) \leftrightarrow (\psi \cdot \mu_2)$.

(U7) If ψ is complete, then $\models (\psi \cdot \mu_1) \wedge (\psi \cdot \mu_2) \to (\psi \cdot (\mu_1 \vee \mu_2))$.

(U8) $\models ((\psi_1 \vee \psi_2) \cdot \mu) \leftrightarrow (\psi_1 \cdot \mu) \vee (\psi_2 \cdot \mu)$.

The following is a slight reformulation of the AGM postulates in the spirit of Katsuno-Mendelzon, taken from [LMS01], see also Table 4.3. We consider here a symmetrical version, in the sense that K and α can both be theories, and simplify by considering only consistent theories.

($*0$) If $\models T \leftrightarrow S$, $\models T' \leftrightarrow S'$, then $T * T' = S * S'$,

($*1$) $T * T'$ is a consistent, deductively closed theory,

($*2$) $T' \subseteq T * T'$,

($*3$) If $T \cup T'$ is consistent, then $T * T' = Cn(T \cup T')$,

($*4$) If $T * T'$ is consistent with T'', then $T * (T' \cup T'') = Cn((T * T') \cup T'')$.

Finally, we recall the definition of a pseudo-distance from [LMS01], see Definition 4.2.2.

Definition 1.8.13

$d : U \times U \to Z$ is called a pseudo-distance on U iff Z is totally ordered by a relation $<$.

1.8.6.2 Some Important Logical Properties of Updates

A number of logical properties of the operator [ ] will now be described and discussed. The reasons why those properties hold are varied: some depend on very little of our assumptions, some on almost all of them. We shall try to make the appropriate distinctions.

Lemma 1.8.13

For any σ, $[\sigma]$ is a theory.

This property is analogous to AGM's ($K*1$) and is implicit in Katsuno-Mendelzon's presentation [KM90].

This depends only on the fact that Definition 1.8.12 defines $[\sigma]$ as the set of all formulas that hold for all the models in a given set. Indeed, this is a property that is expected to hold by the structure of belief sets, not by the definition of explanation or certain properties of the preference relation.

The following properties hold by the definition of explanation, i.e., Definition 1.8.10.

Lemma 1.8.14

If α and α' are logically equivalent, then for any sequences σ, τ: $[\sigma \cdot \alpha \cdot \tau] = [\sigma \cdot \alpha' \cdot \tau]$.

This property is analogous to AGM's ($K * 6$) but notice that, there, it is needed only for the second argument of the revision operation, since it is implicit for the first, a theory. It parallels (U4) in Katsuno-Mendelzon's [KM90].

Lemma 1.8.14 follows from Definition 1.8.10, that implies that the histories that explain $\sigma \cdot \alpha \cdot \tau$ are exactly those that explain $\sigma \cdot \alpha' \cdot \tau$. The preferred histories are therefore the same. The next property is more original.

1.8 Various Results and Approaches

Lemma 1.8.15

If $\beta \models \alpha$, then for any sequences σ, τ: $[\sigma \cdot \alpha \cdot \beta \cdot \tau] = [\sigma \cdot \beta \cdot \tau] = [\sigma \cdot \beta \cdot \alpha \cdot \tau]$.

This property has no clear analogue in the AGM or KM frameworks, but is closely related to (U2) of [KM90]. The first equation is property (C1) of Darwiche-Pearl's [DP94] and (I5) of [Leh95]. The second equation is a weakening of (I4) of [Leh95]. Here we request α to be a logical consequence of β, there we only asked that α be in $[\sigma \cdot \beta]$. Lemma 1.8.15 is a consequence of the fact that the histories that explain $\sigma \cdot \beta \cdot \tau$, $\sigma \cdot \alpha \cdot \beta \cdot \tau$ and $\sigma \cdot \beta \cdot \alpha \cdot \tau$ are the same.

Corollary 1.8.16

For any sequence σ, $[\sigma \cdot \mathbf{true}] = [\sigma]$.

The next property deals with disjunction.

Lemma 1.8.17

If γ is a member both of $[\sigma \cdot \alpha \cdot \tau]$ and $[\sigma \cdot \beta \cdot \tau]$, then it is a member of $[\sigma \cdot \alpha \vee \beta \cdot \tau]$. In other words $[\sigma \cdot \alpha \cdot \tau] \cap [\sigma \cdot \beta \cdot \tau] \subseteq [\sigma \cdot \alpha \vee \beta \cdot \tau]$.

This property is similar to one half of (U8) of [KM90], and to a consequence of AGM's $(K * 7)$, as pointed out in [Gar88], property

(3.14): $(K * A) \cap (K * B) \subseteq K * (A \vee B)$.

Lemma 1.8.17 depends only on Definitions 1.8.10 and 1.8.11, but does not depend on any properties of the preference relation. A history h that explains $\sigma \cdot \alpha \vee \beta \cdot \tau$ explains $\sigma \cdot \alpha \cdot \tau$ or $\sigma \cdot \beta \cdot \tau$. A preferred history for $\sigma \cdot \alpha \vee \beta \cdot \tau$ must therefore either be a preferred history for $\sigma \cdot \alpha \cdot \tau$ (since any history explaining the latter explains the former) or preferred history for $\sigma \cdot \beta \cdot \tau$. The next lemma is a strengthening of Lemma 1.8.17, and it depends on the modularity of the preference relation.

Lemma 1.8.18

The theory $[\sigma \cdot \alpha \vee \beta \cdot \tau]$ is equal to $[\sigma \cdot \alpha \cdot \tau]$, equal to $[\sigma \cdot \beta \cdot \tau]$ or is the intersection of the two theories above.

This property is a weakening of (U8) of [KM90], compare also to property (3.16) in [Gar88]:

$K * (A \vee B) = K * A$ or $K * (A \vee B) = K * B$ or $K * (A \vee B) = (K * A) \cap (K * B)$.

Proof

Exercise, solution in the Appendix.

□

The next properties follow from the assumption that sub-histories are preferred to more complete histories. Remark first that, if a history h explains a non-empty sequence σ of observations but the last model of h does not satisfy the last observation of σ, then there is a shorter (initial) sub-history of h that explains σ. The history h cannot be, in this case, a preferred history for σ.

Lemma 1.8.19

For any sequence σ of observations and any formula α: $\alpha \in [\sigma \cdot \alpha]$.

This property is similar to AGM's $(K * 2)$ and (U1) of [KM90].

In [FH96a], Friedman and Halpern question this postulate. Here, it finds a justification, grounded in our preference for shorter explanations.

Lemma 1.8.20

For any sequence σ of observations and any formulas α and β, if $\neg\beta \notin [\sigma \cdot \alpha]$, then $[\sigma \cdot \alpha \cdot \beta] = [\sigma \cdot \alpha \wedge \beta] = \mathcal{C}n([\sigma \cdot \alpha], \beta)$.

This property is analogous to AGM's $(K * 7)$ and $(K * 8)$.

Proof We show that the preferred histories for $\sigma \cdot \alpha \cdot \beta$ are exactly those preferred histories of $\sigma \cdot \alpha$ whose last element satisfies β. First, clearly, any preferred history for $\sigma \cdot \alpha$ whose last element satisfies β explains $\sigma \cdot \alpha \cdot \beta$ and is a preferred history for it. Secondly, since $\neg\beta \notin [\sigma \cdot \alpha]$, there is a preferred history h for $\sigma \cdot \alpha$ whose last element satisfies β. As we have just seen h is a preferred history for $\sigma \cdot \alpha \cdot \beta$. Let f be a preferred history for $\sigma \cdot \alpha \cdot \beta$. It explains $\sigma \cdot \alpha$. If it were not a preferred history for $\sigma \cdot \alpha$, there would be a history f', $f' < f$ that explains $\sigma \cdot \alpha$. By modularity, we would have $f' < h$ or $h < f$, which are both impossible. We conclude that f is a preferred history for $\sigma \cdot \alpha$. But its last element satisfies β, by Lemma 1.8.19. We have shown that $[\sigma \cdot \alpha \cdot \beta] = \mathcal{C}n([\sigma \cdot \alpha], \beta)$.

To conclude the proof, notice that, by the above, any preferred history for $\sigma \cdot \alpha \cdot \beta$ explains $\sigma \cdot \alpha \wedge \beta$ and that any history explaining $\sigma \cdot \alpha \wedge \beta$ also explains $\sigma \cdot \alpha \cdot \beta$.

□

Corollary 1.8.21

If $\neg\alpha \notin [\sigma]$, then $[\sigma \cdot \alpha] = \mathcal{C}n([\sigma], \alpha)$.

This parallels AGM's $(K * 3)$ and $(K * 4)$.

Proof By Corollary 1.8.16, $[\sigma] = [\sigma \cdot \mathbf{true}]$. By Lemma 1.8.20, $[\sigma \cdot \mathbf{true} \cdot \alpha] = \mathcal{C}n([\sigma \cdot \mathbf{true}], \alpha) = [\sigma \cdot \mathbf{true} \wedge \alpha]$. By Corollary 1.8.16 and Lemma 1.8.15 $\mathcal{C}n([\sigma], \alpha) = [\sigma \cdot \alpha]$.

□

Our last property depends on well-foundedness, which is trivial in a finite setting.

1.8 Various Results and Approaches

Lemma 1.8.22

For any sequence σ, $[\sigma]$ is consistent.

This parallels AGM's $(K * 5)$ and KM's (U3). This property depends on two assumptions. First we assumed observations were consistent formulas. It follows that any sequence of observations is explained by some history. By finiteness, if h explains σ and h is not a preferred history for σ, then there is a preferred history h' for σ (in fact one such that $h' < h$). Therefore $[\sigma]$ is consistent.

1.8.6.3 A Representation Theorem

Introduction

In Section 1.8.6.2, we have presented several logical properties of operators [ ] based on history ranking. In this Section, we will give a full characterization. We generalize here results about revision reported in [LMS01].

We will first show that a straightforward generalization of the not necessarily symmetrical revision case fails already for sequences of length 3, and will then give a characterization for sequences of arbitrary finite length in Theorem 1.8.25. A technical problem for the latter is that we have to work with "illegal" sets of histories, which do not correspond to sequences of sets of models. E.g. the set of sequences $\{\langle 0, 0, 0\rangle, \langle 0, 1, 1\rangle\}$ is not the product of any sequence of sets - $\{0\} \times \{0, 1\} \times \{0, 1\}$ contains too many sequences. Our operator is, however, only defined for such sequences of sets. We use the idea of a patch, a cover of such sets of sequences by products of sequences of sets, to show our result.

As mentioned before, we will want to work mainly with sequences of sets of models. Such sets are freely interchangeable with observations, and every sequence thereof also defines a set of histories. We would like the set of all explaining histories to be representable as a sequence of sets of models. Definition 1.8.10 allows the set of explaining histories to be infinite, and we will limit the sets of histories dealt with by an assumption strengthening Lemma 1.8.19. While we do not assume full sub-history preference, we assume that a history explains a sequence of observations if they are of the same length, and the ith model in the history models the ith observation in the sequence. This assumption, like the Lemma, is justified by sub-history preference, with an implicit agreement that consecutive repeats of a model in a history are merely another way to write that the model explains several observations. In other words, to make the formal phrasing and proof a little easier, we will write for each observation in the sequence σ the model that explains it in the history h. A history h' containing h as a sub-history will not be preferable to h, and sub-histories of h are just represented as longer than they are. Intuitively, an assumption is made here that e.g. $\langle m_1, m_1, m_2\rangle$ and $\langle m_1, m_2, m_2\rangle$, both being representations of $\langle m_1, m_2\rangle$, are equally preferred, and are both considered better than $\langle m_1, m_2, m_3\rangle$. This assumption is neither used nor needed in the theorem or its proof. We make no further assumption on the history-preference relation.

Histories and sequences of observations of length or dimension 2 are closely parallel to the not necessarily symmetric case of [LMS01].

We first recall the corresponding representation result in Section "2-D Representation". Then, we show that a simple generalization of this result fails already in the case of length 3 (Section "Simple Generalization"). Finally, in Section "n-Dimensional Representation", we prove a valid representation theorem for the general, n-dimensional case.

The 2-D Representation Theorem

First, let us quote a theorem characterizing the revision operators representable by a pseudo-distance function (Proposition 2.5 of [LMS01]), see also Proposition 4.3.5.

The pseudo-distance mentioned here is actually no more than a preference relation over pairs of models (or histories of length 2). The theorem deals with an operator $|$ which revises a belief set by an observation, i.e., $A \mid B$ is the belief set held by an agent who has held a belief set A, after observing B. Now, let X be a finite and complete universe (the set of possible models of the language). In such a universe, A and B are interchangeably formulas, theories and sets of models. Let $\mathcal{P}(X)$ designate the set of all *non-empty* subsets of X.

Definition 1.8.14

An operation $| : \mathcal{P}(X) \times \mathcal{P}(X) \to \mathcal{P}(X)$ is *representable* iff there is a pseudo-distance $d : X \times X \to \mathcal{Z}$ such that $A \mid B = \{b \in B \mid \exists a \in A \text{ such that } \forall a' \in A, b' \in B, d(a,b) \leq d(a',b')\}$.

Thus, intuitively, if A and B are sets, $A \mid B$ is the set of those elements of B, which are closest to the set A. By abuse of notation, if A and B are formulas, $A \mid B$ is the set of formulas valid in the set of those models of B, which are closest to the set of models of A.

For this theorem, Lehmann, Magidor and Schlechta define a relation $R_|$ on pairs from $\mathcal{P}(X) \times \mathcal{P}(X)$, which intuitively means "provably closer" or "provably preferable", i.e., assuming the underlying pseudo-distance exists, this relation represents information about it that may be deduced by examining the revision operation. For instance, if $(A \mid (B \cup C)) \cap B \neq \emptyset$, B is provably at least as close to A, as C is to A. This information can only apply to the best-preferred pairs of models in the pairs of sets, so $(A, B) R_|(A', B')$ actually means we have evidence that the best pair of models in $A \times B$ is at least as preferable as the best pair in $A' \times B'$.

1.8 Various Results and Approaches

Definition 1.8.15

Given an operation $|$, define a relation $R_|$ on pairs from $\mathcal{P}(X) \times \mathcal{P}(X)$ by: $(A, B) R_|(A', B')$ iff one of the following two cases obtains:

(1) $A = A'$ and $(A \mid (B \cup B')) \cap B \neq \emptyset$,
(2) $B = B'$ and $((A \cup A') \mid B) \neq (A' \mid B)$.

In the rest of this subsection we shall write R instead of $R_|$. As usual, we shall denote by R^* the transitive closure of R.

Now, we can quote the representation theorem:

Theorem 1.8.23

An operation $|$ is representable iff it satisfies the four conditions below for any non-empty sets $A, A', B, B' \subseteq X$:

(1) $(A \mid B) \subseteq B$,
(2) $((A \cup A') \mid B) \subseteq (A \mid B) \cup (A' \mid B)$,
(3) If $(A, B) R^*(A, B')$, then $(A \mid B) \subseteq (A \mid (B \cup B'))$,
(4) If $(A, B) R^*(A', B)$, then $(A \mid B) \subseteq ((A \cup A') \mid B)$

This is the strongest version of this theorem proven. Fixing the conditions of the theorem, the characterization grows stronger as the definition of R becomes narrower, as it then (possibly) puts less constraints on $|$. A weaker characterization (which is also valid) has R defined to be wider, as follows:

Definition 1.8.16

We say the relation $R_|$ holds iff at least one of the following cases obtains:

(1) $A \supseteq A', B \supseteq B' \Rightarrow (A, B) R(A', B')$
(2) $(A \mid (B \cup B')) \cap B \neq \emptyset \Rightarrow (A, B) R(A, (B \cup B'))$
(3) $((A \cup A') \mid B) \neq (A' \mid B) \Rightarrow (A, B) R((A \cup A'), B)$

Ultimate Goal: The n-Dimensional Case

We want to prove a theorem analogous to Theorem 1.8.23 that relates to strings of observations of length n (instead of length 2, if we ignore the difference in role between a previous observation and a previous belief set), that is, we want to characterize representable operations $[\,] : \mathcal{P}(X)^n \to \mathcal{P}(X)$:

Definition 1.8.17

An operation $[\,] : \mathcal{P}(X)^n \to \mathcal{P}(X)$ is *representable* iff there is a totally ordered set $\mathcal{Z}$ (the order is $<$) and a function $r : X^n \to \mathcal{Z}$ (that will be intuitively understood as a history ranking), such that, for any non-empty subsets $A_1, \ldots A_n \subseteq X$,

$[A_1 \cdots A_n] =$
$\{a_n \in A_n \mid \exists a_1 \in A_1 \ldots a_{n-1} \in A_{n-1} \forall a'_1 \in A_1 \ldots a'_n \in A_n \; r(a_1, \ldots, a_n) \leq r(a'_1, \ldots, a'_n)\}$

We will now see that this may not be achieved by straightforward generalization of the tight 2-dimensional characterization, even for just three dimensions.

Simple Generalization Is Not Valid

The first attempt at generalizing this theorem is held short at $n = 3$. Let us phrase the suggested theorem and disprove it, starting with a new definition for R:

Definition 1.8.18

Given an operation $[\,]$, one defines a relation $R_{[\,]}$ on triplets of non-empty subsets of X by: $(A, B, C) R_{[\,]} (A', B', C')$ iff one of the following cases obtains:

(1) $A = A'$, $B = B'$ and $[A \cdot B \cdot (C \cup C')] \cap C \neq \emptyset$.
(2) $A = A'$, $C = C'$ and $[A \cdot (B \cup B') \cdot C] \neq [A \cdot B' \cdot C]$.
(3) $B = B'$, $C = C'$ and $[(A \cup A') \cdot B \cdot C] \neq [A' \cdot B \cdot C]$.

From here on, unless otherwise stated, R stands for $R_{[\,]}$. Now, the suggested theorem - which is WRONG, as we will see:

Theorem 1.8.24

An operation $[\,]$ is representable iff it satisfies the six conditions below for any non-empty sets $A, A', B, B', C, C' \subseteq X$:

(1) $[A \cdot B \cdot C] \subseteq C$.
(2) $[A \cdot (B \cup B') \cdot C] \subseteq [A \cdot B \cdot C] \cup [A \cdot B' \cdot C]$.
(3) $[(A \cup A') \cdot B \cdot C] \subseteq [A \cdot B \cdot C] \cup [A' \cdot B \cdot C]$.
(4) $(A, B, C) R^*(A', B, C) \Rightarrow [A \cdot B \cdot C] \subseteq [(A \cup A') \cdot B \cdot C]$.
(5) $(A, B, C) R^*(A, B', C) \Rightarrow [A \cdot B \cdot C] \subseteq [A \cdot (B \cup B') \cdot C]$.
(6) $(A, B, C) R^*(A, B, C') \Rightarrow [A \cdot B \cdot C] \subseteq [A \cdot B \cdot (C \cup C')]$.

This theorem is not valid.

Proof

There is a counter-example, as follows: Let $X = \{0, 1\}$ and $n = 3$. This seems to be the simplest n-dimensional case for $n > 2$. For convenience, we will write 0 for $\{0\}$, 1 for $\{1\}$, X for $\{0, 1\}$, and $\star$ for any of them. Define $[\,]$ as follows:

$[\star \cdot \star \cdot 0] := 0$

$[\star \cdot \star \cdot 1] := 1$

$[0 \cdot 0 \cdot X] := 0$

$[0 \cdot 1 \cdot X] := 0$

$[1 \cdot 0 \cdot X] := 1$

$[1 \cdot 1 \cdot X] := X$

$[0 \cdot X \cdot X] := 0$

$[1 \cdot X \cdot X] := 1$

$[X \cdot 0 \cdot X] := 0$

$[X \cdot 1 \cdot X] := X$

$[X \cdot X \cdot X] := X.$

The cases where $C = \{0\}$ and $C = \{1\}$ are forced by condition 1 of the theorem, and are not very interesting. As for the case $C = X$, one may check for each two triplets that fall under the conditions of the theorem that it holds. The check is simplified by the fact that Definition 1.8.18 gives no way to show that $(A, B, C) R(A', B', C)$ when $A \subseteq A'$ and $B \subseteq B'$, and the fact that there is no valid union of sets of sequences of different cardinalities, unless one is contained within the other. The operator complies with all the conditions of Suggested Theorem 1.8.24. But suppose there is a ranking that defines $[\,]$, we reach the following conclusions (using $x \preceq y$ for xRy, and $x \prec y$ for xRy but provably $\neg y R^* x$, and remembering that the conditions of the Suggested Theorem hold):

$(1 \cdot 1 \cdot X) \preceq (0 \cdot 1 \cdot X)$

by Definition 1.8.18, case 3,

$(1 \cdot 0 \cdot X) \prec (1 \cdot 1 \cdot X)$

by case 2 of Definition 1.8.18, and negation of condition 5 of the Suggested Theorem, and

$(0 \cdot 0 \cdot X) \prec (1 \cdot 0 \cdot X)$

by case 3 of the definition of R and negation of condition 4 of the Suggested Theorem.

Thus, the best history in $\{0\} \cdot \{0\} \cdot X$ (which we know to be $0 \cdot 0 \cdot 0$) would also be the single best history in $X \cdot X \cdot X$, but we find $[X \cdot X \cdot X] = X$ instead, which means there really is no underlying ranking.

□

The next natural step is to use a definition more like Definition 1.8.16 in the generalized theorem. If R was defined as follows:

Definition 1.8.19

We say the relation $R_{[\,]}$ holds iff any of the following cases obtains:

(1) $A \supseteq A', B \supseteq B', C \supseteq C' \Rightarrow (A,B,C)R(A',B',C')$
(2) $[A \cdot B \cdot (C \cup C')] \cap C \neq \emptyset \Rightarrow (A,B,C)R(A,B,C \cup C')$
(3) $[A \cdot (B \cup B') \cdot C] \neq [A \cdot B' \cdot C] \Rightarrow (A,B,C)R(A,B \cup B',C)$
(4) $[(A \cup A') \cdot B \cdot C] \neq [A' \cdot B \cdot C] \Rightarrow (A,B,C)R(A \cup A',B,C)$

Then, for the operator defined in the counter-example, we could show that
$(1,0,X)R(1,X,X)R(X,X,X)R(X,0,X)$
but
$[1 \cdot 0 \cdot X] \not\subseteq [X \cdot 0 \cdot X]$

so the operator violates condition 4, and is thus not a counter-example. We have not been able to prove a representation theorem using Definition 1.8.19, and believe it not to be valid. Yet, we have not found any counter-example. We have discovered (by computerized enumeration of the operators) that there are no counter-examples with $n = 3$ and $|X| = 2$, and the question whether it is valid for all n and $|X|$, and if not, for which n and $|X|$ it is valid, remains open at this stage.

An n-Dimensional Representation Theorem

Patches

We believe that for this theorem, we need a way to work around "illegal" sets of histories, i.e. such that are not sequences of sets of models. We call such sets "illegal" because, not being interchangeable with sequences of observations, our operator cannot be applied to them. For the technique used in this proof, it is especially regretful that for most sequences σ, if we try to present σ as a disjoint union of a "legal" sequence (say, a singleton) τ with the rest of the histories in the sequence $\sigma \setminus \tau$ then the latter will be illegal. Instead, we use a family of legal sequences whose union is $\sigma \setminus \tau$. As it complements τ to σ, we call this family a *patch*.

Definition 1.8.20

Given a sequence $A_1 \cdots A_n$ and a sequence $A'_1 \cdots A'_n$ such that $\forall i A'_i \subseteq A_i$, we define the *patch to* $A_1 \cdots A_n$ *from* $A'_1 \cdots A'_n$ to be the set of sequences comprised of all the sequences $B^i_1 \cdots B^i_n$ for each i such that $A'_i \neq A_i$, where

(1) $\forall j \neq i, B^i_j = A_j$

1.8 Various Results and Approaches

(2) $B_i^i = A_i \setminus A_i'$

Let, for instance, $n = 3$, $A := \{0,1\} \cdot \{0,1\} \cdot \{0,1\}$, $A' := \{0\} \cdot \{0,1\} \cdot \{0\}$, then the patch to A from A' is $\{B^1 := \{1\} \cdot \{0,1\} \cdot \{0,1\}, B^3 := \{0,1\} \cdot \{0,1\} \cdot \{1\}\}$. Note that the (disjoint) union of A' with $\bigcup_i B^i$ is equal to A.

As with the definition of $R_|$, this definition is made as tight as possible to make the theorem stronger. We will see later that if we relax the definition a little, allowing many patches from A' to A by letting B_i^i be any set such that $A_i' \cup B_i^i = A_i$, our theorem will become the generalization of the weaker form of Theorem 1.8.23 (the one using Definition 1.8.16). Notice also that Definition 1.8.20 may be easily written in the language of formulae; we do not wish our basic concepts, or the phrasing of the representation theorem, to stray too far from our original problem domain.

The Theorem

With the new tool, the patch, we may define a new, more powerful relation of "provable preferability".

Definition 1.8.21

Given an operation $[\ ]$, define a relation $R_{[\ ]}$ on sequences of length n of non-empty subsets of X by: $(A_1', \ldots, A_{n-1}', C') R_{[\ ]}(A_1, \ldots, A_{n-1}, C)$ iff one of the following cases obtains:

(1) $\forall i A_i \subseteq A_i'$ and $C \subseteq C'$.
(2) $\forall i A_i = A_i'$, $C' \subseteq C$ and $[A_1 \cdots A_{n-1} \cdot C] \cap C' \neq \emptyset$.
(3) $\forall i A_i \supseteq A_i'$, $C = C'$, $\{B_1^i \cdots B_{n-1}^i\}$ is the patch to $A_1 \cdots A_{n-1}$ from $A_1' \cdots A_{n-1}'$, and one of the following holds:

(3.1) $\bigcap_i [B_1^i \cdots B_{n-1}^i \cdot C] \not\subseteq [A_1 \cdots A_{n-1} \cdot C]$
(3.2) $[A_1 \cdots A_{n-1} \cdot C] \not\subseteq \bigcup_i [B_1^i \cdots B_{n-1}^i \cdot C]$

Now we are ready to phrase the representation theorem for the n-dimensional case:

Theorem 1.8.25

An operation $[\ ]$ is representable iff it satisfies the three conditions below for any non-empty sets $A_1, \ldots A_{n-1}, C \subseteq X$:

(1) $[A_1 \cdots A_{n-1} \cdot C] \subseteq C$,
(2) If $\forall i A_i' \subseteq A_i$ and $\{B_1^i \cdots B_{n-1}^i\}$ is the patch to $A_1 \cdots A_{n-1}$ from $A_1' \cdots A_{n-1}'$ then $[A_1 \cdots A_{n-1} \cdot C] \subseteq [A_1' \cdots A_{n-1}' \cdot C] \cup \bigcup_i [B_1^i \cdots B_{n-1}^i \cdot C]$

(3) If $\forall i A'_i \subseteq A_i, C' \subseteq C$ and $(A'_1 \cdots A'_{n-1} \cdot C')R^\star(A_1 \cdots A_{n-1} \cdot C)$, then $[A'_1 \cdots A'_{n-1} \cdot C'] \subseteq [A_1 \cdots A_{n-1} \cdot C]$.

Before proving this theorem, we note that when taking the more relaxed version (as mentioned above) of Definition 1.8.20, and taking Definition 1.8.16 to define $R_|$, Theorem 1.8.23 becomes the special case of Theorem 1.8.25 with $n = 2$: Under these definitions, for any sets A and A', A' (or rather, the set comprised of it as a single sequence of length 1) is a patch to $A \cup A'$ from A, so the concept of the patch coincides with union. When the patch has only one sequence (of one set) in it, then $\bigcap [B_1^1 \cdot C] = \bigcup [B_1^1 \cdot C] = [B_1^1 \cdot C]$, and so the cases (3.1) and (3.2) of Definition 1.8.21 together become case 3 of Definition 1.8.16. And finally, the conditions 3 and 4 of Theorem 1.8.23 are expressed together as condition 3 of Theorem 1.8.25. Note that condition 3 in Theorem 1.8.25 is essentially a loop condition. Let us now prove the more general theorem.

Proof For the proof of Theorem 1.8.25, a number of lemmas will be needed. These lemmas will be presented when needed, and their proof inserted in the midst of the main proof. First, we shall deal with the *soundness* of the theorem, and then with the more challenging *completeness*.

Suppose, then, that $[\,]$ is representable. The ranking r of histories may be extended to sequences of sets in the usual way, by taking the minimum over the sets: $r(A_1, \ldots, A_n) = min_{a_i \in A_i}\{r(a_1, \ldots, a_n)\}$. One may then write the equation defining representability as $[A_1 \cdots A_{n-1} \cdot C] = \{c \in C \mid r(A_1, \ldots, A_{n-1}, \{c\}) = r(A_1, \ldots, A_{n-1}, C)\}$ Let us now show that the three conditions of Theorem 1.8.25 hold:

Condition 1 is obvious. For Condition 2, notice that, on one hand, $A_1 \cdots A_{n-1} = \bigcup_i \{B_1^i \cdots B_{n-1}^i\} \cup (A'_1 \cdots A'_{n-1})$, and on the other hand, if $r(a_1, \ldots, a_{n-1}, c) = r(A_1, \ldots, A_{n-1}, C)$, $\langle a_1, \ldots, a_n, c \rangle \in A''_1 \cdots A''_{n-1} \cdot C''$ for some A''_i, C'', and $A''_1 \cdots A''_{n-1} \cdot C'' \subseteq A_1 \cdots A_{n-1} \cdot C$ then also $r(a_1, \ldots, a_{n-1}, c) = r(A''_1, \ldots, A''_{n-1}, C'')$ and $c \in [A''_1 \cdots A''_{n-1} \cdot C'']$. For Condition 3 we need a little lemma:

Lemma 1.8.26

For any non-empty sets $A_i, A'_i, C, C' \subseteq X$, $(A'_1, \ldots, A'_{n-1}, C')R(A_1, \ldots, A_{n-1}, C) \Rightarrow r(A'_1, \ldots, A'_{n-1}, C') \leq r(A_1, \ldots, A_{n-1}, C)$

Proof

Exercise, solution in the Appendix.

□

We conclude that Condition 3 holds and the characterization is sound.
For the completeness part, assume the operator $[\,]$ complies with the conditions of Theorem 1.8.25. Using the Generalized Abstract Nonsense Lemma 2.1 of [LMS01], see also Lemma 1.2.3. extend R to a total preorder S satisfying

1.8 Various Results and Approaches

(A) $xSy, ySx \Rightarrow xR^\star y$.

Let $\mathcal{Z}$ be the totally ordered set of equivalence classes of $\mathcal{P}(X)^n$ defined by the total pre-order S. Define a function $d\ \mathcal{P}(X)^n \to \mathcal{Z}$ to send a sequence of subsets $A_1, \ldots, A_n$ to its equivalence class under S. We shall define $r\ X^n \to \mathcal{Z}$ by $r(a_1, \ldots, a_n) := d(\{a_1\}, \ldots, \{a_n\})$. While we aim to prove that r represents [ ], we will have to use d for the proof. d has the following obvious properties:

(B) $(A_1, \ldots, A_{n-1}, C)R(A'_1, \ldots, A'_{n-1}, C') \Rightarrow d(A_1, \ldots, A_{n-1}, C) \leq d(A'_1, \ldots, A'_{n-1}, C')$,

and, from (A),

(C) $d(A_1, \ldots, A_{n-1}, C) = d(A'_1, \ldots, A'_{n-1}, C') \Rightarrow (A_1, \ldots, A_{n-1}, C)R^\star(A'_1, \ldots, A'_{n-1}, C')$.

d also has the less obvious property shown by the next lemma, which means it approximates representation as far as the last argument is concerned:

Lemma 1.8.27

For any $A_1, \ldots, A_{n-1}, C$, $d(A_1, \ldots, A_{n-1}, C) = min_{c \in C}\{d(A_1, \ldots, A_{n-1}, \{c\})\}$ and

(D) $[A_1 \cdots A_{n-1} \cdot C] = \{c \in C \mid d(A_1, \ldots, A_{n-1}, \{c\}) = d(A_1, \ldots, A_{n-1}, C)\}$

Proof

Suppose $c \in C$, then $(A_1, \ldots, A_{n-1}, C)R(A_1, \ldots, A_{n-1}, \{c\})$ and from (B) we get $d(A_1, \ldots, A_{n-1}, C) \leq min_{c \in C}\{d(A_1, \ldots, A_{n-1}, c)\}$. If moreover $c \in [A_1 \cdots A_{n-1} \cdot C]$, then $[A_1 \cdots A_{n-1} \cdot C] \cap \{c\} \neq \emptyset$, and by Definition 1.8.21, part 2, $(A_1, \ldots, A_{n-1}, \{c\})R(A_1, \ldots, A_{n-1}, C)$ and therefore $d(A_1, \ldots, A_{n-1}, c) = d(A_1, \ldots, A_{n-1}, C)$. We have shown that the left hand side of (D) is included in the right hand side.
Since $[A_1 \cdots A_{n-1} \cdot C]$ is not empty, $\exists c \in [A_1 \cdots A_{n-1} \cdot C]$ and, by the previous remark, $d(A_1, \ldots, A_{n-1}, C) = d(A_1, \ldots, A_{n-1}, c)$ and therefore we conclude that $d(A_1, \ldots, A_{n-1}, C) = min_{c \in C}\{d(A_1, \ldots, A_{n-1}, c)\}$.
To see the converse inclusion, notice that $d(A_1, \ldots, A_{n-1}, C) = d(A_1, \ldots, A_{n-1}, c)$ implies $(A_1, \ldots, A_{n-1}, c)R^\star(A_1, \ldots, A_{n-1}, C)$ and, by Property 3 of Theorem 1.8.25, $[A_1 \cdots A_{n-1} \cdot \{c\}] \subseteq [A_1 \cdots A_{n-1} \cdot C]$, so $c \in [A_1 \cdots A_{n-1} \cdot C]$ by Property 1 of the theorem.
□

We now have to show that

(E) $[A_1 \cdots A_{n-1} \cdot C] = \{c \in C \mid \exists a_1, \ldots, a_{n-1}, \text{ s.t. } \forall a'_1, \ldots, a'_{n-1}, c' \in C, r(a_1, \ldots, a_{n-1}, c) \leq r(a'_1, \ldots, a'_{n-1}, c')\}$ (where $\forall i\ a_i, a'_i \in A_i$).

To see that the right hand side is a subset of the left hand side, assume that $a_i \in A_i, c \in C$ are such that for all $a'_i \in A_i, c' \in C$, $r(a_1, \ldots, a_{n-1}, c) \leq r(a'_1, \ldots, a'_{n-1}, c')$. We have to show that

$c \in [A_1 \cdots A_{n-1} \cdot C]$. We will show by induction on the size of $A'_1 \cdots A'_{n-1}$, where $a_i \in A'_i \subseteq A_i$ for all i, that $c \in [A'_1 \cdots A'_{n-1} \cdot C]$.

For the base of the induction, if $A'_1 \cdots A'_{n-1}$ is a singleton, then $A'_i = \{a_i\}$ and by Lemma 1.8.27, remembering that in this case r coincides with d, we find $c \in [A'_1 \cdots A'_{n-1} \cdot C]$.

Otherwise, i.e. if not all A'_i are singletons, we may choose for each i an $a'_i \in A'_i$ such that $a'_i \neq a_i$ if $A'_i \neq \{a_i\}$. Since $A'_1 \cdots A'_{n-1}$ is not a singleton, at least one of the set inequalities holds and $\langle a'_1, \ldots, a'_{n-1}\rangle \neq \langle a_1, \ldots, a_{n-1}\rangle$.

Let $\{B^i_1 \cdots B^i_{n-1}\}$ be the patch to $A'_1 \cdots A'_{n-1}$ from $\{a'_1\} \cdots \{a'_{n-1}\}$. For all i we have, by definition of the patch, $|B^i_1 \cdots B^i_{n-1}| < |A'_1 \cdots A'_{n-1}|$. By choice of a'_i, $\langle a_1, \ldots, a_{n-1}\rangle \in B^i_1 \cdots B^i_{n-1}$ for all i and hence by the induction hypothesis, $c \in [B^i_1 \cdots B^i_{n-1} \cdot C]$. Assume now that $c \notin [A'_1 \cdots A'_{n-1} \cdot C]$, then by Property (3.1) of Definition 1.8.21, $(a'_1, \ldots, a'_{n-1}, C) R(A'_1, \ldots, A'_{n-1}, C)$. There is some $c' \in [a'_1 \cdots a'_{n-1} \cdot C]$ and by Definition 1.8.21 part 2, $(a'_1, \ldots, a'_{n-1}, c') R(a'_1, \ldots, a'_{n-1}, C)$. Also, by Definition 1.8.21 part 1, $(A'_1, \ldots, A'_{n-1}, C) R(a_1, \ldots, a_{n-1}, c)$. We have established that $(a'_1, \ldots, a'_{n-1}, c') R^\star(a_1, \ldots, a_{n-1}, c)$ so by our original assumption we have $d(a'_1, \ldots, a'_{n-1}, c') = d(a_1, \ldots, a_{n-1}, c)$. By (C) this implies also $(a_1, \ldots, a_{n-1}, c) R^\star(a'_1, \ldots, a'_{n-1}, c')$ and hence $(a_1, \ldots, a_{n-1}, c) R^\star(A'_1, \ldots, A'_{n-1}, C)$. But then by Conditions 3 and 1 of Theorem 1.8.25, we get $c \in [A'_1 \cdots A'_{n-1} \cdot C]$ which is a contradiction. We conclude that the right hand side of (E) is a subset of the left hand side.

For the converse assume that $c \in C$ is such that for all $a_1, \ldots, a_{n-1}$ ($a_i \in A_i$), there exist $a'_1, \ldots, a'_{n-1}$ and $c' \in C$ such that $c \neq c'$ and $r(a_1, \ldots, a_{n-1}, c) \not\leq r(a'_1, \ldots, a'_{n-1}, c')$. We need to prove that $c \notin [A_1 \cdots A_{n-1} \cdot C]$. Since X is finite, we may change the order of quantifiers in the assumption to $\exists a'_1, \ldots, a'_{n-1}, c' \in C, \forall a_1, \ldots, a_{n-1}, r(a_1, \ldots, a_{n-1}, c) \not\leq r(a'_1, \ldots, a'_{n-1}, c')$ Note that $r(a_1, \ldots, a_{n-1}, c) \not\leq r(a'_1, \ldots, a'_{n-1}, c') \Rightarrow \neg(a_1, \ldots, a_{n-1}, c) R^\star(a'_1, \ldots, a'_{n-1}, c')$. As above, but switching the roles of a_i with those of a'_i, let $A'_1, \ldots, A'_{n-1}$ be such that $a'_i \in A'_i \subseteq A_i$, and prove by induction on $|A'_1 \cdots A'_{n-1}|$, i.e. the cardinality of $A'_1 \cdots A'_{n-1}$, that $c \notin [A'_1 \cdots A'_{n-1} \cdot C]$. If $A'_1 \cdots A'_{n-1}$ is a singleton, then $A'_i = \{a'_i\}$, and by Lemma 1.8.27 $c \notin [A'_1 \cdots A'_{n-1} \cdot C]$. Otherwise choose for each i an $a_i \in A'_i$ with $a_i \neq a'_i$ if possible, and let $\{B^i_1 \cdots B^i_{n-1}\}$ be the patch to $A'_1 \cdots A'_{n-1}$ from $\{a_1\} \cdots \{a_{n-1}\}$. Again we have for all i, $\langle a'_1, \ldots, a'_{n-1}\rangle \in B^i_1 \cdots B^i_{n-1}$ and $|B^i_1 \cdots B^i_{n-1}| < |A'_1 \cdots A'_{n-1}|$ and hence by induction $c \notin \bigcup_i [B^i_1 \cdots B^i_{n-1} \cdot C]$. Now if $c \in [A'_1 \cdots A'_{n-1} \cdot C]$ we get two consequences. First, by Definition 1.8.21, part (3.2), $(a_1, \ldots, a_{n-1}, C) R(A'_1, \ldots, A'_{n-1}, C)$. By the same definition, part 1, we have $(A'_1, \ldots, A'_{n-1}, C) R(a'_1, \ldots, a'_{n-1}, c')$. Second, from condition 2 of Theorem 1.8.25, we derive $c \in [a_1 \cdots a_{n-1} \cdot C]$ so that $(a_1, \ldots, a_{n-1}, c) R(a_1, \ldots, a_{n-1}, C)$. The above together give us $(a_1, \ldots, a_{n-1}, c) R^\star(a'_1, \ldots, a'_{n-1}, c')$ which contradicts our assumption. We have thus shown the converse inclusion.

□

1.8.7 Orderings on $\mathcal{L}$ and Completeness Results

1.8.7.1 Introduction

For rigorous completeness proofs the reader is referred to [KLM90] and [LM92]. In those proofs, several orderings between formulas are used, which arise more or less naturally from the logic $\mathrel|\!\sim$. We now slightly modify one such order (defined in [LM92]), and examine its properties and connections to other orderings in detail. For convenience, we use the above soundness and completeness results. We then reconstruct the completeness proofs, emphasizing the main ideas and the central role of the orderings. This gives us also an opportunity to incorporate results by Gardenfors and Makinson [GM94], who show that, given a partial order $\leq$ on formulas, satisfying EE1-EE3 of the epistemic entrenchment axioms (see Chapter 5), there is a natural way to define a rational entailment relation $\mathrel|\!\sim_{\leq}$, and vice versa - we will come to this later in a short summary. Gardenfors and Makinson also give a completeness proof directly from $\leq$, which we will present in outline along with the other completeness results.

In this section, taken from [Sch97-2], we first motivate, define and examine a natural ordering constructible from a logic (see Section 1.8.7.2). We then proceed to relate our ordering to others to be found in the literature (see Section 1.8.7.3, Section 1.8.7.4). In Section 1.8.7.4, we also take the opportunity to note some marginalia to the results of [GM94]. In Section 1.8.7.5, we reconstruct the completeness proofs of [KLM90], [LM92], [GM94], emphasizing the role of such orderings as above, focussing on intuition and the main ideas. All three proofs proceed by showing the existence of a canonical model. Again, the results presented in [GM94] are slightly augmented. The reader will notice that the completeness proof for R via the [GM94]-technique makes extensive use of rational montony, which is needed in the [LM92]-approach only at two points. We finish in Section 1.8.7.6 by relating our ordering to the rank as defined in [LM92].

So let us define and discuss the promised order.

1.8.7.2 A Natural Ordering

We discuss a relation almost identical to (the inverse of) $<_{LM}$, defined at the beginning of Section 5.3 in [LM92]. So all credit should go to [LM92]. The systematic investigation as done in Fact 1.8.28 is, however, our own - though several results summarized there can be found explicitely or implicitely in the literature.

If we read $\alpha \mathrel|\!\sim \beta$ as "most α's are β's", we can compare sizes: $\alpha \mathrel|\!\sim \beta$ says that α and $\alpha \wedge \beta$ have approximately the same size, and $\alpha \wedge \neg\beta$ is much smaller - provided α is consistent : if $\alpha \mathrel|\!\sim \bot$ ($= false$), then any subset of α has identical size 0.

So we define

$\alpha > \alpha \wedge \beta :\leftrightarrow \alpha \mathrel{|\!\sim} \neg\beta$ and $\alpha \mathrel{|\!\not\sim} \beta \leftrightarrow \alpha \mathrel{|\!\sim} \neg\beta$ and $\alpha \mathrel{|\!\not\sim} \bot$

(We use a little logic in the last equivalence, of course, but this will hold in all systems that interest us.)

Next, we would like to compare any two sets (formulas), not only those where one is contained in the other. If we use the same approach, there are two extreme possibilities:

- $\alpha > \beta :\leftrightarrow true \mathrel{|\!\not\sim} \neg\alpha$, but $true \mathrel{|\!\sim} \neg\beta$
- $\alpha > \beta :\leftrightarrow \alpha \vee \beta \mathrel{|\!\sim} \neg\beta$

The first extreme should be rejected for the following reason: "From the point of view of God, dwarfs and giants are alike" - the measure is too coarse. So, the most cautious approach, taking the least set containing both, i.e. $\alpha \vee \beta$, gives the finest discrimination.

We make this official now:

Definition 1.8.22

$\alpha > \beta :\leftrightarrow \alpha \vee \beta \mathrel{|\!\sim} \neg\beta$ and $\alpha \vee \beta \mathrel{|\!\not\sim} \bot$

$\alpha \sim \beta :\leftrightarrow \alpha \not< \beta \wedge \beta \not< \alpha \leftrightarrow (\alpha \vee \beta \mathrel{|\!\sim} \bot)$ or $(\alpha \vee \beta \mathrel{|\!\not\sim} \neg\alpha, \alpha \vee \beta \mathrel{|\!\not\sim} \neg\beta)$

$\alpha \geq \beta :\leftrightarrow \alpha > \beta$ or $\alpha \sim \beta \leftrightarrow \alpha \not< \beta$

To show the last equivalence: "$\to$": $\alpha > \beta \to \alpha \not< \beta$ (by Fact 1.8.28, 2) c) below), $\alpha \sim \beta \to \alpha \not< \beta$ "$\leftarrow$": $\alpha \not< \beta \to (\alpha \not< \beta \wedge \beta \not< \alpha)$ or $(\alpha \not< \beta \wedge \beta < \alpha)$, the first is equivalent to $\alpha \sim \beta$, and from the second, conclude $\beta < \alpha$.

To make our motivation a little more formal, we can argue $\alpha \vee \beta \mathrel{|\!\sim} \neg\beta \leftrightarrow \alpha \vee \beta \mathrel{|\!\sim} \alpha \wedge \neg\beta \leftrightarrow \pi(\alpha \vee \beta) \approx \pi(\alpha) \approx \pi(\alpha - \beta) \leftrightarrow \pi(\alpha) >> \pi(\beta)$ (π a measure)

The systems P and R were defined in Definition 1.2.10.

Note the following facts:

Fact 1.8.28

In P holds:

1)

    a) $\alpha \vee \beta \mathrel{|\!\sim} \neg\beta \leftrightarrow \alpha \vee \beta \mathrel{|\!\sim} \alpha \wedge \neg\beta$,

    b) $\alpha > \beta \to m(\alpha) \neq \emptyset$,

2)

    a) $\alpha \not< \alpha$,

    b) $\alpha < \beta < \gamma \to \alpha < \gamma$,

1.8 Various Results and Approaches

c) $\alpha < \beta \rightarrow \beta \not< \alpha$,
d) $\alpha < \beta \rightarrow \alpha \wedge \gamma < \beta$,
e) $\alpha < \beta \rightarrow \alpha < \beta \vee \gamma$,
f) $\alpha < \beta \wedge \gamma \mathrel{\vert\!\sim} \alpha \rightarrow \gamma < \beta$,
g) $\alpha < \beta \wedge \beta \mathrel{\vert\!\sim} \gamma \rightarrow \alpha < \gamma$,

3)

a) $\alpha < \beta \rightarrow \alpha \leq \beta$,
b) $\alpha < \beta \ \beta < \alpha \ \alpha \sim \beta$,
c) $\alpha \leq \beta \ \beta \leq \alpha$,

4)

a) $\alpha \sim \alpha$,
b) $\alpha \leq \alpha$,
c) $\alpha \leq \beta \rightarrow \alpha \wedge \gamma \leq \beta$,
d) $\alpha \leq \beta \rightarrow \alpha \leq \beta \vee \gamma$,
e) $\alpha \leq \beta \wedge \gamma \mathrel{\vert\!\sim} \alpha \rightarrow \gamma \leq \beta$,
f) $\alpha \leq \beta \wedge \beta \mathrel{\vert\!\sim} \gamma \rightarrow \alpha \leq \gamma$,
g) $\bot \leq \phi \leq true$ for any ϕ,
h) $true \mathrel{\vert\!\not\sim} \bot \rightarrow \bot < true$,

5)

a) $\alpha < \beta \wedge \gamma < \beta \rightarrow \alpha \vee \gamma < \beta$,
b) $\alpha < \beta \wedge \gamma < \beta \rightarrow \alpha < \beta \wedge \neg\gamma$,

6)

a) $\alpha \sim \beta \rightarrow \alpha \vee \beta \sim \alpha$,
b) $\alpha < \beta \rightarrow \beta \wedge \neg\alpha \sim \beta$.

Not in P, but in R holds:

7)

a) $\alpha \sim \beta \sim \gamma \rightarrow \alpha \sim \gamma$,
b) $\alpha < \beta \sim \gamma \rightarrow \alpha < \gamma$,
c) $\alpha \sim \beta < \gamma \rightarrow \alpha < \gamma$,

8)

a) $\alpha < \beta \leq \gamma \to \alpha < \gamma$,

b) $\alpha \leq \beta < \gamma \to \alpha < \gamma$,

c) $\alpha \leq \beta \leq \gamma \to \alpha \leq \gamma$,

9)

a) $\gamma \sim \alpha < \beta \to \alpha \vee \gamma < \beta$,

b) $\gamma \sim \alpha < \beta \to \alpha < \beta \wedge \neg \gamma$,

10) $<$ is modular.

Proof

We will almost always argue semantically, building on the completeness and correctness proofs in [KLM90] and [LM92].

1)

a) $\alpha \vee \beta \mathrel{|\!\sim} \neg \beta \leftrightarrow \alpha \vee \beta \mathrel{|\!\sim} \alpha \wedge \neg \beta$

"$\leftarrow$": trivial

"$\to$": Case 1: $\alpha \vee \beta \mathrel{|\!\sim} \bot$: trivial. Case 2: $\alpha \vee \beta \mathrel{|\!\not\sim} \bot$: Let $m \in \mu(\alpha \vee \beta) \subseteq m(\alpha \vee \beta) \to m \models \neg \beta \to m \models \alpha$.

b) $\alpha > \beta \to m(\alpha) \neq \emptyset$ $\alpha > \beta \to$ (by 1) a) $\alpha \vee \beta \mathrel{|\!\sim} \alpha \wedge \neg \beta \to$ (by $\alpha \vee \beta \mathrel{|\!\not\sim} \bot$) $m(\alpha) \neq \emptyset$.

2)

Exercise, solution in the Appendix.

3)

a) $\alpha < \beta \to \alpha \leq \beta$ By Definition.

b) $\alpha < \beta$ $\beta < \alpha$ $\alpha \sim \beta$ By Definition.

c) $\alpha \leq \beta$ $\beta \leq \alpha$ By Definition.

4)

a) $\alpha \sim \alpha$ By 2) a).

b) $\alpha \leq \alpha$ By 2) a).

c) $\alpha \leq \beta \to \alpha \wedge \gamma \leq \beta$ Equivalently, $\beta \not< \alpha \to \beta \not< \alpha \wedge \gamma$, or $\beta < \alpha \wedge \gamma \to \beta < \alpha$. But $\beta < \alpha \wedge \gamma \to$ (by 2) e) $\beta < (\alpha \wedge \gamma) \vee \alpha = \alpha$

1.8 Various Results and Approaches

d) $\alpha \leq \beta \to \alpha \leq \beta \vee \gamma$ Equivalently, $\beta \not< \alpha \to \beta \vee \gamma \not< \alpha$, or $\beta \vee \gamma < \alpha \to \beta < \alpha$. But $\beta \vee \gamma < \alpha \to$ (by 2) d) $\beta = (\beta \vee \gamma) \wedge \beta < \alpha$

e) $\alpha \leq \beta \wedge \gamma \mathrel{|\!\sim} \alpha \to \gamma \leq \beta$ Equivalently, $\gamma \mathrel{|\!\sim} \alpha \to (\alpha \leq \beta \to \gamma \leq \beta)$, or $\gamma \mathrel{|\!\sim} \alpha \to (\beta < \gamma \to \beta < \alpha)$, which holds by 2) g).

f) $\alpha \leq \beta \wedge \beta \mathrel{|\!\sim} \gamma \to \alpha \leq \gamma$ Equivalently, $\beta \mathrel{|\!\sim} \gamma \to (\gamma < \alpha \to \beta < \alpha)$, which holds by 2) f).

g) $\bot \leq \phi \leq true$ for any ϕ Suppose $\phi < \bot$, then $\phi = \phi \vee \bot \mathrel{|\!\sim} \neg \phi$, so $\phi \vee \bot \mathrel{|\!\sim} \bot$, contradiction. $\phi \leq true$: Suppose $true < \phi$, then $true \vee \phi \mathrel{|\!\sim} \neg true = \bot$, contradiction.

h) $true \mathrel{|\!\not\sim} \bot \to \bot < true$ $true \mathrel{|\!\not\sim} \bot \to true \vee \bot \mathrel{|\!\not\sim} \bot$. But $true \vee \bot \sim | true = \neg \bot$.

5)

Exercise, solution in the Appendix.

6)

a) $\alpha \sim \beta \to \alpha \vee \beta \sim \alpha$ Case 1 (of $\alpha \sim \beta$): $\alpha \vee \beta \mathrel{|\!\sim} \bot$: trivial Case 2: $\alpha \vee \beta \mathrel{|\!\not\sim} \neg \alpha, \neg \beta$ and $\alpha \vee \beta \mathrel{|\!\not\sim} \bot$: Then $\alpha \vee \beta \mathrel{|\!\not\sim} \neg \alpha$, and $\alpha \vee \beta \mathrel{|\!\not\sim} \neg(\alpha \vee \beta)$ by $\alpha \vee \beta \mathrel{|\!\not\sim} \bot$.

b) $\alpha < \beta \to \beta \wedge \neg \alpha \sim \beta$ By 4b and 4c $\beta \wedge \neg \alpha \leq \beta$. We show $\beta \wedge \neg \alpha \not< \beta$ by proving $\beta = \beta \vee (\beta \wedge \neg \alpha) \mathrel{|\!\not\sim} \neg(\beta \wedge \neg \alpha)$. Suppose $\beta \mathrel{|\!\sim} \neg \beta \vee \alpha$. As $\alpha < \beta$, $m(\beta) \neq \emptyset$. Let $m \in \mu(\beta)$. So $m \models \neg \beta$ or $m \models \alpha$. $m \models \neg \beta$ is impossible. By $\alpha < \beta$ there is $m' \prec m$, $m' \models \beta \wedge \neg \alpha$, contradiction.

7)

Exercise, solution in the Appendix.

8)

a) $\alpha < \beta \leq \gamma \to \alpha < \gamma$ The counterexample of 7) b) shows it does not hold for P. In R: If $\beta < \gamma$, it is 2) b), if $\beta \sim \gamma$, it is 7) b).

b) $\alpha \leq \beta < \gamma \to \alpha < \gamma$ The counterexample for P is in 7) c). In R: If $\alpha < \beta$: 2) b), if $\alpha \sim \beta$: 7) c).

c) $\alpha \leq \beta \leq \gamma \to \alpha \leq \gamma$ The counterexample for P:

Consider $m_0 \models \neg \alpha, \beta, \gamma$, $m_1 \models \alpha, \neg \beta, \neg \gamma$, $m_2 \models \alpha, \beta, \neg \gamma$ with $m_1 \prec m_0$.

Then $\alpha \vee \beta \mathrel{|\!\not\sim} \neg \beta$, $\gamma \vee \beta \mathrel{|\!\not\sim} \neg \gamma$, $\alpha \vee \gamma \mathrel{|\!\sim} \neg \gamma$, $\alpha \vee \gamma \mathrel{|\!\not\sim} \bot$. So $\beta \not< \alpha$, $\gamma \not< \beta$, $\gamma < \alpha$, and thus $\alpha \leq \beta \leq \gamma$, but $\alpha \not\leq \gamma$.

In R: The cases follow from 2) b), 7) a)-c).

9)

a) $\gamma \sim \alpha < \beta \rightarrow \alpha \vee \gamma < \beta$ The counterexample for P is in 7) c): There $\alpha \sim \beta < \gamma$, but $\alpha \vee \beta \vee \gamma \not\mathrel{|\!\sim} \neg(\alpha \vee \beta) = \neg\alpha \wedge \neg\beta$, so $\alpha \vee \beta \not< \gamma$. In R: $\alpha \sim \gamma \rightarrow$ (by 6) a) $\alpha \vee \gamma \sim \alpha$. $\alpha < \beta$, $\alpha \vee \gamma \sim \alpha \rightarrow$ (by 7) c) $\alpha \vee \gamma < \beta$.

b) $\gamma \sim \alpha < \beta \rightarrow \alpha < \beta \wedge \neg\gamma$ A Counterexample for P:

$m_0 \models \alpha, \beta, \gamma$, $m_1 \models \neg\alpha, \beta, \gamma$, $m_2 \models \alpha, \beta, \gamma$, $m_3 \models \neg\alpha, \beta, \neg\gamma$ with $m_1 \prec m_0, m_3 \prec m_2 \ldots$ Then $\alpha \vee \beta \not\mathrel{|\!\sim} \bot$, $\alpha \vee \beta \mathrel{|\!\sim} \neg\alpha$, $\alpha \vee \gamma \not\mathrel{|\!\sim} \neg\alpha, \neg\gamma$. But $\alpha \vee (\beta \wedge \neg\gamma) \not\mathrel{|\!\sim} \neg\alpha$.

In R: $\alpha < \beta$, $\alpha \sim \gamma \rightarrow \gamma < \beta$ by 7) c). By 6) b) then $\beta \wedge \neg\gamma \sim \beta$, so $\alpha < \beta \wedge \neg\gamma$ by 7) b).

10) $<$ is modular $<$ is not necessarily modular in P:

Consider $m_0 \models \alpha, \beta, \gamma$, $m_1 \models \alpha, \neg\beta, \neg\gamma$, $m_3 \models \alpha, \neg\beta, \gamma$ with $m_1 \prec m_0$.

Then $\beta < \alpha$: $\alpha \vee \beta \mathrel{|\!\sim} \neg\beta$. But neither $\beta < \gamma$ nor $\gamma < \alpha$: $\beta \vee \gamma \not\mathrel{|\!\sim} \neg\beta$, $\gamma \vee \alpha \not\mathrel{|\!\sim} \neg\gamma$.

We now work in R: Let $\beta < \alpha$, $\beta \not< \gamma$, we show $\gamma < \alpha$. Case 1: $\gamma < \beta$, then $\gamma < \alpha$ by 2) b), Case 2: $\gamma \not< \beta$, so $\beta \sim \gamma$. So $\gamma < \alpha$ by 7) c).

□

Next, we compare our order with other relations in the literature: $\leq$ in [KLM90], $<$ and R in [LM92], $\leq$ in [GM94].

1.8.7.3 Comparison to Orders in [KLM90] and [LM92]

Definition 1.8.23

([KLM90]) $\alpha \leq_{KLM} \beta :\leftrightarrow \alpha \vee \beta \mathrel{|\!\sim} \alpha$

Fact 1.8.29

(in P)

a) $\alpha \vee \beta \not\mathrel{|\!\sim} \bot \rightarrow (\alpha \leq_{KLM} \beta \leftrightarrow \alpha > \beta \wedge \neg\alpha)$

$\alpha \vee \beta \mathrel{|\!\sim} \bot \rightarrow \alpha \leq_{KLM} \beta$, $\beta \leq_{KLM} \alpha$, $\beta \wedge \neg\alpha \not< \alpha$

b) conversely, $\alpha \leq_{KLM} \beta \leq_{KLM} \alpha \rightarrow \alpha \mathrel{|\!\sim} \beta$, $\beta \mathrel{|\!\sim} \alpha$

c) $\beta \mathrel{|\!\sim} \alpha \rightarrow \alpha \leq_{KLM} \beta$, but not vice versa

d) $\alpha \leq_{KLM} \beta \rightarrow \overline{\overline{\alpha}} = \overline{\alpha \vee \beta}$

1.8 Various Results and Approaches

Proof

a) $\vdash \alpha \vee \beta \leftrightarrow \alpha \vee (\beta \wedge \neg \alpha)$. Moreover, $\alpha \vee \beta \mathrel{\mid\!\sim} \alpha \leftrightarrow \alpha \vee \beta \mathrel{\mid\!\sim} \neg \beta \vee \alpha$: "$\to$" is trivial. "$\leftarrow$": Let $m \in \mu(\alpha \vee \beta)$, $m \models \neg \beta \vee \alpha$. As $m \in m(\alpha \vee \beta)$, $m \models \alpha$. So $\alpha \vee \beta \mathrel{\mid\!\sim} \alpha \leftrightarrow \alpha \vee (\beta \wedge \neg \alpha) \mathrel{\mid\!\sim} \neg(\beta \wedge \neg \alpha)$. $\alpha \vee \beta \mathrel{\mid\!\not\sim} \bot \to (\alpha \leq_{KLM} \beta \leftrightarrow \alpha > \beta \wedge \neg \alpha)$: "$\to$": 1. $\alpha \vee \beta \mathrel{\mid\!\not\sim} \bot \to \alpha \vee (\beta \wedge \neg \alpha) \mathrel{\mid\!\not\sim} \bot$ 2. $\alpha \vee \beta \mathrel{\mid\!\sim} \alpha \to \alpha \vee (\beta \wedge \neg \alpha) \mathrel{\mid\!\sim} \neg(\beta \wedge \neg \alpha)$ "$\leftarrow$" is trivial by the above. $\alpha \vee \beta \mathrel{\mid\!\sim} \bot \to \alpha \leq_{KLM} \beta, \beta \leq_{KLM} \alpha, \beta \wedge \neg \alpha \not< \alpha$: $\alpha \vee \beta \mathrel{\mid\!\sim} \bot \to \alpha \vee \beta \mathrel{\mid\!\sim} \alpha, \beta$. Moreover, by $\alpha \vee \beta \mathrel{\mid\!\sim} \bot$, $\alpha \vee (\beta \wedge \neg \alpha) \mathrel{\mid\!\sim} \bot$, so $\beta \wedge \neg \alpha \not< \alpha$.

b) $\alpha \vee \beta \mathrel{\mid\!\sim} \alpha, \beta \to \alpha \mathrel{\mid\!\sim} \beta, \beta \mathrel{\mid\!\sim} \alpha$: $\alpha \mathrel{\mid\!\sim} \beta$: Let $m \in \mu(\alpha)$. If $m \notin \mu(\alpha \vee \beta)$, then there is $m' \prec m$, $m' \models \alpha, \beta$, contradiction. So $m \models \beta$. Likewise for $\beta \mathrel{\mid\!\sim} \alpha$.

c) "$\to$" is trivial, the following structure $\mathcal{M}$ shows the failure of the converse:

$m \models \beta, \neg \alpha$, $m' \models \alpha, \neg \beta$, $m' \prec m$.

Then $\alpha \vee \beta \mathrel{\mid\!\sim}_{\mathcal{M}} \alpha$, but $\beta \mathrel{\mid\!\not\sim}_{\mathcal{M}} \alpha$.

d) trivial, by cumulativity.

$\square$

Definition 1.8.24

([LM92])

$\alpha <_{LM} \beta :\leftrightarrow \alpha \vee \beta \mathrel{\mid\!\sim} \neg \beta$

$\alpha \leq_{LM} \beta :\leftrightarrow \beta \not<_{LM} \alpha$

Fact 1.8.30

(in P)

a) $\alpha \vee \beta \mathrel{\mid\!\not\sim} \bot \to (\alpha <_{LM} \beta \leftrightarrow \alpha > \beta)$
   $\alpha \vee \beta \mathrel{\mid\!\sim} \bot \to \alpha <_{LM} \beta, \beta <_{LM} \alpha, \beta \not< \alpha, \alpha \not< \beta$

b) conversely, $\alpha <_{LM} \beta <_{LM} \alpha \to \alpha \vee \beta \mathrel{\mid\!\sim} \bot$

c) $\alpha \leq_{LM} \beta \to \alpha \not< \beta \leftrightarrow \alpha \geq \beta$

d) $\alpha <_{LM} \beta \to \alpha \leq_{KLM} \beta$

e) $\alpha \mathrel{\mid\!\sim} \beta \to \alpha <_{LM} \alpha \wedge \neg \beta$

Proof

Exercise, solution in the Appendix.

Remark 1.8.31

Set $\alpha <_{LM'} \beta :\leftrightarrow \alpha \leq_{LM} \beta$, then:

For $\alpha, \beta \mathrel{\mid\!\not\sim} \bot$, we have $\overline{\alpha} <_{LM'} \overline{\beta} \leftrightarrow \alpha <_{LM} \beta \leftrightarrow \beta < \alpha$

Proof

$\overline{\alpha} <_{LM'} \overline{\beta} :\leftrightarrow \overline{\alpha} \leq_{LM'} \overline{\beta} \wedge \alpha \not\sim_{LM'} \beta \leftrightarrow \alpha <_{LM'} \beta \wedge \neg(\beta <_{LM'} \alpha) \leftrightarrow \alpha \vee \beta \not\sim \neg\alpha \wedge \alpha \vee \beta \sim \neg\beta \leftrightarrow \alpha \vee \beta \not\sim \bot \wedge \alpha \vee \beta \sim \neg\beta \leftrightarrow \alpha \vee \beta \sim \neg\beta$ (by $\alpha, \beta \in S$).

$\square$

1.8.7.4 The Results of [GM94]

Some of the results in [GM94] relating orderings and logics can be described - very briefly - and connected to our definition as follows:

[GM94] work with a slight generalization $\vdash'$ of $\vdash$, s.th.

1. $\vdash \subseteq \vdash'$ (as a relation)
2. $\vdash'$ is compact
3. $\overline{A+x}' \cap \overline{A+y}' \subseteq \overline{A+(x \vee y)}'$, (where $\overline{T}' := \{\phi : T \vdash' \phi\}$).
4. $\vdash'$ is a Tarski consequence relation, i.e., $A \subseteq \overline{A}'$, $\overline{(\overline{A}')}' = \overline{A}'$, $A \subseteq B \rightarrow \overline{A}' \subseteq \overline{B}'$.
5. We call $\sim$ consistency preserving wrt. $\vdash'$ iff $A \sim \bot \rightarrow A \vdash' \bot$.

The logic $\vdash'$ is connected to preferential models as follows: Any structure $\mathcal{M}$ gives rise to such a logic $\vdash'$ by "forgetting about $\prec$": $A \vdash' x :\leftrightarrow \forall m \in M_\mathcal{M}(A).m \models x$ (instead of $\mu_\mathcal{M}(A)$ for $\sim$). $M_\mathcal{M}(A)$ is here the set of models of A in the structure $\mathcal{M}$. To simplify the picture, we largely ignore $\vdash'$ - we will not miss any basic ideas and just assume that $\vdash'=\vdash$, i.e. essentially that all models in $M_\mathcal{L}$ occur in the structure.

Fact 1.8.32

1) ([GM94]) Given $\vdash$ and $\sim$, $\sim$ satisfying R, and consistency preserving wrt. $\vdash$, the definition

   $(*)$ $x \leq_{GM_\sim} y :\leftrightarrow (true \vdash x \wedge y$ or $\neg(x \wedge y) \not\sim x)$

   gives an order $\leq_{GM_\sim}$ on $\mathcal{L}$ satisfies the postulates EE1-EE3 of an epistemic entrenchment relation (defined in Chapter 5). ($\alpha <_{GM} \beta :\leftrightarrow (\alpha \leq_{GM} \beta$ and $\beta \not\leq_{GM} \alpha) \leftrightarrow \beta \not\leq_{GM} \alpha$)

2) ([GM94]) Conversely, given such $\leq_{GM}$ satisfying EE1-EE3, $\sim_{\leq_{GM}}$ defined by

   $(**)$ $a \sim_{\leq_{GM}} x :\leftrightarrow (a \vdash x$ or $\exists b(a \wedge b \vdash x$ and $\neg a <_{GM} b))$

   will satisfy R, and be consistency preserving.

1.8 Various Results and Approaches

Remark 1.8.33

(This Remark is our own.)

a) From $(*)$, we conclude: $x \leq_{GM_\sim} y \leftrightarrow \neg y \leq \neg x$ and $\alpha <_{GM_\sim} \beta \leftrightarrow \neg \beta < \neg \alpha$

b) $\leq_{GM_\sim}$ satisfies also EE5.

c) Note that changing $\leq_{GM}$ by lifting true to become the only top element (up to equivalence by $\vdash$) will not change the logic defined by $(**)$. So we can assume wlog. that EE5 holds. The important thing is that then $x \leq_{GM} y \leftrightarrow (true \vdash x \wedge y$ or $\neg(x \wedge y)\ |\!\!\not\sim\ \leq_{GM} x)$, $i.e. (*)$ holds:

d) If $\leq_{GM}$ satisfies EE1-EE3, EE5, and $|\!\!\sim\ =\ |\!\!\sim\ \leq_{GM}$ is defined as in $(**)$, then

1) $x \leq_{GM} y \leftrightarrow (true \vdash x \wedge y$ or $\neg(x \wedge y)\ |\!\!\not\sim\ x)$ and thus

2) $a\ |\!\!\sim\ x \leftrightarrow (a \vdash x$ or $\exists b(a \wedge b \vdash x$ and $\neg b < a))$. (Note that $<_{GM}$ has become $<$ as defined in Section 1.8.7.2. Moreover, $\neg b < a \rightarrow \neg b < true$ (in R) $\rightarrow true\ |\!\!\sim\ b$ by 8) a) in Fact 1.8.28.)

Thus the amount of additional information we can use depends on the size of a: If a is very large, e.g. true, we can use many additional b's. If a is small, e.g. $\bot$, nothing is smaller than $\bot$, so we can't use any additional b's, which gives consistency preservation.

Proof

a) $true \vdash x \wedge y \leftrightarrow \neg x \vee \neg y \vdash \bot$ (by a semantic argument). By consistency preservation, $\neg x \vee \neg y \vdash \bot \leftrightarrow \neg x \vee \neg y\ |\!\!\sim\ \bot$. Thus $x \leq_{GM_\sim} y \leftrightarrow \neg x \vee \neg y\ |\!\!\sim\ \bot$ or $\neg x \vee \neg y\ |\!\!\not\sim\ x \leftrightarrow \neg x \not< \neg y \leftrightarrow \neg y \leq \neg x$. And $\alpha <_{GM_\sim} \beta \leftrightarrow \beta \not\leq_{GM_\sim} \alpha \leftrightarrow \neg \beta < \neg \alpha$

b) $\leq_{GM_\sim}$ also satisfies EE5: It suffices to show $true \leq_{GM_\sim} y \rightarrow true \vdash y$. Suppose there is y, $true \leq_{GM_\sim} y$, $true \not\vdash y$, so $true \not\vdash true \wedge y$. Thus, by definition, $\neg(true \wedge y)\ |\!\!\not\sim\ true$, contradiction.

d) Exercise, solution in the Appendix.

$\square$

We are now in a position for a reconstruction of the completeness results :

1.8.7.5 Completeness Results

Our intention is to give the reader the essential ideas, and we feel free to use semantic arguments (as already above) - in hindsight, so to say.

We closely follow the respective articles.

Completeness Proof for P [KLM90]

Any logic $\vdash$ satisfying P can be represented by a smooth, irreflexive, transitive structure, i.e. there is such $\mathcal{M}$ s.th. $\alpha \mathrel{\vdash} \beta \leftrightarrow \alpha \models_{\mathcal{M}} \beta$. We use the notation of [KLM90].

Recollect that: $\alpha \leq_{KLM} \beta :\leftrightarrow \alpha \vee \beta \mathrel{\vdash} \alpha$. The representing structure $\mathcal{M} = \langle M, \prec \rangle$ is defined by $M := \{\langle m, \alpha \rangle : m$ is a normal world for α, i.e. $m \models \overline{\overline{\alpha}}\}$ and $\langle m, \alpha \rangle \prec \langle n, \beta \rangle :\leftrightarrow \alpha \leq_{KLM} \beta$ and $m \not\models \beta$. (This is the canonical model.)

Lemma 1.8.34

$\alpha \leq_{KLM} \beta, m \models \overline{\overline{\alpha}}, \beta \to m \models \overline{\overline{\beta}}.$

Proof

$\alpha \vee \beta \mathrel{\vdash} \alpha \to m(\beta) \cap \mu(\alpha) \subseteq \mu(\beta) \to (\beta \mathrel{\vdash} \delta \to \alpha \mathrel{\vdash} \beta \to \delta).$

□

Lemma 1.8.35

$\alpha \leq_{KLM} \beta \leq_{KLM} \gamma, m \models \overline{\overline{\alpha}}, \gamma \to m \models \overline{\overline{\beta}}.$

Proof

$\alpha \leq_{KLM} \beta \leq_{KLM} \gamma \to \mu(\alpha) \cap m(\gamma) \subseteq \mu(\beta) \to (\beta \mathrel{\vdash} \delta \to \alpha \mathrel{\vdash} \gamma \to \delta)$
(Suppose $m \in \mu(\alpha)$, $m \models \gamma$, $m \notin \mu(\beta)$. By $\beta \vee \gamma \mathrel{\vdash} \beta$, there is $m' \prec m$, $m' \models \beta$, by $\alpha \vee \beta \mathrel{\vdash} \alpha$, there is $m'' \preceq m'$, $m'' \models \alpha$, contradiction: $m \models \gamma$, $m' \models \beta$, $m'' \models \alpha$ with $m'' \prec m' \prec m$.)

□

Lemma 1.8.36

Transitivity of $\prec$: $\langle m_2, \alpha_2 \rangle$, $\langle m_1, \alpha_1 \rangle \models \neg \alpha_2$, $< m_0, \alpha_0 > \models \neg \alpha_1$ with $\langle m_0, \alpha_0 \rangle \prec < m_1, \alpha_1 > \prec \langle m_2, \alpha_2 \rangle$.

Proof

The crucial thing to show is $\langle m_0, \alpha_0 \rangle \models \neg \alpha_2$. But $\langle m_0, \alpha_0 \rangle \models \alpha_2 \to \langle m_0, \alpha_0 \rangle \models \overline{\overline{\alpha_1}}$ (by Lemma 1.8.35), contradiction.

□

Lemma 1.8.37

$\langle m, \beta \rangle \in \mu(\alpha) \leftrightarrow \langle m, \beta \rangle \models \alpha, \beta \leq_{KLM} \alpha.$

1.8 Various Results and Approaches

Proof

"←": Suppose $\langle m, \beta \rangle \models \alpha$, $\langle n, \gamma \rangle \models \alpha, \neg \beta$ with $< n, \gamma > \prec \langle m, \beta \rangle$. By $\gamma \leq_{KLM} \beta \leq_{KLM} \alpha$ and Lemma 1.8.35, $\langle n, \gamma \rangle \models \beta$, *contradiction* . "→": Suppose $\beta \not\leq_{KLM} \alpha$, i.e. $\alpha \vee \beta \not\hspace{-2pt}\sim \beta$, let $n \models \overline{\overline{\alpha \vee \beta}}$, $n \models \neg \beta$, so $n \models \alpha$ $\langle m, \beta \rangle \models \alpha$, $\langle n, \alpha \vee \beta \rangle \models \alpha, \neg \beta$ with $\langle n, \alpha \vee \beta \rangle \prec \langle m, \beta \rangle$ So $\langle m, \beta \rangle \notin \mu(\alpha)$.

□

Remark 1.8.38

$\alpha \vee \beta \not\hspace{-2pt}\sim \beta$ does not imply (in P) that there is τ, $\alpha \vee \beta \hspace{-2pt}\sim \tau$, $\beta \not\hspace{-2pt}\sim \tau$: $m \models \alpha, \beta$, $m' \models \alpha, \neg \beta$, $m'' \models \alpha, \beta$ with $m' \prec m$. Here $\alpha \vee \beta \not\hspace{-2pt}\sim \beta$. And $\alpha \vee \beta \hspace{-2pt}\sim \tau \leftrightarrow (\alpha \wedge \neg \beta) \vee (\alpha \wedge \beta) = \alpha \wedge (\beta \vee \neg \beta) = \alpha \vdash \tau$, $\beta \hspace{-2pt}\sim \tau \leftrightarrow \alpha \wedge \beta \vdash \tau$. Thus, $\alpha \vee \beta \hspace{-2pt}\sim \tau \rightarrow \alpha \vdash \tau \rightarrow \alpha \wedge \beta \vdash \tau \rightarrow \beta \hspace{-2pt}\sim \tau$.

□

Lemma 1.8.39

$[\alpha]$ is smooth, where $[\alpha] := \{\langle m, \beta \rangle : m \models \alpha\}$.

Proof

Let $\langle m, \beta \rangle \models \alpha$. If $\beta \leq_{KLM} \alpha$, i.e. $\alpha \vee \beta \hspace{-2pt}\sim \beta$, then $\langle m, \beta \rangle \in \mu(\alpha)$ by Lemma 1.8.37. Otherwise, there is $n \models \overline{\overline{\alpha \vee \beta}}, \neg \beta, \alpha$: $\langle m, \beta \rangle \models \alpha$, $\langle n, \alpha \vee \beta \rangle \models \alpha, \neg \beta$ with $\langle n, \alpha \vee \beta \rangle \prec \langle m, \beta \rangle$ But then $\langle n, \alpha \vee \beta \rangle \in \mu(\alpha)$ by Lemma 1.8.37.

□

Crucial for the representation is:

Fact 1.8.40

1) $\langle m, \alpha \rangle \in \mu(\alpha)$
2) $\langle m, \beta \rangle \in \mu(\alpha) \rightarrow \langle m, \beta \rangle \models \overline{\overline{\alpha}}$

Proof

1) is trivial by definition of $\prec$.

2) $\langle m, \beta \rangle \in \mu(\alpha) \rightarrow \beta \leq_{KLM} \alpha$ (by Lemma 1.8.37) $\rightarrow \langle m, \beta \rangle \models \overline{\overline{\alpha}}$ (by Lemma 1.8.34)

□

Completeness Proof for R ([LM92])

Any logic $\mathrel{|\!\sim}$ satisfying R can be represented by a smooth, ranked, irreflexive, transitive structure. We use notation of [LM92]. Recollect that $\alpha <_{LM} \beta :\leftrightarrow \alpha \vee \beta \mathrel{|\!\sim} \neg \beta$. The representing structure $\mathcal{M} = \langle M, \prec \rangle$ is defined by $M := \{\langle m, \alpha \rangle : m$ is a normal world for α, i.e. $m \models \overline{\overline{\alpha}}\}$ and $\langle m, \alpha \rangle \prec \langle n, \beta \rangle :\leftrightarrow \alpha <_{LM} \beta$. This is again the canonical model.

Lemma 1.8.41

In P holds:

$\beta <_{LM} \alpha \to \beta \mathrel{|\!\sim} \neg \alpha$

Proof

$\alpha \vee \beta \mathrel{|\!\sim} \neg \alpha \to \beta \mathrel{|\!\sim} \neg \alpha$.

□

Lemma 1.8.42

Transitivity of $\prec$ holds in P.

Proof

$\alpha <_{LM} \beta <_{LM} \gamma \to \alpha < \gamma$ (by Fact 1.8.28, 2)b), and our above comparison result.) Moreover, $\alpha <_{LM} \gamma \to \alpha \mathrel{|\!\sim} \neg \gamma$ by Lemma 1.8.41.

□

Lemma 1.8.43

In P holds:

$Con(\overline{\overline{\alpha}}, \beta) \to \alpha \vee \beta \mathrel{|\!\not\sim} \neg \beta$

Proof

$Con(\overline{\overline{\alpha}}, \beta) \to \alpha \not\Rightarrow \neg \beta \to \alpha \vee \beta \mathrel{|\!\not\sim} \neg \beta$

□

Lemma 1.8.44

In P holds:

$\langle m, \alpha \rangle \in \mu(\beta) \to \alpha \vee \beta \mathrel{|\!\not\sim} \neg \alpha, \alpha \vee \beta \mathrel{|\!\not\sim} \neg \beta$

Proof

$\langle m, \alpha \rangle \in \mu(\beta) \to \langle m, \alpha \rangle \models \overline{\overline{\alpha}}, \beta \to \alpha \vee \beta \mathrel{|\!\not\sim} \neg \beta$ by Lemma 1.8.43. Suppose $\alpha \vee \beta \mathrel{|\!\sim} \neg \alpha$, thus $\beta <_{LM} \alpha$, and $\beta \mathrel{|\!\sim} \neg \alpha$. Then $\langle m, \alpha \rangle, \langle n, \beta \rangle \models \overline{\overline{\beta}}, \neg \alpha$ with $<n, \beta> \prec \langle m, \alpha \rangle$ so $\langle m, \alpha \rangle \notin \mu(\beta)$.

□

1.8 Various Results and Approaches

Corollary 1.8.45

In P holds:

$\langle m, \alpha \rangle \in \mu(\alpha)$.

Proof

Let $\langle n, \beta \rangle \prec \langle m, \alpha \rangle$, then $\langle n, \beta \rangle \models \neg \alpha$.

□

Lemma 1.8.46

In R holds:

$[\alpha]$ is smooth.

Proof

Let $\langle m, \beta \rangle \models \alpha$. By Lemma 1.8.43, $\alpha \vee \beta \hspace{2pt}|\!\!\!\not\sim \neg \alpha$. Case 1: $\alpha \vee \beta \hspace{2pt}|\!\sim \neg \beta$. Then $\alpha <_{LM} \beta$, $\alpha \hspace{2pt}|\!\sim \neg \beta$, and $\langle m, \beta \rangle \models \alpha$, $\langle n, \alpha \rangle \models \neg \beta$, with $\langle n, \alpha \rangle \prec \langle m, \beta \rangle$, $\langle n, \alpha \rangle$ is minimal in $[\alpha]$ by Corollary 1.8.45. Case 2: $\alpha \vee \beta \hspace{2pt}|\!\!\!\not\sim \neg \beta$. If $\langle m, \beta \rangle \notin \mu(\alpha)$, then $\langle m, \beta \rangle \models \alpha$, $\langle n, \gamma \rangle \models \alpha$ with $\langle n, \gamma \rangle \prec \langle m, \beta \rangle$ for some $\langle n, \gamma \rangle$. Then by Lemma 1.8.43 $\gamma \not<_{LM} \alpha$. By R, $\gamma <_{LM} \beta$, $\gamma \not<_{LM} \alpha \to \alpha <_{LM} \beta \to \alpha \vee \beta \hspace{2pt}|\!\sim \neg \beta$, *contradiction*.

□

Lemma 1.8.47

In R holds:

$\langle m, \alpha \rangle \in \mu(\beta) \to \langle m, \alpha \rangle \models \overline{\overline{\beta}}$.

Proof

By Lemma 1.8.44, $\alpha \vee \beta \hspace{2pt}|\!\!\!\not\sim \neg \alpha$. Now, $\beta \hspace{2pt}|\!\sim \gamma \to \alpha \vee \beta \hspace{2pt}|\!\sim \beta \to \gamma$ (see the completeness proof in [KLM90]). By R, and $\alpha \vee \beta \hspace{2pt}|\!\!\!\not\sim \neg \alpha$, $\alpha \hspace{2pt}|\!\sim \beta \to \gamma$ for $\beta \hspace{2pt}|\!\sim \gamma$. As $\langle m, \alpha \rangle \models \beta$, $\langle m, \alpha \rangle \models \gamma$ for all $\beta \hspace{2pt}|\!\sim \gamma$.

□

Rankedness is a direct consequence of R, see Fact 1.8.28, 10), and $\alpha <_{LM} \beta \leftrightarrow \beta < \alpha$ for $\alpha, \beta \hspace{2pt}|\!\sim$-consistent.

Completeness Proofs for [LM92] and [GM94]

We use the ideas in [GM94] to give a second proof for the [LM92] result, too. The reader will notice that the axiom R (or its consequences) is used here extensively.

Let $\leq$ be total and transitive, and $x < y \leftrightarrow y \not\leq x$.

Let $x^+ := \{y : x \leq y\}$,
$X := \{x\} \cup \{y : \neg x < y\}$,
$M := \{m \subseteq \mathcal{L}: m \text{ is maximal} \vdash -consistent\}$,
and for models $m \prec n :\leftrightarrow$ there is x s.th. $m \models x^+, n \models \neg x$.

If these prerequisites hold, then:

Lemma 1.8.48

1)

    a) $x < y \leq z \to x < z$
    b) $x \leq y < z \to x < z$
    c) $y \in X - \{x\} \to y^+ \subseteq X$
    d) $y \in x^+ \to y^+ \subseteq x^+$

2) $\prec$ is transitive

3) $\prec$ is ranked

4) $Con(x, z^+) \to z^+ \subseteq X$

5) consequently: $n \models X \to n \in \mu(x)$

6) the structure is smooth

7) $m \in \mu(x) \to m \models X$

Proof

Exercise, solution in the Appendix.

To finish [GM94], we just have to note that by 5) and 7) and Definition, the normal worlds for x and $\mu(x)$ coincide, thus $x \mathrel{\vert\!\sim}_{\leq y} \leftrightarrow x \models_M y$.

We turn to the [LM92] completeness result.

Fact 1.8.49

(in P) For x, let $\widehat{x} := \{x\} \cup \{y : \neg y \vee x \mathrel{\vert\!\sim} y\}$ and $\overline{\overline{x}} := \{y : x \mathrel{\vert\!\sim} y\}$. Then $m \models \overline{\overline{x}} \leftrightarrow m \models \widehat{x}$.

Proof

"$\to$": Let $m \models \overline{\overline{x}}$, $y \in \widehat{x}$, so $\neg y \vee x \mathrel{\vert\!\sim} y$, but then $x \mathrel{\vert\!\sim} y$. "$\leftarrow$": Let $x \mathrel{\vert\!\sim} z$, $m \models \widehat{x}$. Consider $y := \neg x \vee z$. Then $\neg y \vee x = \neg(\neg x \vee z) \vee x = (x \wedge \neg z) \vee x = x \mathrel{\vert\!\sim} z$, so $\neg y \vee x \mathrel{\vert\!\sim} \neg x \vee z = y$, so $m \models x, \neg x \vee z$, thus $m \models z$.

□

1.8 Various Results and Approaches

Define now $x \leq' y :\leftrightarrow \neg y \leq \neg x$ and $x <' y :\leftrightarrow \neg y < \neg x$ (both $<, \leq$ in the sense of Definition 1.8.22). By Fact 1.8.28, 3c) and 8c) (in $R!$) $\leq'$ is total and transitive, moreover $x <' y \leftrightarrow y \not\leq' x$.

We conclude by showing $m \models X \leftrightarrow m \models \overline{\overline{x}}$, where $X := \{x\} \cup \{y : \neg x <' y\}$. Case 1: $x \mathrel{\mid\!\sim} \bot$. Then $x \vdash \bot$ by consistency preservation, so $m(x) = m(X) = m(\overline{\overline{x}}) = \emptyset$. Case 2: $x \not\mathrel{\mid\!\sim} \bot$. But now $\neg y < x \leftrightarrow x \vee \neg y \mathrel{\mid\!\sim} y$, thus $\neg x <' y \leftrightarrow x \vee \neg y \mathrel{\mid\!\sim} y$, so $X = \widehat{x}$, and we conclude by above Fact.

□

1.8.7.6 The Rank

This might be the right place to shortly mention another order defined in [LM92], the order by rank of a formula in a database, measuring exceptionality. See also Section 5.7.

Definition 1.8.25

Let a conditional knowledge base K be given, i.e. a set of $\alpha \mathrel{\mid\!\sim} \beta$'s, $\alpha, \beta \in \mathcal{L}$.

a) α is called exceptional for K, iff K preferentially entails $true \mathrel{\mid\!\sim} \neg \alpha$ $\alpha \mathrel{\mid\!\sim} \beta$ is called exceptional for K, iff α is exceptional for K $E(K)$ is the set of $\alpha \mathrel{\mid\!\sim} \beta \in K$ exceptional for K

b) $C_0 := K$

   $C_{\tau+1} := E(C_\tau)$

   $C_\lambda := \bigcap \{C_\rho : \rho < \lambda\}$ for limit λ.

c) If α is exceptional for all C_τ, we set $rk(\alpha) := \infty$ (where $\infty > \rho$ for "all" ordinals ρ). Otherwise, set $rk(\alpha) := \tau$ iff τ is least s.th. α is not exceptional for C_τ.

We note the following result, whose part b) uses notation and a theorem of [LM92].

Fact 1.8.50

a)

   1. If $rk(\alpha \vee \beta) < \infty$, then $\alpha <_{LM} \beta \rightarrow rk(\alpha) < rk(\beta)$
   2. thus $\beta < \alpha \rightarrow rk(\alpha) < rk(\beta)$
   3. Usually, $rk(\alpha) < rk(\beta) \not\rightarrow \alpha <_{LM} \beta$.

b) If K is admissible, then

   $rk(\alpha) < rk(\beta)$ in $K \leftrightarrow \beta < \alpha$ in $\overline{K}$

Proof

a)

1.: $C_{rk(\alpha\vee\beta)} \models \alpha\vee\beta \mathrel{|\!\sim} \neg\beta$ (where $\models$ means preferentially entails). We first show $C_{rk(\alpha\vee\beta)} \models true \mathrel{|\!\sim} \neg\beta$: Let $m \in \mu(true)$. Case 1: $m \models \alpha\vee\beta$, then $m \in \mu(\alpha\vee\beta)$, so $m \models \neg\beta$. Case 2: $m \not\models \alpha\vee\beta$, so $m \models \neg\beta$. Now, $rk(\alpha\vee\beta) = min(rk\alpha, rk\beta)$. If $rk\beta \leq rk\alpha$, then $rk(\alpha\vee\beta) = rk\beta$. But $C_{rk\beta} \not\models true \mathrel{|\!\sim} \neg\beta$, contradiction.

3.: Consider the model

$m'' \models \alpha, \beta$, $m' \models \neg\alpha, \neg\beta$, $m \models \alpha, \neg\beta$ with $m' \prec m''$.

Then $true \mathrel{|\!\sim} \neg\beta$, $true \mathrel{|\!\not\sim} \neg\alpha$, $\alpha\vee\beta \mathrel{|\!\not\sim} \neg\beta$. Thus $rk\alpha = 0$, $rk\beta > 0$, $\alpha \not<_{LM} \beta$.

b) "$\rightarrow$": Let rk be defined in K. $rk\alpha < rk\beta \rightarrow rk(\alpha\vee\beta) = min(rk\alpha, rk\beta) = rk\alpha < rk\beta = rk((\alpha\vee\beta)\wedge\beta) \rightarrow \alpha\vee\beta \mathrel{|\!\sim}_{\overline{K}} \neg\beta \rightarrow \alpha <_{LM} \beta$ in $\overline{K}$. Suppose $\alpha\vee\beta \mathrel{|\!\sim}_{\overline{K}} \bot$. Then either Case 1: $rk(\alpha\vee\beta) < rk((\alpha\vee\beta)\wedge true)$ (in K), contradiction, or Case 2: $\alpha\vee\beta$ has no rank (in K) $\rightarrow \alpha, \beta$ have no rank $\rightarrow rk\alpha \not< rk\beta$ (in K). Thus, $rk\alpha < rk\beta$ (in K) $\rightarrow \alpha\vee\beta \mathrel{|\!\not\sim}_{\overline{K}} \bot$, thus also $\beta < \alpha$ in $\overline{K}$.

"$\leftarrow$": Let $\alpha\vee\beta \mathrel{|\!\not\sim} \bot$, $\alpha\vee\beta \mathrel{|\!\sim} \neg\beta$ in $\overline{K}$. Case 2 of $\alpha\vee\beta \mathrel{|\!\sim} \neg\beta$: $\alpha\vee\beta$ has no rank, but then $\alpha\vee\beta \mathrel{|\!\sim} \bot$, contradiction. So Case 1 holds: $rk(\alpha\vee\beta) < rk((\alpha\vee\beta)\wedge\beta) = rk\beta$ (in K) So by $rk(\alpha\vee\beta) = min(rk\alpha, rk\beta)$ $rk\alpha < rk\beta$.

□

1.8.8 Preferential Choice Representation Theorems for Branching Time Structures

1.8.8.1 Overview

The idea of preferential choice is applied here to dynamic structures in two directions: 1. We show that a deontic choice function of "good" developments can be represented by a ranked, smooth preferential relation on all developments. 2. We generalize the Katsuno/Mendelzon Update Semantics to preferences between developments and obtain a representation theorem for arbitrarily many time points.

In the first part (Theorem 1.8.58), we show that a deontic choice function of "good" developments defined in [Tho84] can be represented by a ranked, smooth relation on all developments. Thus, we do not compare single models, but developments in a branching time structure. Again, the question of acting in a way that those

1.8 Various Results and Approaches 217

preferred developments are reached (or not left), is left open, we only discuss - as R. Thomason does in [Tho84] - the "quality" of the developments, and show that again a local choice by a binary preference relation suffices, and even a very nice one.

In the second part (Theorem 1.8.63), we give a characterization of coupled logics which can be obtained from a preference relation on developments. The Katsuno/Mendelzon Update Semantics [KM90] restricts comparisons to single worlds (points), and does not compare two arbitrary developments, only those originating from the same initial state. Their approach is thus in the spirit of Lewis' Counterfactual Conditional Semantics. We generalize this to compare arbitrary developments through two or arbitrarily many time points by a preferential relation, and characterize the resulting functions. More precisely - for the case of two time points - we suppose that we are given two theories, S and T in a fixed propositional language $\mathcal{L}$ that describe the state of affairs at two times, t and t'. On this approach, worlds are identified with models. A subset Π of $M_\mathcal{L} \times M_\mathcal{L}$ (where $M_\mathcal{L}$ is the set of models of $\mathcal{L}$) represents the possible transitions from worlds to worlds. Thus, where S is an initial and T a final theory, any pair $\langle u, v \rangle \in \Pi$ such that $u \models S$ and $v \models T$ represents a possible transition. The preferential relation $\prec$ on Π tells us which developments are the preferred ones. So, the $\prec$-minimal transitions $\mu(\Pi \cap (M_S \times M_T)) \subseteq \Pi \cap (M_S \times M_T)$ (M_S the set of models of S etc.) are the preferred ones, and we consider first $\{x : \exists \langle u, v \rangle \in \mu(\Pi \cap (M_S \times M_T)).x = u\}$ and $\{y : \exists \langle u, v \rangle \in \mu(\Pi \cap (M_S \times M_T)).y = v\}$, i.e. both projections $\pi_i(\mu(\Pi \cap (M_S \times M_T)))$, and then the theories generated by them: $S' := \{\phi : \forall m \in \pi_0(\mu(\Pi \cap (M_S \times M_T))).m \models \phi\}$ and $T' := \{\phi : \forall m \in \pi_1(\mu(\Pi \cap (M_S \times M_T))).m \models \phi\}$. Obviously, S' and T' are stronger than or equivalent to S and T respectively. Thus, we use $\prec$ on $\Pi \subseteq M_\mathcal{L} \times M_\mathcal{L}$ to map a pair of theories $\langle S, T \rangle$ to another such pair $\langle S', T' \rangle$, and consider a pair of logics, which are connected through a common preferential relation. We present a soundness and completeness theorem, first for the combinatorial functions $\pi_i(\mu(\Pi \cap (M_S \times M_T)))$ (Theorem 1.8.60), and then for the resulting logics (Theorem 1.8.63). The approach generalizes in a straightforward manner to arbitrary cartesian products (Theorem 1.8.61).

The material is taken from [Sch95-3].

1.8.8.2 A Ranked and Smooth Preferential Representation for a Deontic Choice Function

Outline

We show here that Thomason's deontic choice function O of "good" branches (see [Tho84]) can be represented by a ranked, smooth preferential relation on all branches (Theorem 1.8.58). We look for a ranked smooth order on the set of branches $B := \bigcup\{B_t : t \in T\}$, which describes $O_t \subseteq B_t$ as the minimal elements of B_t. So we have to find an order $\prec$ on B such that $O_t = \mu(B_t) := \{b \in B_t :$

$\neg \exists b' \in B_t . b' \prec b\}$, a totally ordered set $(Z, <)$, and a function $h : B \to Z$ such that $b \prec b' \leftrightarrow h(b) < h(b')$. For the construction of $(Z, <)$, we (essentially) rearrange $\{B_t : t \in T\}$ in a suitable way to a total order (Definition 5), and for $b \in B$, we let $h(b)$ be the $\subseteq$-largest B_t, $t \in T$ such that $b \in O_t$. For this to be defined, we close $\{B_t : t \in T\}$ under unions of $\subseteq$-chains, this is done in the first part of the text, up to Definition 3. Finally, we inherit the order $\prec$ on B from $<$ via h. It is easily shown that $\prec$ is smooth and represents the choice function O.

We conclude by remarking that Bartha's choice of "bad" branches by validity of a special formula S at branch/point pairs (see [Bar91]) can also express Thomason's O-function (Fact 1.8.59).

The Details

Definition 1.8.26

Let $\mathcal{X} \subseteq \mathcal{P}(X)$. We say that $\mathcal{X}' \subseteq \mathcal{X}$ is a chain in $\mathcal{X}$ iff $\mathcal{X}'$ is totally ordered under $\subseteq$. We say that $\mathcal{X}$ satisfies

(tp) iff $A, B \in \mathcal{X} \to A \cap B = \emptyset$ or $A \subseteq B$ or $B \subseteq A$.

Thus, if $\mathcal{X}$ has (tp), and is ordered by $A < B :\leftrightarrow B \subset A$, it is almost tree-like.

Fact 1.8.51

Let $\{\mathcal{X}_i : i \in I\}$ be a collection of chains in $\mathcal{X}$, $X_i := \bigcup \mathcal{X}_i$, and let $\mathcal{X}$ satisfy (tp), then so does $\mathcal{X}' := \mathcal{X} \cup \{X_i : i \in I\}$.

Proof

Exercise, solution in the Appendix

Definition 1.8.27

Let $\mathcal{X} \subseteq \mathcal{P}(X)$ and $f : \mathcal{X} \to \mathcal{P}(X)$ be such that for $A, B \in \mathcal{X}$

(1) $f(A) \subseteq A$
(2) $f(A) \neq \emptyset$ if $A \neq \emptyset$
(3) $A \subseteq B \to f(B) \cap A \subseteq f(A)$
(4) $A \subseteq B \wedge f(A) \cap f(B) \neq \emptyset \to f(A) = f(B) \cap A$.

We then say that a chain $\mathcal{X}' \subseteq \mathcal{X}$ satisfies

(ip) iff for $A, B \in \mathcal{X}'$ $f(A) \cap f(B) \neq \emptyset$.

Note that f is then monotone on $\mathcal{X}'$: $A \subseteq B \to f(A) \subseteq f(B)$ by (4) for $A, B \in \mathcal{X}'$. But f is also continuous in the following sense:

1.8 Various Results and Approaches   219

Fact 1.8.52

Let f be as in Definition 1.8.27. Let $\mathcal{X}' \subseteq \mathcal{X}$ satisfy (ip), $\bigcup \mathcal{X}' \in \mathcal{X}$, then $f(\bigcup \mathcal{X}') = \bigcup \{f(A) : A \in \mathcal{X}'\}$.

Proof

The case $\bigcup \mathcal{X}' = \emptyset$ is trivial. Let $X' := \bigcup \mathcal{X}'$. We first show that $f(X') \cap f(B) \neq \emptyset$ for all $B \in \mathcal{X}'$. As $f(X') \neq \emptyset$, there is $a \in f(X') \subseteq X'$, so $a \in A$ for some $A \in \mathcal{X}'$. But then $a \in f(A)$ by (3). So $f(X') \cap f(A) \neq \emptyset$, and $f(A) = A \cap f(X')$ by (4). Let $B \in \mathcal{X}'$, then there is $b \in f(A) \cap f(B)$, so $b \in f(X')$. But now by (4) for each $A \in \mathcal{X}'$ $f(A) = f(X') \cap A$, so by $X' = \bigcup \mathcal{X}'$ and $f(X') \subseteq X'$ $f(X') = \bigcup \{f(A) : A \in \mathcal{X}'\}$.

□

Fact 1.8.53

Let $\mathcal{X}$ satisfy (tp) and let f be as in Definition 1.8.27. Let $\mathcal{A}, \mathcal{B} \subseteq \mathcal{X}$ satisfy (ip), $A := \bigcup \mathcal{A}$, $B := \bigcup \mathcal{B}$, $A \subseteq B$, $A' \in \mathcal{A}$, $B' \in \mathcal{B}$, then $f(B') \cap A' \subseteq f(A')$.

Proof

Let $x \in f(B') \cap A'$, so $A' \cap B' \neq \emptyset$. Consider $\mathcal{B}'' := \{B'' \in \mathcal{B} : B'' \supseteq B'\}$. Suppose there is $B'' \in \mathcal{B}''$ such that $A' \subseteq B''$, then, by $f(B') \subseteq f(B'')$, $x \in f(B'') \cap A'$, so $x \in f(A')$ by (3). Otherwise, as $B' \cap A' \neq \emptyset$, for all $B'' \in \mathcal{B}''$ $B'' \cap A' \neq \emptyset$, so for all $B'' \in \mathcal{B}''$ $B'' \subseteq A'$ by (tp) for $\mathcal{X}$. But, as $\bigcup \mathcal{B}'' = \bigcup \mathcal{B}$, $B \subseteq A' \subseteq A \subseteq B$, so $A' = B$. By Fact 1.8.52 $f(A') = \bigcup \{f(B'') : B'' \in \mathcal{B}\}$, so $x \in f(A')$.

□

Fact 1.8.54

Let $\mathcal{X}$ satisfy (tp) and let f be as in Definition 1.8.27. Let $\mathcal{A}, \mathcal{A}' \subseteq \mathcal{X}$ satisfy (ip), $\bigcup \mathcal{A} = \bigcup \mathcal{A}'$, then $\bigcup \{f(A) : A \in \mathcal{A}\} = \bigcup \{f(A') : A' \in \mathcal{A}'\}$.

Proof

We show "$\subseteq$", "$\supseteq$" is symmetrical. Let $a \in \bigcup \{f(A) : A \in \mathcal{A}\}$, so $a \in f(A) \subseteq A$ for some $A \in \mathcal{A}$. So there is $A' \in \mathcal{A}'$ such that $a \in A'$, thus $a \in f(A) \cap A'$, so $a \in f(A')$ by Fact 1.8.53.

□

Fact 1.8.55

Let $\mathcal{X}$ satisfy (tp), f be as in Definition 1.8.27, $\mathcal{Y} := \{\mathcal{A} \subseteq \mathcal{X} : \mathcal{A} \text{ satisfies } (ip)\}$, and $\mathcal{X}' := \mathcal{X} \cup \{\bigcup \mathcal{A} : \mathcal{A} \in \mathcal{Y}\}$. Extend f to $f' : \mathcal{X}' \to \mathcal{P}(X)$ by defining $f'(\bigcup \mathcal{A}) := \bigcup \{f(A) : A \in \mathcal{A}\}$ for any $\mathcal{A} \in \mathcal{Y}$ with $\bigcup \mathcal{A} \notin \mathcal{X}$. Then f' also satisfies $(1) - (4)$ of Definition 1.8.27.

Proof

Exercise, solution in the Appendix

Definition 1.8.28

(Closure and extension to unions of (ip)-chains):

We can thus extend $\mathcal{X}$ and f to $\mathcal{Y} \subseteq \mathcal{P}(X)$, the closure under all unions which satisfy (ip) as follows:

$\mathcal{X}_0 := \mathcal{X}$, $f_0 := f$.

In the successor step, let $\mathcal{X}_{\alpha+1} := \mathcal{X}'_\alpha$, $f_{\alpha+1} :=$ the extension of f_α to $\mathcal{X}'_\alpha$ as defined and examined in Fact 1.8.51 and Fact 1.8.55.

In the limit step, let $\mathcal{X}_\lambda := \bigcup\{\mathcal{X}_\alpha : \alpha < \lambda\}$, $f_\lambda := \bigcup\{f_\alpha : \alpha < \lambda\}$. Obviously, the limit step preserves (tp) and $(1) - (4)$ of f.

Finally, set $\mathcal{Y} := \bigcup\{\mathcal{X}_\alpha : \alpha < \kappa\}$, $g := \bigcup\{f_\alpha : \alpha < \kappa\}$, where κ is large enough, $(card(\mathcal{P}(X)))^+$ will do.

We now work in $\mathcal{Y}$, with g, as just defined.

Definition 1.8.29

For $\emptyset \neq A \in \mathcal{Y}$, let A^+ be the largest $B \in \mathcal{Y}$ such that $A \subseteq B$, $g(A) \cap g(B) \neq \emptyset$.

This is well-defined: Let $\boldsymbol{B}_A := \{B' \in \mathcal{Y} : A \subseteq B' \wedge g(A) \cap g(B') \neq \emptyset\}$. Obviously, for $B', B'' \in \boldsymbol{B}_A$, $B' \cap B'' \neq \emptyset$, so by (tp) $\boldsymbol{B}_A$ is totally ordered by $\subseteq$. By (4), $g(A) \subseteq g(B')$ for all $B' \in \boldsymbol{B}_A$, so $\boldsymbol{B}_A$ has (ip), and by construction $\bigcup \boldsymbol{B}_A \in \mathcal{Y}$, but $A^+ = \bigcup \boldsymbol{B}_A$.

Moreover, the following holds:

Fact 1.8.56

Let $g(A) \cap g(A') \neq \emptyset$, then $A^+ = A'^+$.

Proof

Let $x \in g(A) \cap g(A')$. By $A \cap A' \neq \emptyset$, $A \subseteq A'$ or $A' \subseteq A$. Consider the case $A \subseteq A'$. Define $\boldsymbol{B}_A$, $\boldsymbol{B}_{A'}$ as above. Let $B' \in \boldsymbol{B}_{A'}$. Then $B' \supseteq A' \supseteq A$, and $g(A) \cap g(B') \neq \emptyset$: By $A' \subseteq B'$ and $g(A') \cap g(B') \neq \emptyset$, $g(A') \subseteq g(B')$, so $x \in g(A') \subseteq g(B')$. Thus $\boldsymbol{B}_{A'} \subseteq \boldsymbol{B}_A$, and $A'^+ \subseteq A^+$. Furthermore, if $B \in \boldsymbol{B}_A$, then $A \subseteq B$, so $A' \cap B \neq \emptyset$, thus $A' \subseteq B$ or $B \subseteq A'$. If $B \subseteq A'$, then there is $B' := A' \in \boldsymbol{B}_{A'}$ with $B \subseteq B'$. If $A' \subseteq B$, then $g(A') \cap g(B) \neq \emptyset$: By $B \in \boldsymbol{B}_A$, $g(A) \subseteq g(B)$, so $x \in g(A') \cap g(B)$. Thus $B \in \boldsymbol{B}_{A'}$. In any case, for $B \in \boldsymbol{B}_A$, there is $B' \in \boldsymbol{B}_{A'}$ with $B \subseteq B'$. Thus $A'^+ = \bigcup \boldsymbol{B}_{A'} \supseteq \bigcup \boldsymbol{B}_A = A^+$.

□

1.8 Various Results and Approaches

Definition 1.8.30

(Total order on $\mathcal{Y}$)

Let $< Y_i : i < \mu >$ be some enumeration of $\mathcal{Y}$. For $\emptyset \neq Y \in \mathcal{Y}$, let $[Y] := \{Y' \in \mathcal{Y} : Y' \supseteq Y\}$. By (tp), [Y] is totally ordered by $\subseteq$.

We note the following obvious Fact:

(a) If $<$ is a strict total order on C, $<'$ a strict total order on C', $C \cap C' = \emptyset$, then "putting C' strictly on top of C" will be a strict total order too. We will denote this ordered set by $C + C'$.

(b) If $(C_i, <_i)$, $i < \lambda$ is an increasing chain of totally, strictly ordered sets, where the $<_i$ agree whenever defined, then $\bigcup \{C_i : i < \lambda\}$ is also totally, strictly ordered by $\bigcup \{<_i : i < \lambda\}$. We will denote this ordered set by $+\{C_i : i < \lambda\}$.

We define now an increasing sequence of total orders X_i:

$Z_0 := X_0 := [Y_0]$ - where $[Y_0]$ is ordered by $Y' < Y'' :\leftrightarrow Y'' \subset Y'$, Y_0 the first element in the above, arbitrary enumeration of $\mathcal{Y}$.

$Z_{i+1} := [Y_{i+1}] - X_i$ (which may be empty), $X_{i+1} := X_i + Z_{i+1}$

for $lim(\lambda)$: $Z_\lambda := [Y_\lambda] - \bigcup\{X_i : i < \lambda\}$, $X_\lambda := (+\{X_i : i < \lambda\}) + Z_\lambda$.

$(\mathcal{Y}, <) := +\{X_i : i < \mu\}$ will be totally, strictly ordered, and if $x < y$ then $\neg(y \supseteq x)$, and thus also $x \subset y \rightarrow y < x$. (This is all we need.)

Finally, we put a (new) element, say t, on top of $\mathcal{Y}$.

Definition 1.8.31

(Ranked order on X)

We define now an order $\prec$ on X, by mapping X by h into $\mathcal{Y} + \{t\}$, and inheriting the order: Let $x \in X$.

Case 1: There is no $A \in \mathcal{Y}$ such that $x \in g(A)$, then $h(x) := t$.

Case 2: There is $A \in \mathcal{Y}$ such that $x \in g(A)$, let then $h(x) := A^+$. By Fact 1.8.56, this is well-defined. Finally, set $x \prec x' :\leftrightarrow h(x) < h(x')$.

It remains to show that this order $\prec$ represents g, and thus f on $\mathcal{X} \subseteq \mathcal{Y}$:

Fact 1.8.57

For $Y \in \mathcal{Y}$, $g(Y) = \mu(Y)$ and $\prec$ is $\mathcal{Y}$-smooth.

Proof

Exercise, solution in the Appendix

We thus have shown:

Theorem 1.8.58

The "good" branches of Thomason's [Tho84] can be chosen by a ranked, smooth preferential order on the set of branches.

More precisely (see Definition 12 in [Tho84]):

Let $\langle T, \prec, O \rangle$ be a tree-like frame for deliberative deontic tense logic, i.e.

1. $T \neq \emptyset$
2. $\prec$ is a transitive irreflexive binary relation such that $t_1 \prec t$, $t_2 \prec t$ implies $t_1 = t_2$ or $t_1 \prec t_2$ or $t_2 \prec t_1$
3. For $t \in T$, let $B_t := \{b : t \in b \text{ and } b \text{ is a maximal subset of } T, \text{ totally ordered under } \prec\}$ (i.e. B_t is the set of (maximal) branches through t) and $O : T \to \bigcup \{B_t : t \in T\}$ is a "choice function of good branches" such that

    (a) $O_t \neq \emptyset$
    (b) $O_t \subseteq B_t$
    (c) $t \prec t'$, $t' \in b \in O_t \to O_{t'} = O_t \cap B_{t'}$.

Then there is a ranked, smooth partial order on $\bigcup \{B_t : t \in T\}$, which represents O.

Proof

Obviously, if $B_t \cap B_{t'} \neq \emptyset$, then $B_t \subseteq B_{t'}$ or $B_{t'} \subseteq B_t$. So $\mathcal{X} := \{B_t : t \in T\}$ satisfies (tp). Properties (1) and (2) of Definition 1.8.27 are obvious for $f := O$. (3) and (4): $B_{t'} \subseteq B_t$ implies $t \prec t'$. If $O_t \cap B_{t'} \neq \emptyset$, then there is $b \in O_t \cap B_{t'}$, so $t' \in b$, but then $O_{t'} = O_t \cap B_{t'}$ by (c). We extend $\mathcal{X}$ and f to $\mathcal{Y}$ and g as in Definition 1.8.28, and represent g as in Definition 1.8.30 and Definition 1.8.31.

We conclude by remarking that Bartha's choice of "bad" branches can express Thomason's choice of "good" branches too. Familiarity with [Bar91] is assumed.

Fact 1.8.59

Bartha, on pages 95-96 of [Bar91], defines "bad" choices by validity of a special constant "proposition" S - which is left unexplained there. Let us introduce a new propositional variable, say $*$, and set $S := F \neg *$, expressing that, sometimes in the future, $*$ will not hold. So the good end-segments of branches, seen from moment m, are those where $\langle b, m \rangle \models \neg F \neg *$ - i.e., everywhere above m on b, $*$ holds. But now, if m' is below m, and b is a branch good above m' and which passes through m, then it is good above m. And if there is another branch b' good above m, then we can "cut and paste" b' onto the intial segment of b, thus $O_m = O_{m'} \cap B_m$.

□

1.8.8.3 An Extension of the Katsuno/Mendelzon Update Semantics

Introduction

The subject of Katsuno/Mendelzon's Update Semantics [KM90] is to describe/characterize reasonable modifications in a database, when learning that the world changes.

Let me begin by emphasizing the following obvious differences:

1) Learning that the world changes (or has changed) is not the same as learning new facts about a static world.

2) Reasoning about possible developments is not the same as reasoning about "normal" developments. E.g., earthquakes are geologically possible at Stuttgart in southern Germany - due to proximity of the Hohenzollerngraben -, but are certainly an unnormal event. For reasons of motivation/simplicity we shall sometimes use the word "probable" as synonym for "normal".

3) The treatment of real time databases is not the same as the treatment of databases on developments: In a real time database, it makes no sense and is impossible to "correct" past beliefs. In a database on developments, say an archeological database on ancient mediterranean cultures, it makes sense to change beliefs about the past. As a matter of fact, the aim of archeological inquiry is to maintain a reconstruction of past events. This database of archeological beliefs is not constructed in the order in which events occurred, and in general will use inferences from later to earlier events as well as inferences from earlier to later ones. As in most areas of empirical inquiry, the process will be nonmonotonic, cases can arise in which beliefs need to be retracted in light of new evidence.

We think that the example of an archeologist trying to make "the best" of the data at hand - and, in an extension, to guide his further research - is a good example of reasoning about developments. He uses information about several time points and "laws" of nature and history to reason about possible developments fitting his information, and uses hypotheses about the normal developments to choose among these the most likely ones, and to conclude back on the probable state of the world at those time points. In the general case, the reasoning processes might be even more complicated, but this exceeds the aims of this Section. A short discussion of such possible extensions is given at the end of this Section.

Criticism of the Katsuno/Mendelzon Approach

Our critique of their approach focusses on two points:

1. We do not see why corrections of past beliefs should be excluded. Katsuno/Mendelzon say in [KM90] on page 2, 3. paragraph, last sentence: "The fact that the real worlds has changed gives us no grounds to conclude that some of the old worlds were actually not possible." Look at the following situation: I have heard that a neighbourhood dog has died. There are two candidates, dog d and e, and I am not sure which. So I think two worlds are possible, one where d is dead, the other where e is (assume that not both are dead). Now I hear dog d barking. No one reasonable would uphold the possibility that d was dead a minute ago. So even in the relatively simple context of possible developments, to be rational, we should revise our belief about the past in the light of new information about the present (or future).

2. Katsuno/Mendelzon's "adding up normality" seems to be too strong a simplification in some situations. In essence, it says that if, from world v, a development to world x is normal, and from world w a development to world y is normal, then, if we believe that v or w holds, we should believe that both developments to x or y are normal. But, why can't we compare developments from different origins? Suppose we live at Naples, and read a book on Pompei last night. Moreover, our neighbour has a truck which makes a horrible noise. Now we are peacefully sitting at my morning's coffee, and hear a terrible rumbling noise. We have two options: Vesuvius has erupted again, or our neighbour has started his truck. The next moment, we feel the house shaking. If we thought only the truck possible, then we would suppose that he has smashed into my house - though a little astonished by the effect on the very stable house. But believing an eruption of Vesuvius possible, this seems to provide a far better explanation of the dramatic developments. So we shall believe in an earthquake, and neglect the second possibility as less normal. Note, however, that if we speak about the possible developments, we are fully entitled to add up all possibilities.

How Can We Patch Things Up?

As basic entity, assume a directed graph Δ (for developments), whose nodes are possible (static) worlds, i.e. propositional models. $w \to v$ means that (in one time step) world w can develop to world v - "can" in the sense of "it is possible". Moreover, we assume given a partial order on the set of arrows, $(w \to v) \prec (x \to y)$ means that the development from w to v is more normal than that from x to y.

Suppose we start reasoning at time t, and have a database $K(t)$ of propositional formulas. This singles out a set possible worlds (nodes) in Δ, where $K(t)$ holds, say $W(t)$. I now get information $K(t+1)$ about the state of affairs at time $t+1$. If I reason about the possible, I shall consider all worlds v as possible, which can be reached in one step from $W(t)$, and where $K(t+1)$ holds (this is as in [KM90]). But, if some world $w \in W(t)$ has no arrow into $K(t+1)$-worlds, then this world should be excluded from $W(t)$ (it has become impossible in the light of the

1.8 Various Results and Approaches

new information), and $K(t)$ be modified (strengthened) accordingly. Analogously, getting information $K(t-1)$, singles out those worlds where $K(t-1)$ holds, and which have an arrow leading into a $K(t)$-world. Again, $K(t)$-worlds, which cannot be reached from $K(t-1)$-worlds, should be eliminated (propagating to $W(t+1)$).

This seems to be a rational approach to the possible. We also note that we can do all three steps simultaneously, by looking at the $K(\tau)$-worlds, and seeing what can be connected by suitable threads, eliminating the rest.

We turn to the probable or normal. Suppose again that we have information $K(t-1)$, $K(t)$, $K(t+1)$. Again, we can single out simultaneously $W(t-1)$, $W(t)$, $W(t+1)$. First, as in the "possible"-case, we eliminate all worlds which do not lie on some complete path (of length 2). But now, we use our partial order on arrows, which we extend suitably to paths: e.g. $\pi < \rho$ iff for all $x \to y$ in π there is $x' \to y'$ in ρ such that $x \to y < x' \to y'$, or, if the order on arrows is ranked, i.e. given by a rotating scale, we can do with one arrow better, and the rest incomparable or better, etc... We now choose the best paths, and accept only those worlds in $W(\tau)$ which lie on some preferred path. (Of course, if we like, we may introduce some order on worlds too, but this will not tell us anything really new.)

Again, this seems to be a rational approach.

The full picture will thus be given by a graph Δ as above, preferences on its arrows, and we consider paths of length e.g. $\mathcal{Z}$ (integers), which we can compare in some reasonable extension of the order on arrows, we have information $K(t)$, $K(t')$, ..., which give us "bottlenecks" $W(t)$, $W(t')$, ..., and we further restrict $W(t)$ to $V(t)$ by considering only those nodes which lie on full paths or even on preferred paths. We generalize even further below by comparing arbitrary functions of the Cartesian product, not necessarily through comparing single arrows.

Note: Directed graphs correspond to binary relations. In the formal part below, we will use the relation language.

Turning back to our archeological example, suppose we are interested in the extinction of the Minoan culture. Then we will try to find information temporarily surrounding the event, to narrow the possible or probable conjectures about the event itself. E.g. decline of other cultures before or after, geological evidence, etc.

We would also like to point out that we do not seek to justify or reason about our choice relation. It is a primitive which we assume given. Neither do we impose any (perhaps plausibel) restrictions like: If the development $a \to b$ is preferred over the development $a' \to b'$, and $b \to c$ over $b' \to c'$, then $a \to b \to c$ is preferred over $a' \to b' \to c'$. We do not speak about subjective versus objective preferences either. In a probabilistic setting, D. Lewis [Lew81] addresses such questions, and an investigation of these problems about preferences could follow Lewis's arguments in a qualitative framework.

Formalizing the Problem, Soundness, and Completeness

We start with two time points; our considerations generalize trivially to the arbitrary case.

So we are given $X \times Y$, and a preference on pairs $\langle x, y \rangle \prec\!\!\prec \langle x', y' \rangle$. We are given $Z \subseteq X \times Y$ as input, where

(1) Z may be $A \times B$, $A \subseteq X$, $B \subseteq Y$

(2) Z may e.g. be an arbitrary union of such $A \times B$'s

(3) Z may be an arbitrary subset of $X \times Y$. The "central machinery" will give the preferred elements $\mu(Z)$ of Z. We can ask the following questions:

(4) What are the preferred elements of Z?

(5) What are the projections of $\mu(Z)$ - i.e. those states in X (Y) through which preferred paths (pairs) pass? More precisely, we look for $\mu(Z)_X := \{x \in X : \exists y \in Y.\langle x, y \rangle \in \mu(Z)\}$ and $\mu(Z)_Y := \{y \in Y : \exists x \in X.\langle x, y \rangle \in \mu(Z)\}$. (Warning: Of course, $x \in \mu(Z)_X$, $y \in \mu(Z)_Y$ does not imply $\langle x, y \rangle \in \mu(Z)$.)

We consider here the case (1) + (5).

Our approach is a generalization of the Katsuno/Mendelzon one: Forget about the first coordinate, assume that all arrows (pairs) of different origin are incomparable, and that for any point x, the arrow $x \to x$ exists and is "the best".

Definition 1.8.32

Fix X, Y. Let $\mathcal{Y}(X) \subseteq \mathcal{P}(X)$, $\mathcal{Y}(Y) \subseteq \mathcal{P}(Y)$.

For $A \subseteq X$, $B \subseteq Y$, let $\mid A \times B \mid := \{A' \times B' : A' \subseteq A, B' \subseteq B, A' \in \mathcal{Y}(X), B' \in \mathcal{Y}(Y)\}$. Note that $A'' \times B'' \subseteq A \times B \to \mid A'' \times B'' \mid \subseteq \mid A \times B \mid$.

For $Z \subseteq X \times Y$, let the projections be $\pi_X(Z) := \{x : \exists y.\langle x, y \rangle \in Z\}$, and $\pi_Y(Z) := \{y : \exists x.\langle x, y \rangle \in Z\}$.

Given $f_X :\mid X \times Y \mid \to \mathcal{P}(X)$, $f_Y :\mid X \times Y \mid \to \mathcal{P}(Y)$, and $A \times B \subseteq X \times Y$, $a \in X$, $b \in Y$, we use the abbreviation $\phi(\langle a, b \rangle, A \times B) := \forall A' \times B' \in \mid A \times B \mid (\langle a, b \rangle \in A' \times B' \to a \in f_X(A' \times B') \land b \in f_Y(A' \times B'))$.

Note that $A'' \times B'' \subseteq A \times B$, $\phi(\langle a, b \rangle, A \times B) \to \phi(\langle a, b \rangle, A'' \times B'')$. (This is trivial by $\mid A'' \times B'' \mid \subseteq \mid A \times B \mid$.)

Recall Proposition 1.3.1:

Let $\mathcal{U} \subseteq \mathcal{P}(U)$, then $f : \mathcal{U} \to \mathcal{P}(U)$ can be represented by a preferential structure iff for all $V, W \in \mathcal{U}$

(f1) $f(V) \subseteq V$

(f2) $V \subseteq W \to f(W) \cap V \subseteq f(V)$.

1.8 Various Results and Approaches

Theorem 1.8.60

a) Let $f : | X \times Y | \to \mathcal{P}(X \times Y)$ be determined by a preferential structure, let $f_X = \pi_X \circ f$, $f_Y = \pi_Y \circ f$ ($\circ$ the composition of functions), then for all $A \times B \in | X \times Y |$

(1X) $f_X(A \times B) \subseteq A$
(1Y) $f_Y(A \times B) \subseteq B$
(2X) For $a \in A$ ($a \in f_X(A \times B) \leftrightarrow \exists b \in B.\phi(\langle a, b \rangle, A \times B))$
(2Y) For $b \in B$ ($b \in f_Y(A \times B) \leftrightarrow \exists a \in A.\phi(\langle a, b \rangle, A \times B))$.

b) Conversely, let $f_X : | X \times Y | \to \mathcal{P}(X)$, $f_Y : | X \times Y | \to \mathcal{P}(Y)$, and

(1X) $f_X(A \times B) \subseteq A$
(1Y) $f_Y(A \times B) \subseteq B$
(2X) For $a \in A$ ($a \in f_X(A \times B) \leftrightarrow \exists b \in B.\phi(\langle a, b \rangle, A \times B))$
(2Y) For $b \in B$ ($b \in f_Y(A \times B) \leftrightarrow \exists a \in A.\phi(\langle a, b \rangle, A \times B))$.

then there is $f : | X \times Y | \to \mathcal{P}(X \times Y)$ determined by a preferential structure, such that $f_X = \pi_X \circ f$, $f_Y = \pi_Y \circ f$.

Proof

We use Proposition 1.3.1.

a)

Exercise, solution in the Appendix

b) For $A \times B \in | X \times Y |$, define $f(A \times B) := \{\langle a, b \rangle \in A \times B : \phi(\langle a, b \rangle, A \times B)\}$. We have to show (f1), (f2) and $f_X = \pi_X \circ f$, $f_Y = \pi_Y \circ f$ (on $| X \times Y |$).

(f1) is trivial.

(f2) Let $A' \times B', A \times B \in | X \times Y |$, $A' \times B' \subseteq A \times B$ and $\langle a, b \rangle \in f(A \times B) \cap A' \times B'$. We have to show $\phi(\langle a, b \rangle, A' \times B')$. But this is trivial by $\phi(\langle a, b \rangle, A \times B)$, $A' \times B' \subseteq A \times B$, and above remark. $f_X = \pi_X \circ f$: Let $A \times B \in | X \times Y |$. "$\subseteq$" Let $a \in f_X(A \times B)$, then by (2X), $\exists b \in B.\phi(\langle a, b \rangle, A \times B)$, thus $\langle a, b \rangle \in f(A \times B)$, so $a \in \pi_X \circ f(A \times B)$. "$\supseteq$" Let $a \in \pi_X \circ f(A \times B)$, so there is $b \in B$ such that $\langle a, b \rangle \in f(A \times B)$, so $\phi(\langle a, b \rangle, A \times B)$, and $a \in A, b \in B$, so by (2X) $a \in f_X(A \times B)$. $f_Y = \pi_Y \circ f$ is analogous.

$\square$

We generalize to the arbitrary Cartesian product (this is straightforward):

Definition 1.8.33

ΠS will denote the Cartesian product. Fix I, $\mathcal{X} := \{X_i : i \in I\}$ and for $i \in I$ let $\mathcal{Y}_i \subseteq \mathcal{P}(X_i)$. For $\mathcal{A} = \{A_i : i \in I\}$ with $A_i \subseteq X_i$ for all $i \in I$, let $|\mathcal{A}| := \{B = \{B_i : i \in I\} : \forall i \in I(B_i \subseteq A_i \land B_i \in \mathcal{Y}_i)\}$. Note again that for $\mathcal{E} = \{E_i : i \in I\}$ and $\Pi\mathcal{E} \subseteq \Pi\mathcal{A} \mid \mathcal{E} \mid \subseteq \mid \mathcal{A} \mid$. Given $f_i :\mid \mathcal{X} \mid \to \mathcal{P}(X_i)$ for all $i \in I$, $\mathcal{A} = \{A_i : i \in I\}$ with $A_i \subseteq X_i$, $g \in \Pi\mathcal{X}$, let $\phi(g, \mathcal{A}) := \forall B \in \mid \mathcal{A} \mid (g \in \Pi B \to \forall i \in I. g(i) \in f_i(\Pi B))$.

Note again that $\Pi B \subseteq \Pi\mathcal{A}, \phi(g, \mathcal{A}) \to \phi(g, B)$. For any $Z \subseteq \Pi\mathcal{X}$ let $\pi_j(Z) := \{x_j \in X_j : \exists z \in Z. z(j) = x_j\}$, the projection to the j-coordinate.

Theorem 1.8.61

a) Let $f : \{\Pi\mathcal{A} : \mathcal{A} \in \mid \mathcal{X} \mid\} \to \mathcal{P}(\Pi\mathcal{X})$ be defined by a preferential structure on $\Pi\mathcal{X}$ and $\forall i \in I \; f_i = \pi_i \circ f$, then for all $\mathcal{A} \in \mid \mathcal{X} \mid$

(1) $\forall i \in I. f_i(\Pi\mathcal{A}) \subseteq A_i$

(2) $\forall i \in I \forall a_i \in A_i \; (a_i \in f_i(\Pi\mathcal{A}) \leftrightarrow \exists g \in \Pi\mathcal{A}(g(i) = a_i \land \phi(g, \mathcal{A})))$

b) Conversely let for all $i \in I \; f_i : \{\Pi\mathcal{A}. \mathcal{A} \in \mid \mathcal{X} \mid\} \to \mathcal{P}(X_i)$ and for all $\mathcal{A} \in \mid \mathcal{X} \mid$

(1) $\forall i \in I. f_i(\Pi\mathcal{A}) \subseteq A_i$

(2) $\forall i \in I \forall a_i \in A_i \; (a_i \in f_i(\Pi\mathcal{A}) \leftrightarrow \exists g \in \Pi\mathcal{A}(g(i) = a_i \land \phi(g, \mathcal{A})))$

then there is $f : \{\Pi\mathcal{A} : \mathcal{A} \in \mid \mathcal{X} \mid\} \to \mathcal{P}(\Pi\mathcal{X})$ determined by a preferential structure such that $\forall i \in I \; f_i = \pi_i \circ f$.

Proof

We use again Proposition 1.3.1.

a) Fix $i \in I$.

(1) is trivial by $f(Z) \subseteq Z$.

(2) "$\to$" Let $a_i \in f_i(\Pi\mathcal{A}) = \pi_i(f(\Pi\mathcal{A}))$. So there is $g \in f(\Pi\mathcal{A})$, $a_i = g(i)$. We show $\phi(g, \mathcal{A})$. Let $B \in \mid \mathcal{A} \mid$, $g \in \Pi B$, so $g \in f(\Pi B)$ by (f2), so $\forall i \in I. g(i) \in \pi_i(f(\Pi B)) = f_i(\Pi B)$. "$\leftarrow$" Let $a_i \in A_i$, $g \in \Pi\mathcal{A}$ such that $g(i) = a_i \land \phi(g, \mathcal{A})$, but then by $\mathcal{A} \in \mid \mathcal{A} \mid a_i = g(i) \in f_i(\Pi\mathcal{A})$.

b) Suppose all f_i satisfy (1) and (2). For $\mathcal{A} \in \mid \mathcal{X} \mid$, define $f(\Pi\mathcal{A}) := \{g \in \Pi\mathcal{A} : \phi(g, \mathcal{A})\}$. We have to show (f1), (f2) and $\forall i \in I. f_i = \pi_i \circ f$.

(f1) is trivial.

(f2) Let $\mathcal{A}, B \in \mid \mathcal{X} \mid$, $\Pi B \subseteq \Pi\mathcal{A}$ and $g \in f(\Pi\mathcal{A}) \cap \Pi B$. We have to show $\phi(g, B)$, but this is trivial by $\Pi B \subseteq \Pi\mathcal{A}$ and $\phi(g, \mathcal{A})$. $f_i = \pi_i \circ f$: "$\subseteq$" Let $a_i \in f_i(\Pi\mathcal{A}) \subseteq A_i$, so by (2) $\exists g \in \Pi\mathcal{A}. g(i) = a_i \land \phi(g, \mathcal{A})$, so

$g \in f(\Pi\mathcal{A})$ and $a_i = g(i) \in \pi_i(f(\Pi\mathcal{A}))$. "$\supseteq$" Let $a_i \in \pi_i(f(\Pi\mathcal{A})) \to \exists g \in f(\Pi\mathcal{A}).g(i) = a_i \to \exists g \in \Pi\mathcal{A}.g(i) = a_i \wedge \phi(g, \mathcal{A}) \to a_i = g(i) \in f_i(\Pi\mathcal{A})$.

$\square$

We formulate the logical counterpart of Theorem 1.8.60 - the proof is easy.

Fix a propositional language $\mathcal{L}$.

Definition 1.8.34

We consider two logics, $.^i : \mathcal{P}(\mathcal{L}) \times \mathcal{P}(\mathcal{L}) \to \mathcal{P}(\mathcal{L})$, $\langle S, T \rangle \mapsto \langle S, T \rangle^i \subseteq \mathcal{L}$.

We say that both logics $\langle S, T \rangle^i$ are given by a function $f : \boldsymbol{D} \times \boldsymbol{D} \to \mathcal{P}(M_\mathcal{L} \times M_\mathcal{L})$ iff for all theories S, T $\langle S, T \rangle^i = \{\phi : \forall m \in \pi_i(f(M_S \times M_T)).m \models \phi\}$, where π_i is the projection on the i-th coordinate.

Analogously, we say that they are given by a preferential structure on $M_\mathcal{L} \times M_\mathcal{L}$ iff f is given by such a structure.

We call a function f on pairs of models definability preserving (dp) iff for all theories S, T $\pi_1(f(M_S \times M_T)) = M_U$ and $\pi_2(f(M_S \times M_T)) = M_V$ for some theories U, V, where M_S is the set of S-models etc. Note that then $\pi_i(f(M_S \times M_T)) = M_{\langle S, T\rangle^i}$.

For U, V complete, consistent theories (cct), S, T any theories, we abbreviate $\psi(U, V, S, T) := \forall S', T'(U \vdash S' \vdash S \wedge V \vdash T' \vdash T \to U \vdash \langle S', T'\rangle^1 \wedge V \vdash \langle S', T'\rangle^2)$.

Fact 1.8.62

Let $\mathcal{Y}(X) = \mathcal{Y}(Y) = \boldsymbol{D}$, $f_X(M_S \times M_T) = M_{\langle S,T\rangle^1}$, $f_Y(M_S \times M_T) = M_{\langle S,T\rangle^2}$, U, V cct, then $\psi(U, V, S, T) \leftrightarrow \phi(\langle m_U, m_V \rangle, M_S \times M_T)$.

Proof

This fact follows immediately from: $\langle m_U, m_V \rangle \in M_{S'} \times M_{T'} \subseteq M_S \times M_T \leftrightarrow U \vdash S' \vdash S, V \vdash T' \vdash T$, $m_U \in f_X(M_{S'} \times M_{T'}) = M_{\langle S',T'\rangle^1} \leftrightarrow U \vdash < S', T' >^1$, $m_V \in f_Y(M_{S'} \times M_{T'}) = M_{\langle S',T'\rangle^2} \leftrightarrow V \vdash < S', T' >^2$.

$\square$

Theorem 1.8.63

Let $\langle S, T \rangle^1$, $\langle S, T \rangle^2$ be two logics on pairs of theories. Then $\langle S, T \rangle^i$ are given by a dp preferential structure iff

(1) $\overline{S} = \overline{S'} \wedge \overline{T} = \overline{T'} \to \langle S, T \rangle^1 = \langle S', T' \rangle^1 \wedge \langle S, T \rangle^2 = \langle S', T' \rangle^2$

(2) $\langle S, T \rangle^i$ is classically closed

(3) $\langle S,T\rangle^1 \vdash S$, $\langle S,T\rangle^2 \vdash T$

(4S) If U is a cct with $U \vdash S$, then $U \vdash \langle S,T\rangle^1$ iff there is a cct V such that $V \vdash T$ and $\psi(U,V,S,T)$

(4T) If V is a cct with $V \vdash T$, then $V \vdash \langle S,T\rangle^2$ iff there is a cct U such that $U \vdash S$ and $\psi(U,V,S,T)$.

Proof

We use Theorem 1.8.60. For a cct U let m_U denote its model. Note that if f is dp, U a cct, and $U \vdash \langle S,T\rangle^1$, then $m_U \in \pi_X(f(M_S \times M_T))$: $m_U \in M_{\langle S,T\rangle^1} = \pi_X(f(M_S \times M_T))$, likewise for $V \vdash \langle S,T\rangle^2$.

"$\rightarrow$" Note that $\pi_X \circ f = f_X$.

(1) and (2) are trivial.

(3) By (1X) of Theorem 1.8.60, $\pi_X(f(M_S \times M_T)) \subseteq M_S$, so $\langle S,T\rangle^1 \vdash S$.

(4S) Let U be cct, $U \vdash S$. $U \vdash \langle S,T\rangle^1$ iff $m_U \in M_{\langle S,T\rangle^1} = \pi_X(f(M_S \times M_T)) = f_X(M_S \times M_T)$ iff (by (2X) of Theorem 1.8.60) there is $m_V \in M_T.\phi(\langle m_U, m_V\rangle, M_S \times M_T)$ iff there is a cct V such that $V \vdash T$ and $\psi(U,V,S,T)$ (by Fact 1.8.62).

(4T) is analogous.

"$\leftarrow$" Let $\langle S,T\rangle^i$ satisfy (1) – (4T), define $f_X, f_Y : D \times D \rightarrow D$ as follows: For $M_S \times M_T$, let $f_X(M_S \times M_T) := M_{\langle S,T\rangle^1}$, $f_Y(M_S \times M_T) := M_{\langle S,T\rangle^2}$. By (1), this is well-defined. We have to show (iX), (iY), of Theorem 1.8.60, note that then $f_X = \pi_X \circ f$, $f_Y = \pi_Y \circ f$. Moreover, we have to show $\langle S,T\rangle^1 = \{\phi : \forall m \in f_X(M_S \times M_T)).m \models \phi\}$ and $\langle S,T\rangle^2 = \{\phi : \forall m \in f_Y(M_S \times M_T)).m \models \phi\}$. We will only show the properties for the first coordinate, the second being analogous.

(1X): Let $m \in f_X(M_S \times M_T) = M_{\langle S,T\rangle^1}$, then, by (3) $m \in M_S$.

(2X): Let $m_U \in M_S$, so $U \vdash S$. Then $m_U \in f_X(M_S \times M_T) = M_{\langle S,T\rangle^1}$ iff $U \vdash \langle S,T\rangle^1$ iff (by (4S)) there is cct V, $V \vdash T$ and $\psi(U,V,S,T)$ iff (by Fact 1.8.62) there is $m_V \in M_T$ and $\phi(\langle m_U, m_V\rangle, M_S \times M_T)$.

It remains to show $\langle S,T\rangle^1 = \{\phi : \forall m \in f_X(M_S \times M_T)).m \models \phi\}$. By (2), it suffices to show $M_{\langle S,T\rangle^1} = f_X(M_S \times M_T)$, but this was the definition.

□

1.8 Various Results and Approaches

Extensions

The First Order Case:

As already pointed out in the introduction, our basic tool, Proposition 1.3.1 is an algebraic characterization and independent of logic. To adapt our result to the first order case (and still other logics, if we wish), we thus have to look into the material from Definition 1.8.34 onward only. We see that the only properties we need of the base logic $\vdash$ and its models $M_\mathcal{L}$ are soundness and completeness (in the strong sense, $T \vdash T'$ iff $M(T) \subseteq M(T')$), and that complete and consistent theories have exactly one model.

The latter property obviously is not satisfied by the first order case, but it does not seem unreasonable to make the additional assumption that all models, which satisfy the same first order formulas, are treated in the same manner. With this assumption, we can work with equivalence classes of models, or one representative in each case, and the proofs go through.

Further Extensions:

If we look back at our motivating scenario of an archeologist, we see that we have covered only part of his reasoning. For instance, he might have discovered a fact, e.g. a burial, but is unable to date it, so he knows that T holds at some time, but is uncertain about the exact time point. (This example is due to a referee.) Obviously, our results do not cover this case. For a formal result in the style of Theorem 1.8.63, we first need a language to speak about developments, allowing to quantify over time points, and then a base logic to reason about such developments - which would be the base models of our language. In a second step, we would work with a preference relation on developments, and use our Proposition 1.3.1 again to work for a formal result.

Chapter 2
Higher Preferential Structures

Abstract Recall from Chapter 1 that the basic property of "normal" preferential structures is that minimization is upward absolute. If $x, y \in X$, $x \prec y$, i.e., y is a non-minimal element in X, and $X \subseteq Y$, then y will be a non-minimal element in Y, too - as $x \in Y$. This results in the fundamental "algebraic" property that for $X \subseteq Y$ $\mu(Y) \cap X \subseteq \mu(X)$, where $\mu(X)$ is the set of minimal elements of X.

We can read $x \prec y$ as "x attacks y". In higher preferential structures, attacks can be attacked themselves. There might be z, and an attack from z against the attack $x \prec y$. If $z \in Y - X$, then $x \prec y$ is valid in X, but not in Y any more. Consequently, above fundamental algebraic property need not hold any more, see Section 2.3.1.

A representation result for the general, not necessarily smooth case is given in Section 2.3.2.

We distinguish two cases of smoothness, the usual one, called total smoothness here, and "essential smoothness", presented in Definition 2.3.5, with results in Section 2.3.4, in particular the representation result Proposition 2.3.13, using $(\mu \subseteq)$, $(\cap)$, and (μCUM).

2.1 Introduction

IBRS, information bearing systems, were introduced by Dov Gabbay, their interpretation as higher preferential structures leads to interesting new semantics. They are "higher" in the sense that relations are not only between points (models), but also between points and pairs of the basic relation, in arrow notation, between nodes and arrows.

Preferentially speaking, the basic property

$$X \subseteq Y \rightarrow \mu(Y) \cap X \subseteq \mu(X)$$

will not hold any more, as $x \prec x'$ can now be attacked by a new element in $Y\text{-}X$, an attack "invisible" in X.

It is evident that this idea introduces a whole new class of semantics.

2.2 IBRS

2.2.1 Definition and Comments

Definition 2.2.1

(1) An information bearing binary relation frame IBR, has the form $(S, \Re)$, where S is a nonempty set and $\Re$ is a subset of S_ω, where S_ω is defined by induction as follows:

(1.1) $S_0 := S$

(1.2) $S_{n+1} := S_n \cup (S_n \times S_n)$

(1.3) $S_\omega = \bigcup \{S_n : n \in \omega\}$.

We call elements from S points or nodes, and elements from $\Re$ arrows. Given $(S, \Re)$, we also set $\boldsymbol{P}((S, \Re)) := S$ and $\boldsymbol{A}((S, \Re)) := \Re$.

If α is an arrow, the origin and destination of α are defined as usual, and we write $\alpha : x \to y$ when x is the origin and y the destination of the arrow α. We also write $o(\alpha)$ and $d(\alpha)$ for the origin and destination of α.

(2) Let Q be a set of atoms, and $\boldsymbol{L}$ be a set of labels (usually $\{0, 1\}$ or $[0, 1]$). An information assignment h on $(S, \Re)$ is a function $h : Q \times \Re \to \boldsymbol{L}$.

(3) An information bearing system IBRS has the form $(S, \Re, h, Q, \boldsymbol{L})$, where $S, \Re, h, Q, \boldsymbol{L}$ are as above.

See Figure 2.1 for an illustration.

We have here:

$$S = \{a, b, c, d, e\},$$
$$\Re = S \cup \{(a, b), (a, c), (d, c), (d, e)\} \cup \{((a, b), (d, c)), (d, (a, c))\},$$
$$Q = \{p, q\}.$$

The values of h for p and q are as indicated in the figure. For example $h(p, (d, (a, c))) = 1$.

2.2 IBRS

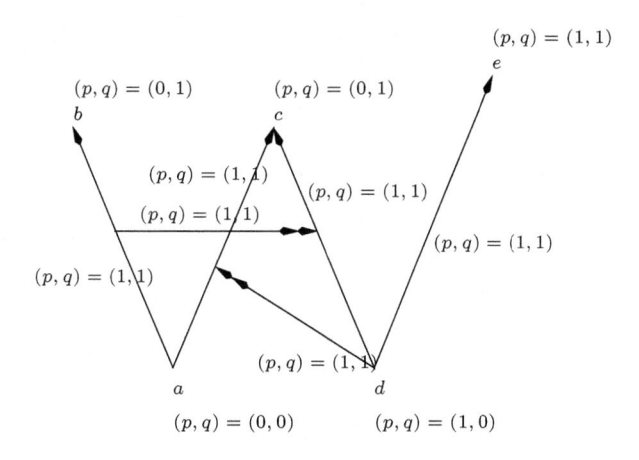

Fig. 2.1 A simple example of an information bearing system

Comment

The elements in Figure 2.1 can be interpreted in many ways, depending on the area of application.

(1) The points in S can be interpreted as possible worlds, or as nodes in an argumentation network or nodes in a neural net or states.
(2) The direct arrows from nodes to nodes can be interpreted as accessibility relations, attack or support arrows in an argumentation networks, connections in a neural net, a preferential ordering in a non-monotonic model, etc.
(3) The labels on the nodes and arrows can be interpreted as fuzzy values in the accessibility relation, weights in a neural net, strength of arguments and their attack in an argumentation net, distances in a counterfactual model, etc.
(4) The double arrows can be interpreted as feedback loops to nodes or to connections, or as reactive links changing the system which are activated as we pass between the nodes.

Thus, IBRS can be used as a source of information for various logics based on the atoms in Q. We now illustrate by listing several such logics.

Modal Logic

One can consider the figure as giving rise to two modal logic models. One with actual world a and one with d, these being the two minimal points of the relation. Consider a language with $\Box q$. how do we evaluate $a \models \Box q$?

The modal logic will have to give an algorithm for calculating the values.

Say we choose algorithm $\mathcal{A}_1$ for $a \models \Box q$, namely:

[$\mathcal{A}_1(a, \Box q) = 1$] iff for all $x \in S$ such that $a = x$ or $(a, x) \in \Re$ we have $h(q, x) = 1$.

According to $\mathcal{A}_1$ we get that $\Box q$ is false at a. $\mathcal{A}_1$ gives rise to a T-modal logic. Note that the reflexivity is not anchored at the relation $\Re$ of the network but in the algorithm $\mathcal{A}_1$ in the way we evaluate. We say $(S, \Re, \ldots) \models \Box q$ iff $\Box q$ holds in all minimal points of $(S, \Re)$.

For orderings without minimal points we may choose a subset of distinguished points.

Non-monotonic Deduction

We can ask whether $p \mid\!\sim q$ according to algorithm $\mathcal{A}_2$ defined below. $\mathcal{A}_2$ says that $p \mid\!\sim q$ holds iff q holds in all minimal models of p. Let us check the value of $\mathcal{A}_2$ in this case:

Let $S_p = \{s \in S \mid h(p, s) = 1\}$. Thus $S_p = \{d, e\}$.

The minimal points of S_p are $\{d\}$. Since $h(q, d) = 0$, we have that $p \not\mid\!\sim q$.

Note that in the cases of modal logic and non-monotonic logic we ignored the arrows $(d, (a, c))$ (i.e., the double arrow from d to the arrow (a, c)) and the h values to arrows. These values do not play a part in the traditional modal or non-monotonic logic. They do play a part in other logics. The attentive reader may already suspect that we have here an opportunity for generalisation of, say, non-monotonic logic, by giving a role to arrow annotations.

Argumentation Nets

Here the nodes of S are interpreted as arguments. The atoms $\{p, q\}$ can be interpreted as types of arguments and the arrows e.g., $(a, b) \in \Re$ as indicating that the argument a is attacking the argument b.

So, for example, let

   a = We must win votes.
   b = Death sentence for murderers.
   c = We must allow abortion for teenagers.
   d = Bible forbids taking of life.
   q = The argument is a social argument.
   p = The argument is a religious argument.

2.2 IBRS

$(d, (a, c))$ = There should be no connection between winning votes and abortion.

$((a, b), (d, c))$ = If we attack the death sentence in order to win votes then we must stress (attack) that there should be no connection between religion (Bible) and social issues.

Thus we have according to this model that supporting abortion can lose votes. The argument for abortion is a social one and the argument from the Bible against it is a religious one.

We can extract information from this IBRS using two algorithms. The modal logic one can check whether for example every social argument is attacked by a religious argument. The answer is no, since the social argument b is attacked only by a which is not a religious argument.

We can also use algorithm $\mathcal{A}_3$ (following Dung) to extract the winning arguments of this system. The arguments a and d are winning since they are not attacked. d attacks the connection between a and c (i.e., stops a attacking c).

The attack of a on b is successful and so b is out. However the arrow (a, b) attacks the arrow (d, c). So c is not attacked at all as both arrrows leading into it are successfully eliminated. So c is in. e is out because it is attacked by d.

So the winning arguments are $\{a, c, d\}$.

In this model we ignore the annotations on arrows. To be consistent in our mathematics we need to say that h is a partial function on $\Re$. The best way is to give more specific definition on IBRS to make it suitable for each logic.

See also [Gab08b] and [BGW05].

Counterfactuals

The traditional semantics for counterfactuals involves closeness of worlds. The clause $y \models p \hookrightarrow q$, where $\hookrightarrow$ is a counterfactual implication is that q holds in all worlds y' "near enough" to y in which p holds. So if we interpret the annotation on arrows as distances then we can define "near" as distance ≤ 2, we get: $a \models p \hookrightarrow q$ iff in all worlds of p-distance ≤ 2 if p holds so does q. Note that the distance depends on p.

In this case we get that $a \models p \hookrightarrow q$ holds. The distance function can also use the arrows from arrows to arrows, etc. There are many opportunities for generalisation in our IBRS set up.

Intuitionistic Persistence

We can get an intuitionistic Kripke model out of this IBRS by letting, for $t, s \in S$, $t\rho_0 s$ iff $t = s$ or $[tRs \wedge \forall q \in Q(h(q, t) \leq h(q, s))]$. We get that

[$r_0 = \{(y, y) \mid y \in S\} \cup \{(a, b), (a, c), (d, e)\}$.]

Let ρ be the transitive closure of ρ_0. Algorithm $\mathcal{A}_4$ evaluates $p \Rightarrow q$ in this model, where $\Rightarrow$ is intuitionistic implication.

$\mathcal{A}_4 : p \Rightarrow q$ holds at the IBRS iff $p \Rightarrow q$ holds intuitionistically at every ρ-minimal point of (S, ρ).

2.2.2 The Power of IBRS

We show now how a number of logics fit into our general picture of IBRS.

(1) Nonmonotonic logics in the form of preferential logics:

There are only arrows from nodes to nodes, and they are unlabelled. The nodes are classical models, and as such all propositional variables of the base language are given a value from $\{0, 1\}$.

The structure is used as described above, i.e., the R-minimal models of a formula or theory are considered.

(2) Theory revision

In the full case, i.e., where the left hand side can change, nodes are again classical models, arrows exist only between nodes, and express by their label the distance between nodes. Thus, there is just one (dummy) p, and a real value as label. In the AGM situation, where the left hand side is fixed, nodes are classical models (on the right) or sets thereof (on the left), arrows go from sets of models to models, and express again distance from a set (the K-models in AGM notation) to a model (of the new formula ϕ).

The structure is used by considering the closest ϕ-models.

The framework is sufficiently general to express revision also differently: Nodes are pairs of classical models, and arrows express that in pair (a, b) the distance from a to b is smaller than the distance in the pair (a', b').

(3) Theory update

As developments of length 2 can be expressed by a binary relation and the distance associated, we can - at least in the simple case - proceed analogously to the first revision situation. It seems, however, more natural to consider as nodes threads of developments, i.e., sequences of classical models, as arrows comparisons between such threads, i.e., unlabelled simple arrows only, expressing that one thread is more natural or likely than another.

The evaluation is then by considering the "best" threads under above comparison, and taking a projection on the desired coordinate (i.e., classical model). The result is then the theory defined by these projections.

(4) Deontic logic

Just as for preferential logics.

2.2 IBRS   239

(5) The logic of counterfactual conditionals

Again, we can compare pairs (with same left element) as above, or, alternatively, compare single models with respect to distance from a fixed other model. This would give arrows with indices, which stand for this other model.

Evaluation will then be as usual, taking the closest ϕ-models, and examining whether ψ holds in them.

(6) Modal logic

Nodes are classical models, and thus have the usual labels, arrows are unlabelled, and only between nodes, and express reachability.

For evaluation, starting from some point, we collect all reachable other models, perhaps adding the point of departure.

(7) Intuitionistic logic

Just as for modal logic.

(8) Inheritance systems

Nodes are properties (or sets of models), arrows come in two flavours, positive and negative, and exist between nodes only.

The evaluation is relatively complicated, and the subject of ongoing discussion.

(9) Argumentation theory

There is no unique description of an argumentation system as an IBRS. For instance, an inheritance system is an argumentation system, so we can describe such a system as detailed above. But an argument can also be a deontic statement, as we saw in the first part of this introduction, and a deontic statement can be described as an IBRS itself. Thus, a node can be, under finer granularity, itself an IBRS. Labels can describe the type of argument (social, etc.) or its validity, etc.

2.2.3 Abstract Semantics for IBRS and Its Engineering Realization

2.2.3.1 Introduction

We give here a rough outline of a formal semantics for IBRS. It consists more of some hints where difficulties are, than of a finished construction. Still, the authors feel that this is sufficient to complete the work, and, on the other hand, our remarks might be useful to those who intend to finalize the construction.

(1) Nodes and arrows

As we may have counterarguments not only against nodes, but also against arrows, they must be treated basically the same way, i.e., in some way there has to be a positive, but also a negative influence on both. So arrows cannot just be concatenation between the contents of nodes.

We will differentiate between nodes and arrows by labelling arrows in addition with a time delay. We see nodes as situations, where the output is computed instantenously from the input, whereas arrows describe some "force" or "mechanism" which may need some time to "compute" the result from the input.

Consequently, if α is an arrow, and β an arrow pointing to α, then it should point to the input of α, i.e., before the time lapse. Conversely, any arrow originating in α should originate after the time lapse.

Apart from this distinction, we will treat nodes and arrows the same way, so the following discussion will apply to both - which we call just "objects".

(2) Defeasibility

The general idea is to code each object, say X, by $I(X) : U(X) \to C(X)$: If $I(X)$ holds then, unless $U(X)$ holds, consequence $C(X)$ will hold. (We adopted Reiter's notation for defaults, as IBRS have common points with the former.)

The situation is slightly more complicated, as there can be several counterarguments, so $U(X)$ really is an "or". Likewise, there can be several supporting arguments, so $I(X)$ also is an "or".

A counterargument must not always be an argument against a specific supporting argument, but it can be. Thus, we should admit both possibilties. As we can use arrows to arrows, the second case is easy to treat (as is the dual, a supporting argument can be against a specific counterargument). How do we treat the case of unspecific pro- and counterarguments? Probably the easiest way is to adopt Dung's idea: an object is in, if it has at least one support, and no counterargument - see [Dun95]. Of course, other possibilities may be adopted, counting, use of labels, etc., but we just consider the simple case here.

(3) Labels

In the general case, objects stand for some kind of defeasible transmission. We may in some cases see labels as restricting this transmission to certain values. For instance, if the label is $p = 1$ and $q = 0$, then the p-part may be transmitted and the q-part not.

Thus, a transmission with a label can sometimes be considered as a family of transmissions, which ones are active is indicated by the label.

Example 2.2.1

In fuzzy Kripke models, labels are elements of $[0, 1]$. $p = 0.5$ as label for a node m' which stands for a fuzzy model means that the value of p is 0.5. $p = 0.5$ as label for an arrow from m to m' means that p is transmitted with value 0.5. Thus, when we look from m to m', we see p with value $0.5 * 0.5 = 0.25$. So, we have $\Diamond p$ with value 0.25 at m - if, e.g., m, m' are the only models.

2.2 IBRS

(4) Putting things together

If an arrow leaves an object, the object's output will be connected to the (only) positive input of the arrow. (An arrow has no negative inputs from objects it leaves.) If a positive arrow enters an object, it is connected to one of the positive inputs of the object, analogously for negative arrows and inputs.

When labels are present, they are transmitted through some operation.

In slightly more formal terms, we have:

Definition 2.2.2

In the most general case, objects of IBRS have the form: $(\langle I_1, L_1\rangle, \ldots, \langle I_n, L_n\rangle)$: $(\langle U_1, L'_1\rangle, \ldots, \langle U_n, L'_n\rangle)$, where the L_i, L'_i are labels and the I_i, U_i might be just truth values, but can also be more complicated, a (possibly infinite) sequence of some values. Connected objects have, of course, to have corresponding such sequences. In addition, the object X has a criterion for each input, whether it is valid or not (in the simple case, this will just be the truth value "TRUE"). If there is at least one positive valid input I_i, and no valid negative input U_i, then the output $C(X)$ and its label are calculated on the basis of the valid inputs and their labels. If the object is an arrow, this will take some time, t, otherwise, this is instantaneous.

Evaluating a Diagram

An evaluation is relative to a fixed input, i.e., some objects will be given certain values, and the diagram is left to calculate the others. It may well be that it oscillates, i.e., shows a cyclic behaviour. This may be true for a subset of the diagram, or the whole diagram. If it is restricted to an unimportant part, we might neglect this. Whether it oscillates or not can also depend on the time delays of the arrows (see Example 2.2.2).

We therefore define for a diagram Δ

$\alpha \mathrel{|\!\sim}_\Delta \beta$ iff

(a) α is a (perhaps partial) input - where the other values are set "not valid"

(b) β is a (perhaps partial) output

(c) after some time, β is stable, i.e., all still possible oscillations do not affect β

(d) the other possible input values do not matter, i.e., whatever the input, the result is the same.

In the cases examined here more closely, all input values will be defined.

2.2.3.2 A Circuit Semantics for Simple IBRS Without Labels

It is natural to implement IBRS by (modified) logical circuits. The nodes will be implemented by sub-circuits which can store information, and the arcs by connections between them. As connections can connect connections, connections will not just be simple wires. The objective of the present discussion is to warn the reader that care has to be taken about the temporal behaviour of such circuits, especially when feedback is allowed.

All details are left to those interested in a realization.

Background: It is standard to implement the usual logical connectives by electronic circuits. These components are called gates. Circuits with feedback sometimes show undesirable behaviour when the initial conditions are not specified. (When we switch a circuit on, the outputs of the individual gates can have arbitrary values.) The technical realization of these initial values shows the way to treat defaults. The initial values are set via resistors (in the order of $1\ k\Omega$) between the point in the circuit we want to intialize and the desired tension (say 0 Volt for FALSE, 5 Volt for TRUE). They are called pull-down or pull-up resistors (for default 0 or 5 Volt). When a "real" result comes in, it will override the tension applied via the resistor.

Closer inspection reveals that we have here a 3 level default situation: The initial value will be the weakest, which can be overridden by any "real" signal, but a positive argument can be overridden by a negative one. Thus, the biggest resistor will be for the initialization, the smaller one for the supporting arguments, and the negative arguments have full power. Technical details will be left to the experts.

We give now an example which shows that the delays of the arrows can matter. In one situation, a stable state is reached, in another, the circuit begins to oscillate.

Example 2.2.2

(In engineering terms, this is a variant of a JK flip-flop with $R*S=0$, a circuit with feedback.)

We have 8 measuring points.

$In1, In2$ are the overall input, $Out1, Out2$ the overall output, $A1, A2, A3, A4$ are auxiliary internal points. All points can be TRUE or FALSE.

The logical structure is as follows:

$A1 = In1 \wedge Out1$, $A2 = In2 \wedge Out2$,

$A3 = A1 \vee Out2$, $A4 = A2 \vee Out1$,

$Out1 = \neg A3$, $Out2 = \neg A4$.

Thus, the circuit is symmetrical, with $In1$ corresponding to $In2$, $A1$ to $A2$, $A3$ to $A4$, $Out1$ to $Out2$.

2.2 IBRS

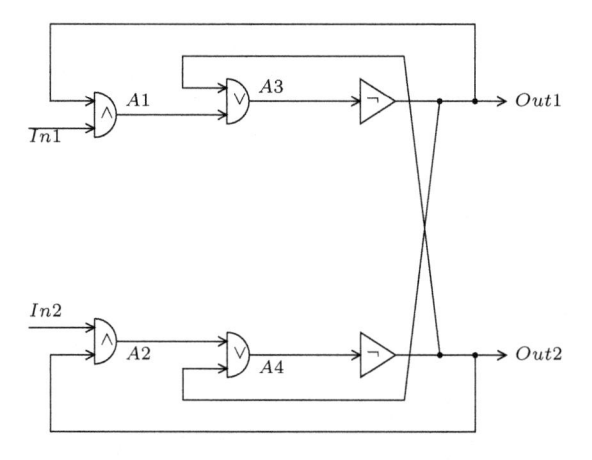

Fig. 2.2 Gate Semantics

The input is held constant. See Figure 2.2.

We suppose that the output of the individual gates is present n time slices after the input was present. n will in the first circuit be equal to 1 for all gates, in the second circuit equal to 1 for all but the AND gates, which will take 2 time slices. Thus, in both cases, e.g., $Out1$ at time t will be the negation of $A3$ at time $t-1$. In the first case, $A1$ at time t will be the conjunction of $In1$ and $Out1$ at time $t-1$, and in the second case the conjunction of $In1$ and $Out1$ at time $t-2$.

We initialize $In1$ as TRUE, all others as FALSE. (The initial value of $A3$ and $A4$ does not matter, the behaviour is essentially the same for all such values.)

The first circuit will oscillate with a period of 4, the second circuit will go to a stable state.

We have the following transition tables (time slice shown at left):

Circuit 1, $delay = 1$ everywhere:

Circuit 2, $delay = 1$ everywhere, except for AND with $delay = 2$:

(Thus, $A1$ and $A2$ are held at their intial value up to time 2, then they are calculated using the values of time $t-2$.)

Note that state 6 of circuit 2 is also stable in circuit 1, but it is never reached in that circuit.

Table 2.1 Oscillating variant

	In1	In2	A1	A2	A3	A4	Out1	Out2	
1:	T	F	F	F	F	F	F	F	
2:	T	F	F	F	F	F	T	T	
3:	T	F	T	F	T	T	T	T	
4:	T	F	T	F	T	T	F	F	
5:	T	F	F	F	T	F	F	F	oscillation starts
6:	T	F	F	F	F	F	F	T	
7:	T	F	F	F	T	F	T	T	
8:	T	F	T	F	T	T	F	T	
9:	T	F	F	F	T	F	F	F	back to start of oscillation

Table 2.2 No oscillation

	In1	In2	A1	A2	A3	A4	Out1	Out2	
1:	T	F	F	F	F	F	F	F	
2:	T	F	F	F	F	F	T	T	
3:	T	F	F	F	T	T	T	T	
4:	T	F	T	F	T	T	F	F	
5:	T	F	T	F	T	F	F	F	
6:	T	F	F	F	T	F	F	T	stable state reached
7:	T	F	F	F	T	F	F	T	
8:	T	F	F	F	T	F	F	T	

2.3 Higher Preferential Structures

2.3.1 Introduction

Definition 2.3.1

An IBR is called a generalized preferential structure iff the origins of all arrows are points. We will usually write x, y, etc. for points, α, β, etc. for arrows.

Definition 2.3.2

Consider a generalized preferential structure $\mathcal{X}$.

(1) Level n arrow:

   Definition by upward induction.

2.3 Higher Preferential Structures

If $\alpha : x \to y$, x, y are points, then α is a level 1 arrow.

If $\alpha : x \to \beta$, x is a point, β a level n arrow, then α is a level $n+1$ arrow. ($o(\alpha)$ is the origin, $d(\alpha)$ is the destination of α.)

$\lambda(\alpha)$ will denote the level of α.

(2) Level n structure:

$\mathcal{X}$ is a level n structure iff all arrows in $\mathcal{X}$ are at most level n arrows.

We consider here only structures of some arbitrary but finite level n.

(3) We define for an arrow α by induction $O(\alpha)$ and $D(\alpha)$.

If $\lambda(\alpha) = 1$, then $O(\alpha) := \{o(\alpha)\}$, $D(\alpha) := \{d(\alpha)\}$.

If $\alpha : x \to \beta$, then $D(\alpha) := D(\beta)$, and $O(\alpha) := \{x\} \cup O(\beta)$.

Thus, for example, if $\alpha : x \to y$, $\beta : z \to \alpha$, then $O(\beta) := \{x, z\}$, $D(\beta) = \{y\}$. Consider also the arrow $\beta := \langle \beta', l' \rangle$ in Figure 2.4. There, $D(\beta) = \{\langle x, i \rangle\}$, $O(\beta) = \{\langle z', m' \rangle, \langle y, j \rangle\}$.

Comment 2.3.1

A counterargument to α is not an argument for $\neg \alpha$ (this is asking for too much), but just showing one case where $\neg \alpha$ holds. In preferential structures, an argument for α is a set of level 1 arrows, eliminating $\neg \alpha$-models. A counterargument is one level 2 arrow, attacking one such level 1 arrow.

Of course, when we have copies, we may need many successful attacks, on all copies, to achieve the goal. As we may have copies of level 1 arrows, we may need many level 2 arrows to destroy them all.

We will not consider here diagrams with arbitrarily high levels. One reason is that diagrams like the following will have an unclear meaning:

Example 2.3.1

$\langle \alpha, 1 \rangle : x \to y$,

$\langle \alpha, n+1 \rangle : x \to \langle \alpha, n \rangle$ ($n \in \omega$).

Is $y \in \mu(X)$?

Definition 2.3.3

Let $\mathcal{X}$ be a generalized preferential structure of (finite) level n.

We define (by downward induction):

(1) Valid $X \mapsto Y$ arrow:

Let $X, Y \subseteq \boldsymbol{P}(\mathcal{X})$.

$\alpha \in \boldsymbol{A}(\mathcal{X})$ is a valid $X \mapsto Y$ arrow iff

(1.1) $O(\alpha) \subseteq X$, $D(\alpha) \subseteq Y$,

(1.2) $\forall \beta : x' \to \alpha.(x' \in X \Rightarrow \exists \gamma : x'' \to \beta.(\gamma$ is a valid $X \mapsto Y$ arrow$))$.

We will also say that α is a valid arrow in X, or just valid in X, iff α is a valid $X \mapsto X$ arrow.

(2) Valid $X \Rightarrow Y$ arrow:

Let $X \subseteq Y \subseteq \boldsymbol{P}(\mathcal{X})$.

$\alpha \in \boldsymbol{A}(\mathcal{X})$ is a valid $X \Rightarrow Y$ arrow iff

(2.1) $o(\alpha) \in X$, $O(\alpha) \subseteq Y$, $D(\alpha) \subseteq Y$,

(2.2) $\forall \beta : x' \to \alpha.(x' \in Y \Rightarrow \exists \gamma : x'' \to \beta.(\gamma$ is a valid $X \Rightarrow Y$ arrow$))$.

Thus, any attack β from Y against α has to be countered by a valid attack on β.

(Note that in particular $o(\gamma) \in X$, and that $o(\beta)$ need not be in X, but can be in the bigger Y.)

Remark 2.3.1

Note that, in the definition of valid $X \mapsto Y$ arrow, X and Y need not be related, but in the definition of valid $X \Rightarrow Y$ arrow, $X \subseteq Y$.

Let us assume now that $X \subseteq Y$, and look at the remaining differences.

In both cases, $D(\alpha) \subseteq Y$.

In the $X \mapsto Y$ case, $O(\alpha) \subseteq X$, and attacks from X are countered.

In the $X \Rightarrow Y$ case, $o(\alpha) \in X$, $O(\alpha) \subseteq Y$, and attacks from Y are countered.

So the first condition is stronger in the $X \mapsto Y$ case, the second in the $X \Rightarrow Y$ case.

Example 2.3.2

(1) Consider the arrow $\beta := \langle \beta', l' \rangle$ in Figure 2.4. $D(\beta) = \{\langle x, i \rangle\}$, $O(\beta) = \{\langle z', m' \rangle, \langle y, j \rangle\}$, and the only arrow attacking β originates outside X, so β is a valid $X \mapsto \mu(X)$ arrow.

(2) Consider the arrows $\langle \alpha', k' \rangle$ and $\langle \gamma', n' \rangle$ in Figure 2.5. Both are valid $\mu(X) \Rightarrow X$ arrows.

Example 2.3.3

See Figure 2.3.

Let $X \subseteq Y$.

Consider the left-hand side of the diagram.

2.3 Higher Preferential Structures

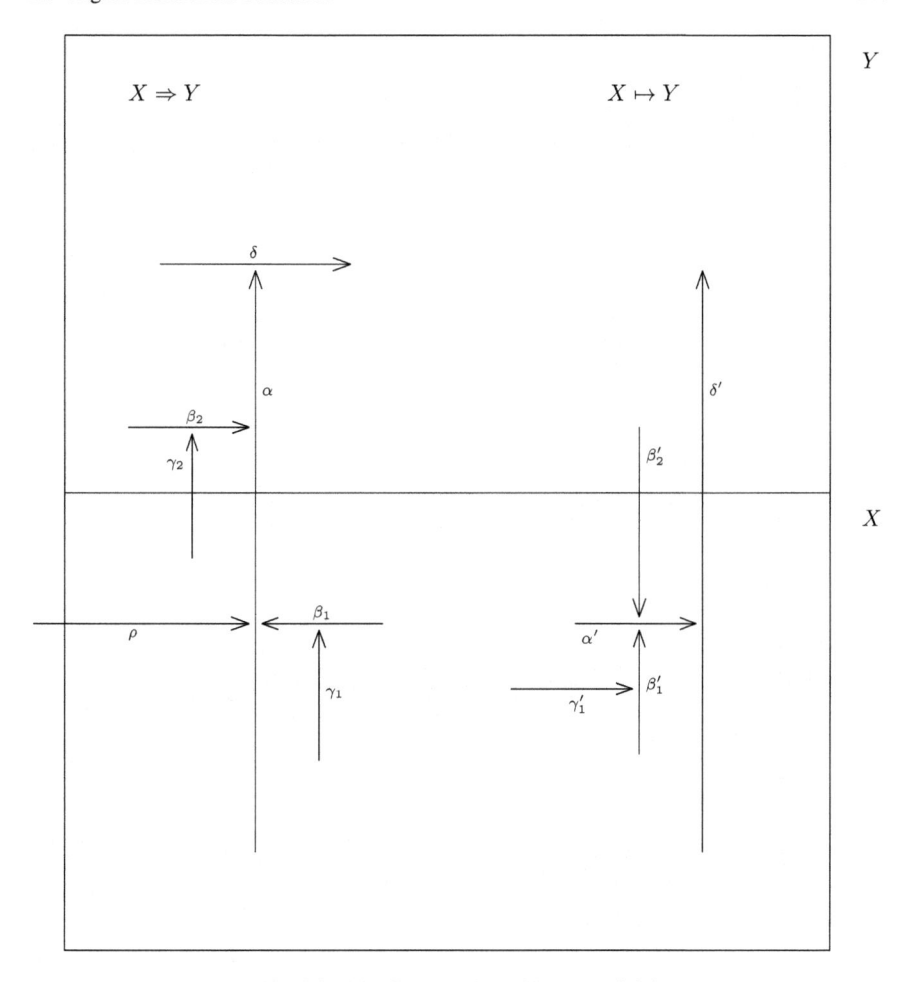

Fig. 2.3 The Construction of Example 2.3.3

The fact that δ originates in Y, but not in X establishes that α is not a valid $X \mapsto Y$ arrow, as the condition $O(\alpha) \subseteq X$ is violated. To be a valid $X \Rightarrow Y$ arrow, we have to show that all attacks on α originating from Y (not only from X) are countered by valid $X \Rightarrow Y$ arrows. This holds, as β_1 is countered by γ_1, β_2 by γ_2. All possible attacks on α, γ_1, γ_2 from outside Y, like ρ, need not be considered.

Consider the right-hand side of the diagram.

The fact that there is no valid counterargument to β'_2 establishes that α' is not a valid $X \Rightarrow Y$ arrow. It is a valid $X \mapsto Y$ arrow, as counterarguments to α' like β'_2, which do not originate in X, are not considered. The counterargument β'_1 is considered, but it is destroyed by the valid $X \mapsto Y$ arrow γ'_1.

Fact 2.3.2

(1) If α is a valid $X \Rightarrow Y$ arrow, then α is a valid $Y \mapsto Y$ arrow.
(2) If $X \subseteq X' \subseteq Y' \subseteq Y \subseteq \boldsymbol{P}(\mathcal{X})$ and $\alpha \in \boldsymbol{A}(\mathcal{X})$ is a valid $X \Rightarrow Y$ arrow, and $O(\alpha) \subseteq Y'$, $D(\alpha) \subseteq Y'$, then α is a valid $X' \Rightarrow Y'$ arrow.

Proof

Exercise, solution in the appendix.

Definition 2.3.4

Let $\mathcal{X}$ be a generalized preferential structure of level n, $X \subseteq \boldsymbol{P}(\mathcal{X})$.
$\mu(X) := \{x \in X : \exists \langle x, i \rangle . \neg \exists \text{ valid } X \mapsto X \text{ arrow } \alpha : x' \to \langle x, i \rangle \}$.

Comment 2.3.2

The purpose of smoothness is to guarantee Cumulativity. Smoothness achieves Cumulativity by mirroring all information present in X in $\mu(X)$. Closer inspection shows that smoothness does more than necessary. This is visible when there are copies (or, equivalently, non-injective labelling functions). Suppose we have two copies of $x \in X$, $\langle x, i \rangle$ and $\langle x, i' \rangle$, and there is a $y \in X$, $\alpha : \langle y, j \rangle \to \langle x, i \rangle$, but there is no $\alpha' : \langle y', j' \rangle \to \langle x, i' \rangle$, $y' \in X$. Then $\alpha : \langle y, j \rangle \to \langle x, i \rangle$ is irrelevant, as $x \in \mu(X)$ anyhow. So mirroring $\alpha : \langle y, j \rangle \to \langle x, i \rangle$ in $\mu(X)$ is not necessary, i.e., it is not necessary to have some $\alpha' : \langle y', j' \rangle \to \langle x, i \rangle$, $y' \in \mu(X)$.

On the other hand, Example 2.3.5 shows that if we want smooth structures to correspond to the property (μCUM), we need at least some valid arrows from $\mu(X)$ for higher-level arrows. This "some" is made precise (essentially) in Definition 2.3.5.

From a more philosophical point of view, when we see the (inverted) arrows of preferential structures as attacks on non-minimal elements, we should see smooth structures as always having attacks from valid (minimal) elements. So, in general structures, attacks from non-valid elements are valid; in smooth structures we always have attacks from valid elements.

The analogue to usual smooth structures on level 2 is then that any successfully attacked level 1 arrow is also attacked from a minimal point.

Definition 2.3.5

Let $\mathcal{X}$ be a generalized preferential structure.
$X \sqsubseteq X'$ iff

2.3 Higher Preferential Structures

(1) $X \subseteq X' \subseteq P(\mathcal{X})$,

(2) $\forall x \in X' - X \ \forall \langle x, i \rangle \ \exists \alpha : x' \to \langle x, i \rangle$ (α is a valid $X \Rightarrow X'$ arrow),

(3) $\forall x \in X \ \exists \langle x, i \rangle$

$(\forall \alpha : x' \to \langle x, i \rangle \ (x' \in X' \Rightarrow \exists \beta : x'' \to \alpha.(\beta$ is a valid $X \Rightarrow X'$ arrow))).

Note that (3) is not simply the negation of (2):

Consider a level 1 structure. Thus all level 1 arrows are valid, but the source of the arrows must not be neglected.

(2) reads now: $\forall x \in X' - X \ \forall \langle x, i \rangle \ \exists \alpha : x' \to \langle x, i \rangle.x' \in X$

(3) reads: $\forall x \in X \ \exists \langle x, i \rangle \ \neg \exists \alpha : x' \to \langle x, i \rangle.x' \in X'$

This is intended: intuitively, read $X = \mu(X')$, and minimal elements must not be attacked at all, but non-minimals must be attacked from X — which is a modified version of smoothness. More precisely: non-minimal elements (i.e., from $X' - X$) have to be validly attacked from X, minimal elements must not be validly attacked at all from X' (only perhaps from the outside).

Remark 2.3.3

We note the special case of Definition 2.3.5 for level 3 structures. We also write it immediately for the intended case $\mu(X) \sqsubseteq X$, and explicitly with copies.

$x \in \mu(X)$ iff

(1) $\exists \langle x, i \rangle \forall \langle \alpha, k \rangle : \langle y, j \rangle \to \langle x, i \rangle$

$(y \in X \Rightarrow \exists \langle \beta', l' \rangle : \langle z', m' \rangle \to \langle \alpha, k \rangle.$

$(z' \in \mu(X) \land \neg \exists \langle \gamma', n' \rangle : \langle u', p' \rangle \to \langle \beta', l' \rangle.u' \in X)).$

See Figure 2.4.

$x \in X - \mu(X)$ iff

(2) $\forall \langle x, i \rangle \exists \langle \alpha', k' \rangle : \langle y', j' \rangle \to \langle x, i \rangle$

$(y' \in \mu(X) \land$

(a) $\neg \exists \langle \beta', l' \rangle : \langle z', m' \rangle \to \langle \alpha', k' \rangle.z' \in X$

or

(b) $\forall \langle \beta', l' \rangle : \langle z', m' \rangle \to \langle \alpha', k' \rangle$

$(z' \in X \Rightarrow \exists \langle \gamma', n' \rangle : \langle u', p' \rangle \to \langle \beta', l' \rangle.u' \in \mu(X)))$

See Figure 2.5.

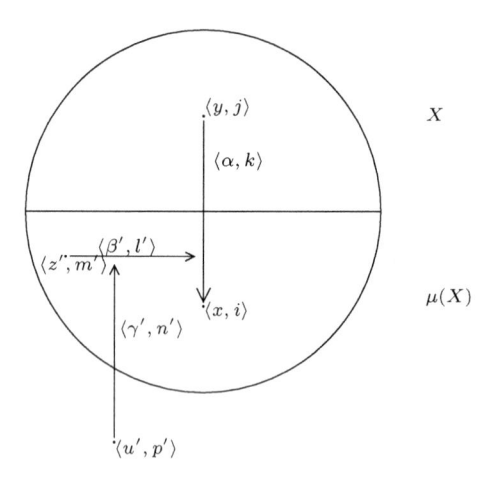

Fig. 2.4 Case 3-1-2

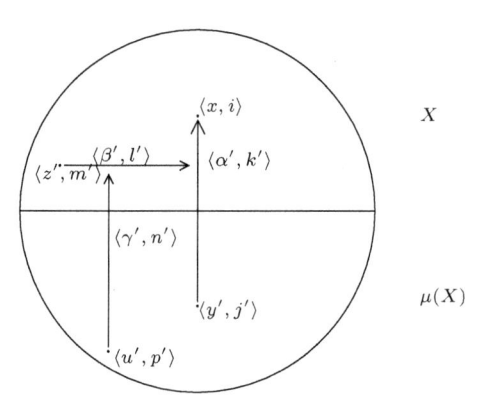

Fig. 2.5 Case 3-2

Fact 2.3.4

(1) If $X \sqsubseteq X'$, then $X = \mu(X')$,
(2) $X \sqsubseteq X', X \subseteq X'' \subseteq X' \Rightarrow X \sqsubseteq X''$. (This corresponds to ($\mu CUM$).)
(3) $X \sqsubseteq X', X \subseteq Y', Y \sqsubseteq Y', Y \subseteq X' \Rightarrow X = Y$. (This corresponds to ($\mu \subseteq \supseteq$).)

Proof

(1) Trivial by Fact 2.3.2 (1).
(2) We have to show

2.3 Higher Preferential Structures

(a) $\forall x \in X'' - X \; \forall \langle x, i \rangle \; \exists \alpha : x' \to \langle x, i \rangle$ (α is a valid $X \Rightarrow X''$ arrow), and

(b) $\forall x \in X \; \exists \langle x, i \rangle \; (\forall \alpha : x' \to \langle x, i \rangle \; (x' \in X'' \Rightarrow \exists \beta : x'' \to \alpha.$($\beta$ is a valid $X \Rightarrow X''$ arrow))).

Both follow from the corresponding condition for $X \Rightarrow X'$, the restriction of the universal quantifier, and Fact 2.3.2 (2).

(3) Exercise, solution in the appendix.

$\Box$

Definition 2.3.6

Let $\mathcal{X}$ be a generalized preferential structure, $X \subseteq P(\mathcal{X})$.

- $\mathcal{X}$ is called totally smooth for X iff

    (1) $\forall \alpha : x \to y \in A(\mathcal{X})(O(\alpha) \cup D(\alpha) \subseteq X \Rightarrow \exists \alpha' : x' \to y.x' \in \mu(X))$

    (2) if α is valid, then there must also exist such α' which is valid

    (y a point or an arrow).

- If $\mathcal{Y} \subseteq P(\mathcal{X})$, then $\mathcal{X}$ is called $\mathcal{Y}$-totally smooth

    iff for all $X \in \mathcal{Y}$ $\mathcal{X}$ is totally smooth for X.

Example 2.3.4

$X := \{\alpha : a \to b, \alpha' : b \to c, \alpha'' : a \to c, \beta : b \to \alpha'\}$ is not totally smooth,

$X := \{\alpha : a \to b, \alpha' : b \to c, \alpha'' : a \to c, \beta : b \to \alpha', \beta' : a \to \alpha'\}$ is totally smooth.

Example 2.3.5

Consider $\alpha' : a \to b, \alpha'' : b \to c, \alpha : a \to c, \beta : a \to \alpha$.

Then $\mu(\{a, b, c\}) = \{a\}, \mu(\{a, c\}) = \{a, c\}$. Thus, (μCUM) does not hold in this structure. Note that there is no valid arrow from $\mu(\{a, b, c\})$ to c.

Definition 2.3.7

Let $\mathcal{X}$ be a generalized preferential structure, $X \subseteq P(\mathcal{X})$.

$\mathcal{X}$ is called essentially smooth for X iff $\mu(X) \sqsubseteq X$. If $\mathcal{Y} \subseteq P(\mathcal{X})$, then $\mathcal{X}$ is called $\mathcal{Y}$-essentially smooth

iff for all $X \in \mathcal{Y}$ $\mu(X) \sqsubseteq X$.

Example 2.3.6

It is easy to see that we can distinguish total and essential smoothness in richer structures, as the following Example shows:

We add an accessibility relation R, and consider only those models which are accessible.

Let e.g., $a \to b \to \langle c, 0 \rangle$, $\langle c, 1 \rangle$, without transitivity. Thus, only c has two copies. This structure is essentially smooth, but of course not totally so.

Let now mRa, mRb, $mR\langle c, 0 \rangle$, $mR\langle c, 1 \rangle$, $m'Ra$, $m'Rb$, $m'R\langle c, 0 \rangle$.

Thus, seen from m, $\mu(\{a, b, c\}) = \{a, c\}$, but seen from m', $\mu(\{a, b, c\}) = \{a\}$, but $\mu(\{a, c\}) = \{a, c\}$, contradicting (CUM).

□

2.3.2 The General Case

The idea to solve the representation problem illustrated by Example 1.2.5 is to use the points c and d as bases for counterarguments against $\alpha : b \to a$ - as is possible in IBRS. We do this now. We will obtain a representation for logics weaker than P by generalized preferential structures.

We will now prove a representation theorem, but will make it more general than for preferential structures only. For this purpose, we will introduce some definitions first.

Definition 2.3.8

Let $\eta, \rho : \mathcal{Y} \to \mathcal{P}(U)$.

(1) If $\mathcal{X}$ is a simple structure

$\mathcal{X}$ is called an attacking structure relative to η representing ρ iff

$\rho(X) = \{x \in \eta(X) : \text{there is no valid } X - to - \eta(X) \text{ arrow } \alpha : x' \to x\}$

for all $X \in \mathcal{Y}$.

(2) If $\mathcal{X}$ is a structure with copies

$\mathcal{X}$ is called an attacking structure relative to η representing ρ iff

$\rho(X) = \{x \in \eta(X) : \text{there is } \langle x, i \rangle \text{ and no valid } X - to - \eta(X) \text{ arrow } \alpha : \langle x', i' \rangle \to \langle x, i \rangle\}$

for all $X \in \mathcal{Y}$.

Obviously, in those cases $\rho(X) \subseteq \eta(X)$ for all $X \in \mathcal{Y}$.

2.3 Higher Preferential Structures

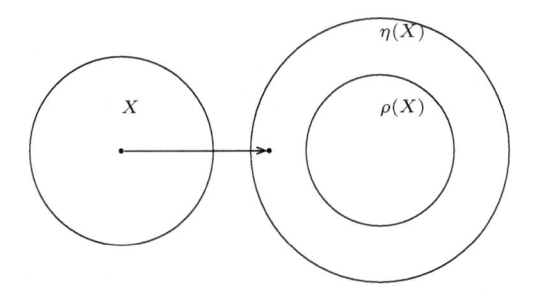

Fig. 2.6 Attacking structure

Thus, $\mathcal{X}$ is a preferential structure iff η is the identity.

See Figure 2.6

(Note that it does not seem very useful to generalize the notion of smoothness from preferential structures to general attacking structures, as, in the general case, the minimizing set X and the result $\rho(X)$ may be disjoint.)

The following result is the first positive representation result here, and shows that we can obtain (almost) anything with level 2 structures.

Proposition 2.3.5

Let $\eta, \rho : \mathcal{Y} \to \mathcal{P}(U)$. Then there is an attacking level 2 structure relative to η representing ρ iff

(1) $\rho(X) \subseteq \eta(X)$ for all $X \in \mathcal{Y}$,
(2) $\rho(\emptyset) = \eta(\emptyset)$ if $\emptyset \in \mathcal{Y}$.

(2) is, of course, void for preferential structures.

Proof

(A) The construction

　We make a two stage construction.

　(A.1) Stage 1.

　　In stage one, consider (almost as usual)

　　$\mathcal{U} := \langle \mathcal{X}, \{\alpha_i : i \in I\} \rangle$ where

　　$\mathcal{X} := \{\langle x, f \rangle : x \in U, f \in \Pi\{X \in \mathcal{Y} : x \in \eta(X) - \rho(X)\}\}$,

　　$\alpha : x' \to \langle x, f \rangle :\Leftrightarrow x' \in ran(f)$. Attention: $x' \in X$, not $x' \in \rho(X)$!

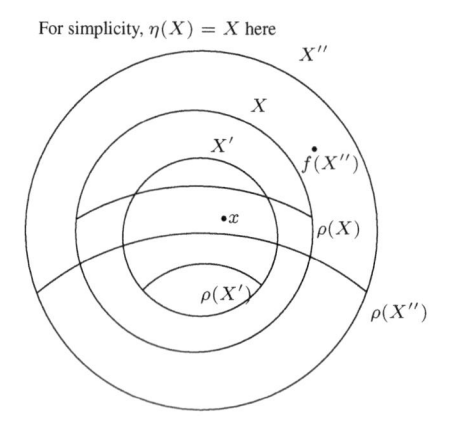

Fig. 2.7 The complicated case

(A.2) Stage 2.

Let $\mathcal{X}'$ be the set of all $\langle x, f, X \rangle$ s.t. $\langle x, f \rangle \in \mathcal{X}$ and
(a) either X is some dummy value, say $*$
or
(b) all of the following (1) – (4) hold:
  (1) $X \in \mathcal{Y}$,
  (2) $x \in \rho(X)$,
  (3) there is $X' \subseteq X$, $x \in \eta(X') - \rho(X')$, $X' \in \mathcal{Y}$, (thus $ran(f) \cap X \neq \emptyset$ by definition),
  (4) $\forall X'' \in \mathcal{Y}.(X \subseteq X'', x \in \eta(X'') - \rho(X'') \Rightarrow (ran(f) \cap X'') - X \neq \emptyset)$.

(Thus, f chooses in (4) for X'' also outside X. If there is no such X'', (4) is void, and only (1) – (3) need to hold, i.e., we may take any f with $\langle x, f \rangle \in \mathcal{X}$.)

See Figure 2.7.

Note: If (1) – (3) are satisfied for x and X, then we will find f s.t. $\langle x, f \rangle \in \mathcal{X}$, and $\langle x, f, X \rangle$ satisfies (1) – (4) : As $X \subset X''$ for X'' as in (4), we find f which chooses for such X'' outside of X.

So for any $\langle x, f \rangle \in \mathcal{X}$, there is $\langle x, f, * \rangle$, and maybe also some $\langle x, f, X \rangle$ in $\mathcal{X}'$.

Let again for any x', $\langle x, f, X \rangle \in \mathcal{X}'$
$$\alpha : x' \to \langle x, f, X \rangle :\Leftrightarrow x' \in ran(f)$$

(A.3) Adding arrows.

Consider x' and $\langle x, f, X \rangle$.

2.3 Higher Preferential Structures

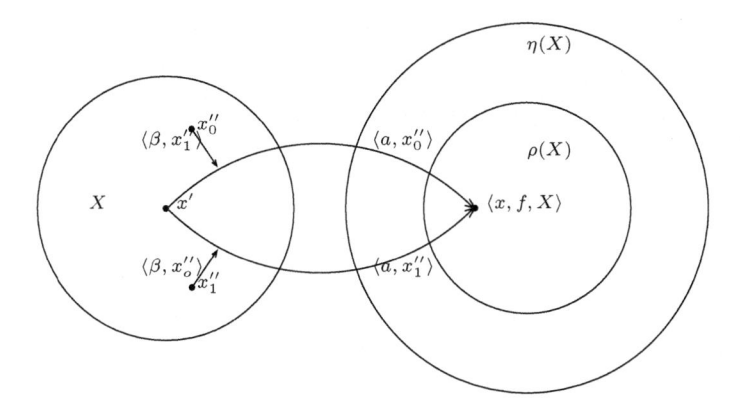

Fig. 2.8 Attacking structure

If $X = *$, or $x' \notin X$, we do nothing, i.e., leave a simple arrow $\alpha : x' \to \langle x, f, X \rangle \Leftrightarrow x' \in ran(f)$.

If $X \in \mathcal{Y}$, and $x' \in X$, and $x' \in ran(f)$, we make X many copies of the attacking arrow and have then: $\langle \alpha, x'' \rangle : x' \to \langle x, f, X \rangle$ for all $x'' \in X$.

In addition, we add attacks on the $\langle \alpha, x'' \rangle : \langle \beta, x'' \rangle : x'' \to \langle \alpha, x'' \rangle$ for all $x'' \in X$.

The full structure $\mathcal{Z}$ is thus:

$\mathcal{X}'$ is the set of elements.

If $x' \in ran(f)$, and $X = *$ or $x' \notin X$ then $\alpha : x' \to \langle x, f, X \rangle$

if $x' \in ran(f)$, and $X \neq *$ and $x' \in X$ then

(a) $\langle \alpha, x'' \rangle : x' \to \langle x, f, X \rangle$ for all $x'' \in X$,

(b) $\langle \beta, x'' \rangle : x'' \to \langle \alpha, x'' \rangle$ for all $x'' \in X$.

See Figure 2.8.

(B) Representation

We have to show that this structure represents ρ relative to η.

Let $y \in \eta(Y), Y \in \mathcal{Y}$.

Case 1. $y \in \rho(Y)$.

We have to show that there is $\langle y, g, Y'' \rangle$ s.t. there is no valid $\alpha : y' \to \langle y, g, Y'' \rangle$, $y' \in Y$. In Case 1.1 below, Y'' will be $*$, in Case 1.2, Y'' will be Y, g will be chosen suitably.

Case 1.1. There is no $Y' \subseteq Y$, $y \in \eta(Y') - \rho(Y'), Y' \in \mathcal{Y}$.

So for all Y' with $y \in \eta(Y') - \rho(Y')$ $Y' - Y \neq \emptyset$. Let $g \in \Pi\{Y' - Y : y \in \eta(Y') - \rho(Y')\}$. Then $ran(g) \cap Y = \emptyset$, and $\langle y, g, * \rangle$ is not attacked from Y. ($\langle y, g \rangle$ was already not attacked in $\mathcal{X}$.)

Case 1.2. There is $Y' \subseteq Y$, $y \in \eta(Y') - \rho(Y')$, $Y' \in \mathcal{Y}$.

Let now $\langle y, g, Y \rangle \in \mathcal{X}'$, s.t. $g(Y'') \not\subseteq Y$ if $Y \subseteq Y''$, $y \in \eta(Y'') - \rho(Y'')$, $Y'' \in \mathcal{Y}$. As noted above, such g and thus $\langle y, g, Y \rangle$ exist. Fix $\langle y, g, Y \rangle$.

Consider any $y' \in ran(g)$. If $y' \notin Y$, y' does not attack $\langle y, g, Y \rangle$ in Y. Suppose $y' \in Y$. We had made Y many copies $\langle \alpha, y'' \rangle$, $y'' \in Y$ with $\langle \alpha, y'' \rangle : y' \to \langle y, g, Y \rangle$ and had added the level 2 arrows $\langle \beta, y'' \rangle : y'' \to \langle \alpha, y'' \rangle$ for $y'' \in Y$. So all copies $\langle \alpha, y'' \rangle$ are destroyed in Y. This was done for all $y' \in Y$, $y' \in ran(g)$, so $\langle y, g, Y \rangle$ is now not (validly) attacked in Y any more.

Case 2. $y \in \eta(Y) - \rho(Y)$.

Let $\langle y, g, Y' \rangle$ (where Y' can be $*$) be any copy of y, we have to show that there is $z \in Y$, $\alpha : z \to \langle y, g, Y' \rangle$, or some $\langle \alpha, z' \rangle : z \to \langle y, g, Y' \rangle$, $z' \in Y'$, which is not destroyed by some level 2 arrow $\langle \beta, z' \rangle : z' \to \langle \alpha, z' \rangle$, $z' \in Y$.

As $y \in \eta(Y) - \rho(Y)$, $ran(g) \cap Y \neq \emptyset$, so there is $z \in ran(g) \cap Y$. Fix such z. (We will modify the choice of z only in Case 2.2.2 below.)

Case 2.1. $Y' = *$.

As $z \in ran(g)$, $\alpha : z \to \langle y, g, * \rangle$. (There were no level 2 arrows introduced for this copy.)

Case 2.2. $Y' \neq *$.

So $\langle y, g, Y' \rangle$ satisfies the conditions (1) – (4) of (b) at the beginning of the proof.

If $z \notin Y'$, we are done, as $\alpha : z \to \langle y, g, Y' \rangle$, and there were no level 2 arrows introduced in this case. If $z \in Y'$, we had made Y' many copies $\langle \alpha, z' \rangle$, $\langle \alpha, z' \rangle : z \to \langle y, g, Y' \rangle$, one for each $z' \in Y'$. Each $\langle \alpha, z' \rangle$ was destroyed by $\langle \beta, z' \rangle : z' \to \langle \alpha, z' \rangle$, $z' \in Y'$.

Case 2.2.1. $Y' \not\subseteq Y$.

Let $z'' \in Y'-Y$, then $\langle \alpha, z'' \rangle : z \to \langle y, g, Y' \rangle$ is destroyed only by $\langle \beta, z'' \rangle : z'' \to \langle \alpha, z'' \rangle$ in Y', but not in Y, as $z'' \notin Y$, so $\langle y, g, Y' \rangle$ is attacked by $\langle \alpha, z'' \rangle : z \to \langle y, g, Y' \rangle$, valid in Y.

Case 2.2.2. $Y' \subset Y$ ($Y = Y'$ is impossible, as $y \in \rho(Y')$, $y \notin \rho(Y)$).

Then there was by definition (condition (b) (4)) some $z' \in (ran(g) \cap Y) - Y'$ and $\alpha : z' \to \langle y, g, Y' \rangle$ is valid, as $z' \notin Y'$. (In this case, there are no copies of α and no level 2 arrows.)

□

2.3 Higher Preferential Structures

Corollary 2.3.6

(1) We cannot distinguish general structures of level 2 from those of higher levels by their ρ-functions relative to η.

(2) Let U be the universe, $\mathcal{Y} \subseteq \mathcal{P}(U)$, $\mu : \mathcal{Y} \to \mathcal{P}(U)$. Then any μ satisfying $(\mu \subseteq)$ can be represented by a level 2 preferential structure. (Choose $\eta = identity$.)

Again, we cannot distinguish general structures of level 2 from those of higher levels by their μ-functions.

□

A remark on the function η :

We can also obtain the function η via arrows. Of course, then we need positive arrows (not only negative arrows against negative arrows, as we first need to have something positive).

If η is the identity, we can make a positive arrow from each point to itself. Otherwise, we can connect every point to every point by a positive arrow, and then choose those we really want in η by a choice function obtained from arrows just as we obtained ρ from arrows.

2.3.3 Discussion of the Totally Smooth Case

Fact 2.3.7

Let $X, Y \in \mathcal{Y}$, $\mathcal{X}$ a level n structure. Let $\langle \alpha, k \rangle : \langle x, i \rangle \to \langle y, j \rangle$, where $\langle y, j \rangle$ may itself be (a copy of) an arrow.

(1) Let $n > 1$, $X \subseteq Y$, $\langle \alpha, k \rangle \in X$ a level $n - 1$ arrow in $\mathcal{X} \restriction X$. If $\langle \alpha, k \rangle$ is valid in $\mathcal{X} \restriction Y$, then it is valid in $\mathcal{X} \restriction X$.

(2) Let $\mathcal{X}$ be totally smooth, $\mu(X) \subseteq Y$, $\mu(Y) \subseteq X$, $\langle \alpha, k \rangle \in \mathcal{X} \restriction X \cap Y$, then $\langle \alpha, k \rangle$ is valid in $\mathcal{X} \restriction X$ iff it is valid in $\mathcal{X} \restriction Y$.

Note that we will also sometimes write X for $\mathcal{X} \restriction X$, when the context is clear.

Proof

(1) If $\langle \alpha, k \rangle$ is not valid in $\mathcal{X} \restriction X$, then there must be a level n arrow $\langle \beta, r \rangle : \langle z, s \rangle \to \langle \alpha, k \rangle$ in $\mathcal{X} \restriction X \subseteq \mathcal{X} \restriction Y$. $\langle \beta, r \rangle$ must be valid in $\mathcal{X} \restriction X$ and $\mathcal{X} \restriction Y$, as there are no level $n+1$ arrows. So $\langle \alpha, k \rangle$ is not valid in $\mathcal{X} \restriction Y$, contradiction.

(2) By downward induction. Case n : $\langle \alpha, k \rangle \in \mathcal{X} \restriction X \cap Y$, so it is valid in both as there are no level $n+1$ arrows. Case $m \to m-1$: Let $\langle \alpha, k \rangle \in \mathcal{X} \restriction X \cap Y$ be

a level $m-1$ arrow valid in $\mathcal{X} \restriction X$, but not in $\mathcal{X} \restriction Y$. So there must be a level m arrow $\langle \beta, r \rangle : \langle z, s \rangle \to \langle \alpha, k \rangle$ valid in $\mathcal{X} \restriction Y$. By total smoothness, we may assume $z \in \mu(Y) \subseteq X$, so $\langle \beta, r \rangle \in \mathcal{X} \restriction X$ is valid by induction hypothesis. So $\langle \alpha, k \rangle$ is not valid in $\mathcal{X} \restriction X$, contradiction.

□

Corollary 2.3.8

Let $X, Y \in \mathcal{Y}$, $\mathcal{X}$ a totally smooth level n structure, $\mu(X) \subseteq Y$, $\mu(Y) \subseteq X$. Then $\mu(X) = \mu(Y)$.

Proof

Exercise, solution in the appendix.

Fact 2.3.9

There are situations satisfying $(\mu \subseteq) + (\mu CUM) + (\cap)$ which cannot be represented by level 2 totally smooth preferential structures.

The proof is given in the following example.

Example 2.3.7

Let $Y := \{x, y, y'\}$, $X := \{x, y\}$, $X' := \{x, y'\}$. Let $\mathcal{Y} := \mathcal{P}(Y)$. Let $\mu(Y) := \{y, y'\}$, $\mu(X) := \mu(X') := \{x\}$, and $\mu(Z) := Z$ for all other sets.

Obviously, this satisfies $(\cap)$, $(\mu \subseteq)$, and (μCUM).

Suppose $\mathcal{X}$ is a totally smooth level 2 structure representing μ.

So $\mu(X) = \mu(X') \subseteq Y - \mu(Y)$, $\mu(Y) \subseteq X \cup X'$. Let $\langle x, i \rangle$ be minimal in $\mathcal{X} \restriction X$. As $\langle x, i \rangle$ cannot be minimal in $\mathcal{X} \restriction Y$, there must be $\alpha : \langle z, j \rangle \to \langle x, i \rangle$, valid in $\mathcal{X} \restriction Y$.

Case 1: $z \in X'$.

So $\alpha \in \mathcal{X} \restriction X'$. If α is valid in $\mathcal{X} \restriction X'$, there must be $\alpha' : \langle x', i' \rangle \to \langle x, i \rangle$, $x' \in \mu(X')$, valid in $\mathcal{X} \restriction X'$, and thus in $\mathcal{X} \restriction X$, by $\mu(X) = \mu(X')$ and Fact 2.3.7 (2). This is impossible, so there must be $\beta : \langle x', i' \rangle \to \alpha$, $x' \in \mu(X')$, valid in $\mathcal{X} \restriction X'$. As β is in $\mathcal{X} \restriction Y$ and $\mathcal{X}$ a level ≤ 2 structure, β is valid in $\mathcal{X} \restriction Y$, so α is not valid in $\mathcal{X} \restriction Y$, contradiction.

Case 2: $z \in X$.

α cannot be valid in $\mathcal{X} \restriction X$, so there must be $\beta : \langle x', i' \rangle \to \alpha$, $x' \in \mu(X)$, valid in $\mathcal{X} \restriction X$. Again, as β is in $\mathcal{X} \restriction Y$ and $\mathcal{X}$ a level ≤ 2 structure, β is valid in $\mathcal{X} \restriction Y$, so α is not valid in $\mathcal{X} \restriction Y$, contradiction.

□

2.3 Higher Preferential Structures

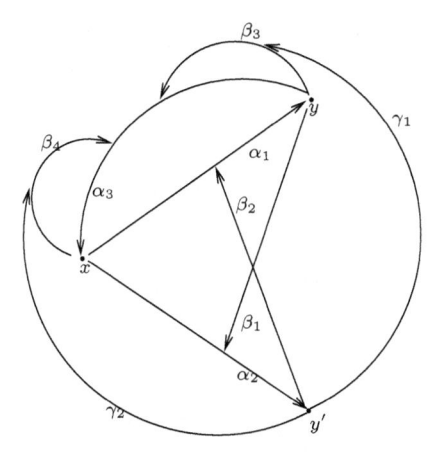

Fig. 2.9 Solution by smooth level 3 structure

It is unknown to the authors whether an analogue is true for essential smoothness, i.e., whether there are examples of such μ function which need at least level 3 essentially smooth structures for representation. Proposition 2.3.13 below shows that such structures suffice, but we do not know whether level 3 is necessary.

Fact 2.3.10

Above Example 2.3.7 can be solved by a totally smooth level 3 structure:

Let $\alpha_1 : x \to y$, $\alpha_2 : x \to y'$, $\alpha_3 : y \to x$, $\beta_1 : y \to \alpha_2$, $\beta_2 : y' \to \alpha_1$, $\beta_3 : y \to \alpha_3$, $\beta_4 : x \to \alpha_3$, $\gamma_1 : y' \to \beta_3$, $\gamma_2 : y' \to \beta_4$.

See Figure 2.9.

The subdiagram generated by X contains α_1, α_3, β_3, β_4. α_1, β_3, β_4 are valid, so $\mu(X) = \{x\}$.

The subdiagram generated by X' contains α_2. α_2 is valid, so $\mu(X') = \{x\}$.

In the full diagram, α_3, β_1, β_2, γ_1, γ_2 are valid, so $\mu(Y) = \{y, y'\}$.

□

Remark 2.3.11

Example 1.2.4 together with Corollary 2.3.8 show that $(\mu \subseteq)$ and (μCUM) without $(\cap)$ do not guarantee representability by a level n totally smooth structure.

2.3.4 The Essentially Smooth Case

Definition 2.3.9

Let $\mu : \mathcal{Y} \to \mathcal{P}(U)$ and $\mathcal{X}$ be given, let $\alpha : \langle y, j \rangle \to \langle x, i \rangle \in \mathcal{X}$.
Define

$$O(\alpha) := \{Y \in \mathcal{Y} : x \in Y - \mu(Y), y \in \mu(Y)\},$$
$$D(\alpha) := \{X \in \mathcal{Y} : x \in \mu(X), y \in X\},$$
$$\Pi(O, \alpha) := \Pi\{\mu(Y) : Y \in O(\alpha)\},$$
$$\Pi(D, \alpha) := \Pi\{\mu(X) : X \in D(\alpha)\}.$$

The following lemma is probably the main technical result of the section.

Lemma 2.3.12

Let U be the universe, $\mu : \mathcal{Y} \to \mathcal{P}(U)$. Let μ satisfy $(\mu \subseteq) + (\mu \subseteq \supseteq)$.
Let $\mathcal{X}$ be a level 1 preferential structure, $\alpha : \langle y, j \rangle \to \langle x, i \rangle$, $O(\alpha) \neq \emptyset$, $D(\alpha) \neq \emptyset$.
We can modify $\mathcal{X}$ to a level 3 structure $\mathcal{X}'$ by introducing level 2 and level 3 arrows s.t. no copy of α is valid in any $X \in D(\alpha)$, and in every $Y \in O(\alpha)$ at least one copy of α is valid. (More precisely, we should write $\mathcal{X}' \upharpoonright X$, etc.)
Thus, in $\mathcal{X}'$,

(1) $\langle x, i \rangle$ will not be minimal in any $Y \in O(\alpha)$,
(2) if α is the only arrow minimizing $\langle x, i \rangle$ in $X \in D(\alpha)$, $\langle x, i \rangle$ will now be minimal in X.

The construction is made independently for all such arrows $\alpha \in \mathcal{X}$.

Proof

The construction

Make $\Pi(D, \alpha)$ many copies of $\alpha : \{\langle \alpha, f \rangle : f \in \Pi(D, \alpha)\}$, all $\langle \alpha, f \rangle : \langle y, j \rangle \to \langle x, i \rangle$. Note that $\langle \alpha, f \rangle \in X$ for all $X \in D(\alpha)$ and $\langle \alpha, f \rangle \in Y$ for all $Y \in O(\alpha)$.

Add to the structure $\langle \beta, f, X_r, g \rangle : \langle f(X_r), i_r \rangle \to \langle \alpha, f \rangle$, for any $X_r \in D(\alpha)$, and $g \in \Pi(O, \alpha)$ (and some or all i_r - this does not matter).

For all $Y_s \in O(\alpha)$:

if $\mu(Y_s) \not\subseteq X_r$ and $f(X_r) \in Y_s$, then add to the structure $\langle \gamma, f, X_r, g, Y_s \rangle : \langle g(Y_s), j_s \rangle \to \langle \beta, f, X_r, g \rangle$ (again for all or some j_s),

2.3 Higher Preferential Structures

[Figure 2.10: ellipse diagram with labels $\mu(X)$, $X \in D(\alpha)$, $\langle f(X_r), i_r \rangle$, $\langle \beta, f, X_r, g \rangle$, $\langle \gamma, f, X_r, g, Y_s \rangle$, $\langle g(Y_s), j_s \rangle$, $\langle x, i \rangle$, $\langle \alpha, f \rangle$, $\langle y, j \rangle$, $\mu(Y)$, $Y \in O(\alpha)$]

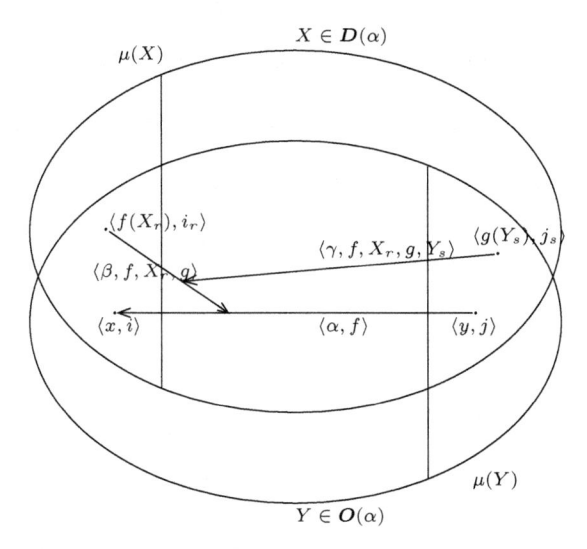

Fig. 2.10 The construction

if $\mu(Y_s) \subseteq X_r$ or $f(X_r) \notin Y_s$, $\langle \gamma, f, X_r, g, Y_s \rangle$ is not added.

See Figure 2.10.

Let $X_r \in D(\alpha)$. We have to show that no $\langle \alpha, f \rangle$ is valid in X_r. Fix f.

$\langle \alpha, f \rangle$ is in X_r, so we have to show that for at least one $g \in \Pi(O, \alpha)$ $\langle \beta, f, X_r, g \rangle$ is valid in X_r, i.e., that for this g, no $\langle \gamma, f, X_r, g, Y_s \rangle$: $\langle g(Y_s), j_s \rangle \to \langle \beta, f, X_r, g \rangle$, $Y_s \in O(\alpha)$ attacks $\langle \beta, f, X_r, g \rangle$ in X_r.

We define g. Take $Y_s \in O(\alpha)$.

Case 1: $\mu(Y_s) \subseteq X_r$ or $f(X_r) \notin Y_s$: choose arbitrary $g(Y_s) \in \mu(Y_s)$.

Case 2: $\mu(Y_s) \not\subseteq X_r$ and $f(X_r) \in Y_s$: Choose $g(Y_s) \in \mu(Y_s) - X_r$.

In Case 1, $\langle \gamma, f, X_r, g, Y_s \rangle$ does not exist, so it cannot attack $\langle \beta, f, X_r, g \rangle$.

In Case 2, $\langle \gamma, f, X_r, g, Y_s \rangle$: $\langle g(Y_s), j_s \rangle \to \langle \beta, f, X_r, g \rangle$ is not in X_r, as $g(Y_s) \notin X_r$.

Thus, no $\langle \gamma, f, X_r, g, Y_s \rangle$: $\langle g(Y_s), j_s \rangle \to \langle \beta, f, X_r, g \rangle$, $Y_s \in O(\alpha)$ attacks $\langle \beta, f, X_r, g \rangle$ in X_r.

So $\forall \langle \alpha, f \rangle : \langle y, j \rangle \to \langle x, i \rangle$

$y \in X_r \Rightarrow \exists \langle \beta, f, X_r, g \rangle : \langle f(X_r), i_r \rangle \to \langle \alpha, f \rangle$

$(f(X_r) \in \mu(X_r) \wedge \neg \exists \langle \gamma, f, X_r, g, Y_s \rangle : \langle g(Y_s), j_s \rangle \to \langle \beta, f, X_r, g \rangle . g(Y_s) \in X_r)$.

But $\langle \beta, f, X_r, g \rangle$ was constructed only for $\langle \alpha, f \rangle$, so was $\langle \gamma, f, X_r, g, Y_s \rangle$, and there was no other $\langle \gamma, i \rangle$ attacking $\langle \beta, f, X_r, g \rangle$, so we are done.

Let $Y_s \in \mathbf{O}(\alpha)$. We have to show that at least one $\langle \alpha, f \rangle$ is valid in Y_s.

We define $f \in \Pi(\mathbf{D}, \alpha)$. Take X_r.

If $\mu(X_r) \not\subseteq Y_s$, choose $f(X_r) \in \mu(X_r) - Y_s$. If $\mu(X_r) \subseteq Y_s$, choose arbitrary $f(X_r) \in \mu(X_r)$.

All attacks on $\langle x, f \rangle$ have the form $\langle \beta, f, X_r, g \rangle : \langle f(X_r), i_r \rangle \to \langle \alpha, f \rangle$, $X_r \in \mathbf{D}(\alpha)$, $g \in \Pi(\mathbf{O}, \alpha)$. We have to show that they are either not in Y_s, or that they are themselves attacked in Y_s.

Case 1: $\mu(X_r) \not\subseteq Y_s$. Then $f(X_r) \not\in Y_s$, so $\langle \beta, f, X_r, g \rangle : \langle f(X_r), i_r \rangle \to \langle \alpha, f \rangle$ is not in Y_s (for no g).

Case 2: $\mu(X_r) \subseteq Y_s$. Then $\mu(Y_s) \not\subseteq X_r$ by $(\mu \subseteq \supseteq)$ and $f(X_r) \in Y_s$, so $\langle \beta, f, X_r, g \rangle : \langle f(X_r), i_r \rangle \to \langle \alpha, f \rangle$ is in Y_s (for all g). Take any $g \in \Pi(\mathbf{O}, \alpha)$. As $\mu(Y_s) \not\subseteq X_r$ and $f(X_r) \in Y_s$, $\langle \gamma, f, X_r, g, Y_s \rangle : \langle g(Y_s), j_s \rangle \to \langle \beta, f, X_r, g \rangle$ is defined, and $g(Y_s) \in \mu(Y_s)$, so it is in Y_s (for all g). Thus, $\langle \beta, f, X_r, g \rangle$ is attacked in Y_s.

Thus, for this f, all $\langle \beta, f, X_r, g \rangle$ are either not in Y_s, or attacked in Y_s, thus for this f, $\langle \alpha, f \rangle$ is valid in Y_s.

So for this $\langle x, i \rangle$

$\exists \langle \alpha, f \rangle : \langle y, j \rangle \to \langle x, i \rangle . y \in \mu(Y_s) \wedge$

(a) $\neg \exists \langle \beta, f, X_r, g \rangle : \langle f(X_r), i \rangle \to \langle \alpha, f \rangle . f(X_r) \in Y_s$

or

(b) $\forall \langle \beta, f, X_r, g \rangle : \langle f(X_r), i \rangle \to \langle \alpha, f \rangle$

$(f(X_r) \in Y_s \Rightarrow$

$\exists \langle \gamma, f, X_r, g, Y_s \rangle : \langle g(Y_s), j_s \rangle \to \langle \beta, f, X_r, g \rangle . g(Y_s) \in \mu(Y_s))$.

As we made copies of α only, introduced only β's attacking the α-copies, and γ's attacking the β's, the construction is independent for different α's.

□

Proposition 2.3.13

Let U be the universe, $\mu : \mathcal{Y} \to \mathcal{P}(U)$.

Then any μ satisfying $(\mu \subseteq)$, $(\cap)$, (μCUM) (or, alternatively, $(\mu \subseteq)$ and $(\mu \subseteq \supseteq)$) can be represented by a level 3 essentially smooth structure.

Proof

In stage one, consider as usual $\mathcal{U} := \langle \mathcal{X}, \{\alpha_i : i \in I\} \rangle$ where $\mathcal{X} := \{\langle x, f \rangle : x \in U, f \in \Pi\{\mu(X) : X \in \mathcal{Y}, x \in X - \mu(X)\}\}$, and set $\alpha : \langle x', f' \rangle \to \langle x, f \rangle :\Leftrightarrow x' \in ran(f)$.

2.3 Higher Preferential Structures   263

For stage two:

Any level 1 arrow $\alpha : \langle y, j \rangle \to \langle x, i \rangle$ was introduced in stage one by some $Y \in \mathcal{Y}$ s.t. $y \in \mu(Y), x \in Y - \mu(Y)$. Do the construction of Lemma 2.3.12 for all level 1 arrows of $\mathcal{X}$ in parallel or successively.

We have to show that the resulting structure represents μ and is essentially smooth. (Level 3 is obvious.)

(1) Representation

Suppose $x \in Y - \mu(Y)$. Then there was in stage 1 for all $\langle x, i \rangle$ some $\alpha : \langle y, j \rangle \to \langle x, i \rangle, y \in \mu(Y)$. We examine the y.

If there is no X s.t. $x \in \mu(X), y \in X$, then there were no β's and γ's introduced for this $\alpha : \langle y, j \rangle \to \langle x, i \rangle$, so α is valid.

If there is X s.t. $x \in \mu(X), y \in X$, consider $\alpha : \langle y, j \rangle \to \langle x, i \rangle$. So $X \in \boldsymbol{D}(\alpha)$, $Y \in \boldsymbol{O}(\alpha)$, so we did the construction of Lemma 2.3.12, and by its result, $\langle x, i \rangle$ is not minimal in Y.

Thus, in both cases, $\langle x, i \rangle$ is successfully attacked in Y, and no $\langle x, i \rangle$ is a minimal element in Y.

Suppose $x \in \mu(X)$ (we change notation to conform to Lemma 2.3.12). Fix $\langle x, i \rangle$.

If there is no $\alpha : \langle y, j \rangle \to \langle x, i \rangle, y \in X$, then $\langle x, i \rangle$ is minimal in X, and we are done.

If there is α or $\langle \alpha, k \rangle : \langle y, j \rangle \to \langle x, i \rangle, y \in X$, then α originated from stage one through some Y s.t. $x \in Y - \mu(Y)$, and $y \in \mu(Y)$. (Note that stage 2 of the construction did not introduce any new level 1 arrows - only copies of existing level 1 arrows.) So $X \in \boldsymbol{D}(\alpha), Y \in \boldsymbol{O}(\alpha)$, so we did the construction of Lemma 2.3.12, and by its result, $\langle x, i \rangle$ is minimal in X, and we are done again.

In both cases, all $\langle x, i \rangle$ are minimal elements in X.

(2) Essential smoothness. We have to show the conditions of Definition 2.3.5. We will, however, work with the reformulation given in Remark 2.3.3.

Case (1), $x \in \mu(X)$.

Case (1.1), there is $\langle x, i \rangle$ with no $\langle \alpha, f \rangle : \langle y, j \rangle \to \langle x, i \rangle, y \in X$. There is nothing to show.

Case (1.2), for all $\langle x, i \rangle$ there is $\langle \alpha, f \rangle : \langle y, j \rangle \to \langle x, i \rangle, y \in X$.

α was introduced in stage 1 by some Y s.t. $x \in Y - \mu(Y), y \in X \cap \mu(Y)$, so $X \in \boldsymbol{D}(\alpha), Y \in \boldsymbol{O}(\alpha)$. In the proof of Lemma 2.3.12, at the end of (2), it was shown that

$$\exists \langle \beta, f, X_r, g \rangle : \langle f(X_r), i_r \rangle \to \langle \alpha, f \rangle$$
$$(f(X_r) \in \mu(X_r) \wedge$$
$$\neg \exists \langle \gamma, f, X_r, g, Y_s \rangle : \langle g(Y_s), j_s \rangle \to \langle \beta, f, X_r, g \rangle . g(Y_s) \in X_r).$$

By $f(X_r) \in \mu(X_r)$, condition (1) of Remark 2.3.3 is true.

Case (2), $x \notin \mu(Y)$. Fix $\langle x, i \rangle$. (We change notation back to Y.)

In stage 1, we constructed $\alpha : \langle y, j \rangle \to \langle x, i \rangle$, $y \in \mu(Y)$, so $Y \in O(\alpha)$.

If $D(\alpha) = \emptyset$, then there is no attack on α, and the condition (2) of Remark 2.3.3 is trivially true.

If $D(\alpha) \neq \emptyset$, we did the construction of Lemma 2.3.12, so

$$\exists \langle \alpha, f \rangle : \langle y, j \rangle \to \langle x, i \rangle . y \in \mu(Y_s) \wedge$$

(a) $\neg \exists \langle \beta, f, X_r, g \rangle : \langle f(X_r), i \rangle \to \langle \alpha, f \rangle . f(X_r) \in Y_s$

or

(b) $\forall \langle \beta, f, X_r, g \rangle : \langle f(X_r), i \rangle \to \langle \alpha, f \rangle$
$$(f(X_r) \in Y_s \Rightarrow$$
$$\exists \langle \gamma, f, X_r, g, Y_s \rangle : \langle g(Y_s), j_s \rangle \to \langle \beta, f, X_r, g \rangle . g(Y_s) \in \mu(Y_s)).$$

As the only attacks on $\langle \alpha, f \rangle$ had the form $\langle \beta, f, X_r, g \rangle$, and $g(Y_s) \in \mu(Y_s)$, condition (2) of Remark 2.3.3 is satisfied.

□

As said after Example 2.3.7, we do not know if level 3 is necessary for representation. We also do not know whether the same can be achieved with level 3, or higher, totally smooth structures.

2.3.5 Translation to Logic

We turn to the translation to logics.

Proposition 2.3.14

Let $\mathrel{|\!\sim}$ be a logic for $\mathcal{L}$. Set $T^{\mathcal{M}} := Th(\mu_{\mathcal{M}}(M(T)))$, where $\mathcal{M}$ is a generalized preferential structure, and $\mu_{\mathcal{M}}$ its choice function. Then

(1) there is a level 2 preferential structure $\mathcal{M}$ s.t. $\overline{\overline{T}} = T^{\mathcal{M}}$ iff (LLE), (CCL), (SC) hold for all $T, T' \subseteq \mathcal{L}$.

(2) there is a level 3 essentially smooth preferential structure $\mathcal{M}$ s.t. $\overline{\overline{T}} = T^{\mathcal{M}}$ iff (LLE), (CCL), (SC), $(\subseteq \supseteq)$ hold for all $T, T' \subseteq \mathcal{L}$.

2.3 Higher Preferential Structures

Proof

The proof is an immediate consequence of Corollary 2.3.6 (2), Fact 2.3.4, and Propositions 2.3.13 and 1.2.18 (10) and (11).

(More precisely, for (2): Let $\mathcal{M}$ be an essentially smooth structure, then by Definition 2.3.7 for all X $\mu(X) \sqsubseteq X$. Consider (μCUM). So by Fact 2.3.4 (2) $\mu(X') \subseteq X'' \subseteq X' \Rightarrow \mu(X') \sqsubseteq X''$, so by Fact 2.3.4 (1) $\mu(X') = \mu(X'')$. $(\mu \subseteq \supseteq)$ is analogous, using Fact 2.3.4 (3).

□

We leave aside the generalization of preferential structures to attacking structures relative to η, as this can cause problems, without giving real insight: It might well be that $\rho(X) \not\subseteq \eta(X)$, but, still, $\rho(X)$ and $\eta(X)$ might define the same theory - due to definability problems.

Chapter 3
Abstract Size

Abstract A natural interpretation of the non-monotonic rule $\phi \mid\sim \psi$ is that the set of exceptional cases, i.e., those where ϕ holds, but not ψ, is a small subset of all the cases where ϕ holds, and the complement, i.e., the set of cases where ϕ and ψ hold, is a big subset of all ϕ-cases.

Section 3.2 shows how one can develop a multitude of rules for nonmonotonic logics from a very small set of principles about reasoning with size. The rules about size can be split into additive and multiplicative rules, and rules about coherence. The tables in Section 3.2.6 summarize definitions and connections.

Section 3.3 develops the same idea in a first order setting, with a "weak filter" on the semantic side, a generalized quantifier ∇ on the syntactic side (with the intuitive meaning "in most cases"), and a representation result connecting syntax and semantics.

3.1 Introduction

We discuss the concept of abstract size in a number of different views:

(1) a multitude of notions of size of different strength, and their immediate connection to non-monotonic logic are discussed in Section 3.2,

(2) multiplicative properties of size and their connections to modular relations are also discussed there, more precisely in Section 3.2.5,

(3) the important connections between multiplicative laws about size and interpolation are discussed in Section 6.5,

(4) we interpret size by a "weak filter" and a generalized quantifier in a first order setting, in Section 3.3,

We conclude the introduction to this chapter by a short remark on the subjects treated in Section 3.2.

A natural interpretation of the non-monotonic rule $\phi \mathrel{\mid\!\sim} \psi$ is that the set of exceptional cases, i.e., those where ϕ holds, but not ψ, is a small subset of all the cases where ϕ holds, and the complement, i.e., the set of cases where ϕ and ψ hold, is a big subset of all ϕ-cases.

This interpretation gives an abstract semantics to non-monotonic logic, in the sense that definitions and rules are translated to rules about model sets without any structural justification of those rules, as they are given, e.g., by preferential structures, which provide structural semantics. Yet, they are extremely useful, as they allow us to concentrate on the essentials, forgetting about syntactical reformulations of semantically equivalent formulas; the laws derived from the standard proof-theoretical rules incite to generalizations and modifications, and reveal deep connections but also differences. One of those insights is the connection between laws about size and (semantical) interpolation for non-monotonic logics.

Section 3.2 shows how one can develop a multitude of rules for nonmonotonic logics from a very small set of principles about reasoning with size.

In our understanding, algebraic semantics describe the abstract properties corresponding model sets have. Structural semantics, on the other hand, give intuitive concepts like accessibility or preference, from which properties of model sets, and thus algebraic semantics, originate.

Varying properties of structural semantics (e.g., transitivity, etc.) result in varying properties of algebraic semantics, and thus of logical rules. We consider operations directly on the algebraic semantics and their logical consequences, and we see that simple manipulations of the size concept result in most rules of nonmonotonic logics. Even more, we show how to generate new rules from those manipulations. The result is one big table, which, in a much more modest scale, can be seen as a "periodic table" of the "elements" of nonmonotonic logic. Some simple underlying principles allow to generate them all.

(See also Section 1.2.5 for structural and algebraic semantics.)

3.1.1 Comparison of Three Abstract Coherent Size Systems

We compare in [Sch04], Section 7.3 three abstract coherent systems based on size:

- The system of S. Ben-David and R. Ben-Eliyahu (see [BB94]),
- the system of the author,
- the system of N. Friedman and J. Halpern (see [FH98]).

For lack of space, we just mention them here, and refer the interested reader to [Sch04] for details.

(1) The system of S. Ben-David and R. Ben-Eliyahu:

Let $\mathcal{N}' := \{\mathcal{N}'(A) : A \subseteq U\}$ be a system of filters for $\mathcal{P}(U)$, i.e. each $\mathcal{N}'(A)$ is a filter over A. The conditions are (in slight modification):

UC': $B \in \mathcal{N}'(A) \to \mathcal{N}'(B) \subseteq \mathcal{N}'(A)$,

DC': $B \in \mathcal{N}'(A) \to \mathcal{N}'(A) \cap \mathcal{P}(B) \subseteq \mathcal{N}'(B)$,

RBC': $X \in \mathcal{N}'(A), Y \in \mathcal{N}'(B) \to X \cup Y \in \mathcal{N}'(A \cup B)$,

SRM': $X \in \mathcal{N}'(A), Y \subseteq A \to A - Y \in \mathcal{N}'(A)$ $X \cap Y \in \mathcal{N}'(Y)$,

GTS': $C \in \mathcal{N}'(A), B \subseteq A \to C \cap B \in \mathcal{N}'(B)$.

The system of the author:

We consider a system of ideals over $\mathcal{P}(U)$, i.e. let for $A \subseteq U$ an ideal $\mathcal{I}(A) \subseteq \mathcal{P}(A)$ be given.

(2) ($\emptyset$) If $A \neq \emptyset$, then $\emptyset \in \mathcal{I}(A)$,

(Coh0) if $B \subseteq C \subseteq D$, $B \in \mathcal{I}(C)$, then $B \in \mathcal{I}(D)$,

(CohCUM) if $A, C \in \mathcal{I}(B)$, then $A - C \in \mathcal{I}(B\text{-}C)$,

(CohRM) if $A \in \mathcal{I}(B), C \subseteq B, B - C \notin \mathcal{I}(B)$, then $A - C \in \mathcal{I}(B\text{-}C)$.

(3) The system of N. Friedman and J. Halpern:

Let U be a set, $<$ a strict partial order on $\mathcal{P}(U)$, (i.e. $<$ is transitive, and contains no cycles). Consider the following conditions for $<$:

(B1) $A' \subseteq A < B \subseteq B' \to A' < B'$,

(B2) if A, B, C are pairwise disjoint, then $C < A \cup B, B < A \cup C \to B \cup C < A$,

(B3) $\emptyset < X$ for all $X \neq \emptyset$,

(B4) $A < B \to A < B\text{-}A$,

(B5) Let $X, Y \subseteq A$. If $A - X < X$, then $Y < A - Y$ or $Y - X < X \cap Y$.

3.2 Basic Definitions and Overview

3.2.1 Introduction

This section is about basic algebraic concepts implicit in many nonmonotonic and related logics. Looking at logics at this level of abstraction will often simplify and clarify seemingly complicated situations. We discuss here the basic ideas and their developments, often into a continuum of variations.

3.2.1.1 Additive and Multiplicative Laws About Size

These laws show connections and how to develop a multitude of logical rules known from non-monotonic logics by combining a small number of principles about size. We can use them as building blocks to construct the rules from.

More precisely, "size" is to be read as "relative size", since it is essential to change the base sets.

The probably easiest way to see a connection between non-monotonic logics and abstract size is by considering preferential structures. Preferential structures define principal filters, generated by the set of minimal elements, as follows: if $\phi \mathrel{|\!\sim} \psi$ holds in such a structure, then $\mu(\phi) \subseteq M(\psi)$, where $\mu(\phi)$ is the set of minimal elements of $M(\phi)$. According to our ideas, we define a principal filter $\mathcal{F}$ over $M(\phi)$ by $X \in \mathcal{F}$ iff $\mu(\phi) \subseteq X \subseteq M(\phi)$. Thus, $M(\phi) \cap M(\neg\psi)$ will be a "small" subset of $M(\phi)$. (Recall that filters contain the "big" sets, and ideals the "small" sets.)

We can now go back and forth between rules on size and logical rules, e.g.:

(For details, see Tables 3.2, 3.3, and 3.4.)

(1) The "AND" rule corresponds to the filter property (finite intersections of big subsets are still big).
(2) "Right weakening" corresponds to the rule that supersets of big sets are still big.
(3) It is natural, but beyond filter properties themselves, to postulate that if X is a small subset of Y, and $Y \subseteq Y'$, then X is also a small subset of Y'. We call such properties "coherence properties" between filters. This property corresponds to the logical rule (wOR).
(4) In the rule (CM_ω) usually called Cautious Monotony, we change the base set a little when going from $M(\alpha)$ to $M(\alpha \wedge \beta)$ (the change is small by the prerequisite $\alpha \mathrel{|\!\sim} \beta$), and still have $\alpha \wedge \beta \mathrel{|\!\sim} \beta'$ if we had $\alpha \mathrel{|\!\sim} \beta'$. We see here a conceptually very different use of "small", as we now change the base set, over which the filter is defined, by a small amount.
(5) The rule of Rational Monotony is the last one in the first table, and somewhat isolated there. It is better seen as a multiplicative law, as described in the third table. It corresponds to the rule that the product of medium (i.e, neither big nor small) sets, has still medium size.

3.2.2 Additive Properties

The definition of a filter or ideal is standard. The definition of a weak filter is due to the author, and permits to treat finite situations like the lottery paradox: the chances that a specific number will win is small, still, one of them will win.

3.2 Basic Definitions and Overview

Definition 3.2.1

Fix a base set X.

A filter (resp. weak filter) on or over X is a set $\mathcal{F} \subseteq \mathcal{P}(X)$ such that $(F1) - (F3)$ (resp. (F1), (F2), $(F3')$) hold:

(F1) $X \in \mathcal{F}$,

(F2) $A \subseteq B \subseteq X, A \in \mathcal{F}$ imply $B \in \mathcal{F}$,

(F3) $A, B \in \mathcal{F}$ imply $A \cap B \in \mathcal{F}$,

$(F3')$ $A, B \in \mathcal{F}$ imply $A \cap B \neq \emptyset$.

We often add:

(F4) $\emptyset \notin \mathcal{F}$.

So a weak filter satisfies $(F3')$ instead of (F3).

"Weak filters" were called $\mathcal{N} - systems$ in [Sch95-1]. We use the terminology $\mathcal{N} - systems$ in Section 3.3, so readers who would like to see more background information, e.g. in [Sch95-1], will not have to change between different notations.

Note that $\mathcal{P}(X)$ is a filter, but not a weak one, but all proper filters are also weak filters.

A filter is called a principal filter iff there is $X' \subseteq X$ s.t. $\mathcal{F} = \{A : X' \subseteq A \subseteq X\}$.

The dual notion is that of a (weak) ideal, i.e. a (weak) ideal on or over X is a set $\mathcal{I} \subseteq \mathcal{P}(X)$ such that $(I1) - (I3)$ (or (I1), (I2), $(I3')$) hold:

(I1) $\emptyset \in \mathcal{I}$,

(I2) $A \subseteq B \subseteq X, B \in \mathcal{I}$ imply $A \in \mathcal{I}$,

(I3) $A, B \in \mathcal{I}$ imply $A \cup B \in \mathcal{I}$,

$(I3')$ $A, B \in \mathcal{I}$ imply $A \cup B \neq X$.

Again, we often add:

(I4) $X \notin \mathcal{I}$.

So a weak ideal satisfies $(I3')$ instead of (I3).

A filter is an abstract notion of size; elements of a filter on X are called big subsets of X; their complements are called small, and the rest have medium size. The dual applies to ideals; this is justified by the following trivial fact - which is just the well known duality between standard filters and ideals.

Fact 3.2.1

If $\mathcal{F}$ is a (weak) filter on X, then $\mathcal{I} := \{X - A : A \in \mathcal{F}\}$ is a (weak) ideal on X; if $\mathcal{I}$ is a (weak) ideal on X, then $\mathcal{F} := \{X - A : A \in \mathcal{F}\}$ is a (weak) filter on X.

□

3.2.2.1 Discussion of the Tables 3.2 and 3.3

The main part of this Section are Table 3.2 and Table 3.3 They show connections and how to develop a multitude of logical rules known from nonmonotonic logics by combining a small number of principles about size. We use them as building blocks to construct the rules from.

These principles are some basic and very natural postulates, (Opt), (iM), $(eM\mathcal{I})$, $(eM\mathcal{F})$, and a continuum of power of the notion of "small", or, dually, "big", from $(1*s)$ to $(<\omega*s)$. From these, we can develop the rest except, essentially, Rational Monotony, and thus an infinity of different rules.

This is a conceptual Section, and it does not contain any more difficult formal results. The interest lies, in our opinion, in the simplicity, paucity, and naturalness of the basic building blocks. We hope that this schema brings more and deeper order into the rich fauna of nonmonotonic and related logics.

This section presents *additive* laws for abstract size, whereas our emphasis in this book is on *multiplicative* laws. We present them mainly to put our work in perspective, to connect this book to more familiar laws about size, and to make the text more self-contained, i.e., as a service to the reader.

We now give the main additive rules for manipulation of abstract size from [GS09a], as presented in Table 3.2 and Table 3.3, "Rules on size".

These laws allow us to give very concise abstract descriptions of laws about nonmonotonic logics, and, by their modularity, to construct new logics by changing some parameters. Moreover, they reveal deep connections between laws like AND and Cumulativity, which are not so obvious in their usual formulations.

(1) "$\mathcal{I}$" evokes "ideal", "$\mathcal{F}$" evokes "filter" though the full strength of both is reached only in $(<\omega*s)$. "s" evokes "small", and "$(x*s)$" stands for "x small sets together are still not everything".

(2) If $A \subseteq X$ is neither in $\mathcal{I}(X)$, nor in $\mathcal{F}(X)$, we say it has medium size, and we define $\mathcal{M}(X) := \mathcal{P}(X) - (\mathcal{I}(X) \cup \mathcal{F}(X))$.

$\mathcal{M}^+(X) := \mathcal{P}(X) - \mathcal{I}(X)$ is the set of subsets which are not small.

(3) $\nabla x \phi$ is a generalized first-order quantifier; it is read "almost all x have property ϕ". $\nabla x(\phi : \psi)$ is the relativized version, read "almost all x with property ϕ have also property ψ". To keep the table "Rules on size" simple, we write mostly only the non-relativized versions.

3.2 Basic Definitions and Overview

Formally, we have $\nabla x\phi :\Leftrightarrow \{x : \phi(x)\} \in \mathcal{F}(U)$ where U is the universe, and $\nabla x(\phi : \psi) :\Leftrightarrow \{x : (\phi \wedge \psi)(x)\} \in \mathcal{F}(\{x : \phi(x)\})$.

See Section 3.3.

(4) Analogously, for propositional logic, we define a logic $\mid\!\sim$ from a filter $\mathcal{F}$ by

$\alpha \mid\!\sim \beta :\Leftrightarrow M(\alpha \wedge \beta) \in \mathcal{F}(M(\alpha))$,

where $M(\phi)$ is the set of models of ϕ.

(5) In preferential structures, $\mu(X) \subseteq X$ is the set of minimal elements of X. This generates a principal filter by $\mathcal{F}(X) := \{A \subseteq X : \mu(X) \subseteq A\}$. Corresponding properties about μ are not listed systematically.

(6) The usual rules (AND) etc. are named here (AND_ω), as they are in a natural ascending line of similar rules, based on strengthening of the filter/ideal properties.

The Groups of Rules

The rules concern properties of $\mathcal{I}(X)$ or $\mathcal{F}(X)$, or dependencies between such properties for different X and Y. All X, Y, etc. will be subsets of some universe, say V. Intuitively, V is the set of all models of some fixed propositional language. It is not necessary to consider all subsets of V; the intention is to consider subsets of V, which are definable by a formula or a theory. So we assume all X, Y, etc. taken from some $\mathcal{Y} \subseteq \mathcal{P}(V)$, which we call the domain. In the former case, $\mathcal{Y}$ is closed under set difference, in the latter case not necessarily so. (We will mention it when we need some particular closure property.)

The rules are divided into five groups:

(1) (Opt), which says that "All" is optimal, i.e., when there are no exceptions, then a soft rule $\mid\!\sim$ holds.

(2) three monotony rules:

   (2.1) (iM) is inner monotony; a subset of a small set is small;

   (2.2) $(eM\mathcal{I})$ is external monotony for ideals: enlarging the base set keeps small sets small;

   (2.3) $(eM\mathcal{F})$ is external monotony for filters: a big subset stays big when the base set shrinks.

   These three rules are very natural if "size" is anything coherent over a change of base sets. In particular, they can be seen as weakening.

(3) $(\approx)$ keeps proportions; it is present here mainly to point the possibility out.

(4) a group of rules $x * s$, which say how many small sets will not yet add to the base set. The notation "$(< \omega * s)$" is an allusion to the full filter property, that filters are closed under $finite$ intersections.

(5) Rational Monotony, which can best be understood as robustness of $\mathcal{M}^+$; see $(\mathcal{M}^{++})$, (3).

We will assume all base sets to be nonempty in order to avoid pathologies and in particular clashes between (Opt) and $(1 * s)$.

Note that the full strength of the usual definitions of a filter and an ideal are reached only in line $(< \omega * s)$.

Regularities

(1) The group of rules $(x * s)$ use ascending strength of $\mathcal{I}/\mathcal{F}$.

(2) The column $(\mathcal{M}^+)$ contains interesting algebraic properties. In particular, they show a strengthening from $(3 * s)$ up to Rationality. They are not necessarily equivalent to the corresponding (I_x) rules, not even in the presence of the basic rules. The examples show that care has to be taken when considering the different variants.

(3) Adding the somewhat superflous (CM_2), we have increasing Cautious Monotony from (wCM) to full (CM_ω).

(4) We have increasing "or" from (wOR) to full (OR_ω).

(5) The line $(2 * s)$ is only there because there seems to be no $(\mathcal{M}_2^+)$; otherwise we could begin $(n * s)$ at $n = 2$.

Direct Correspondences

Several correspondences are trivial and are mentioned now. Somewhat less obvious (in)dependencies are given in Section 3.2.3. Finally, the connections with the μ-rules are given in Section 3.2.3.5. In those rules, (I_ω) is implicit, as they are about principal filters. Still, the μ-rules are written in the table "Rules on size" in their intuitively adequate place.

(1) The columns "Ideal" and "Filter" are mutually dual, when both entries are defined.

(2) The correspondence between the ideal/filter column and the ∇-column is obvious, the latter is added only for completeness' sake, and to point out the trivial translation to (augmented) first order logic. Note that the rules in the ∇-column are object level axioms.

3.2 Basic Definitions and Overview

(3) The ideal/filter and the AND-column correspond directly.

(4) We can construct logical rules from the $\mathcal{M}^+$-column by direct correspondence, e.g. for $(\mathcal{M}_\omega^+)$, (1):

Set $Y := M(\gamma)$, $X := M(\gamma \wedge \beta)$, $A := M(\gamma \wedge \beta \wedge \alpha)$.

- $X \in \mathcal{M}^+(Y)$ will become $\gamma \not\hspace{-2pt}\sim \neg\beta$
- $A \in \mathcal{F}(X)$ will become $\gamma \wedge \beta \hspace{2pt}\mid\hspace{-5pt}\sim \alpha$
- $A \in \mathcal{M}^+(Y)$ will become $\gamma \not\hspace{-2pt}\sim \neg(\alpha \wedge \beta)$.

so we obtain $\gamma \not\hspace{-2pt}\sim \neg\beta, \gamma \wedge \beta \hspace{2pt}\mid\hspace{-5pt}\sim \alpha \Rightarrow \gamma \not\hspace{-2pt}\sim \neg(\alpha \wedge \beta)$.

We did not want to make the table too complicated, so such rules are not listed in the table.

(5) Various direct correspondences:

- In the line (Opt), the filter/ideal entry corresponds to (SC),
- in the line (iM), the filter/ideal entry corresponds to (RW),
- in the line $(eM\mathcal{I})$, the ideal entry corresponds to (PR') and (wOR),
- in the line $(eM\mathcal{F})$, the filter entry corresponds to (wCM),
- in the line $(\approx)$, the filter/ideal entry corresponds to $(disjOR)$ and (NR),
- in the line $(1 * s)$, the filter/ideal entry corresponds to (CP),
- in the line $(2 * s)$, the filter/ideal entry corresponds to $(CM_2) = (OR_2)$.

(6) Note that one can, e.g., write (AND_2) in two flavours:

- $\alpha \hspace{2pt}\mid\hspace{-5pt}\sim \beta, \alpha \hspace{2pt}\mid\hspace{-5pt}\sim \beta' \Rightarrow \alpha \not\hspace{-2pt}\vdash \neg\beta \vee \neg\beta'$, or
- $\alpha \hspace{2pt}\mid\hspace{-5pt}\sim \beta \Rightarrow \alpha \not\hspace{-2pt}\sim \neg\beta$

- this is $(CM_2) = (OR_2)$.

For reasons of simplicity, we mention only one.

Rational Monotony

$(RatM)$ does not fit into adding small sets. We have exhausted the combination of small sets by $(< \omega * s)$, unless we go to languages with infinitary formulas.

The next idea would be to add medium size sets. But, by definition, $2 * medium$ can be all. Adding small and medium sets would not help either: Suppose we have a rule $medium + n * small \neq all$. Taking the complement of the first medium set, which is again medium, we have the rule $2 * n * small \neq all$. So we do not see any meaningful new internal rule. i.e. without changing the base set.

(For a similar reason, rules like $A, B \in \mathcal{I}(X) \Rightarrow A \cup B \notin \mathcal{F}(X)$ will give us nothing new, as then $A \cup B \in \mathcal{I}(X)$ is the full filter/ideal property, and $A \cup B \in \mathcal{M}(X)$ means just that 4 subsets of X may add up to the full X.)

Probably, $(RatM)$ has more to do with independence: by default, all "normalities" are independent, and intersecting with another formula preserves normality as much as possible. One should not forget here either the double use of "small" sets in our context: Non-monotonicity works with "small" exception sets, and many rules concern modifying the base set by a "small" subset. See also Remark 3.2.2 below.

Still, Rational Monotony has its natural place in an ascending chain of conditions, which can be seen by looking at the $\mathcal{M}^+$ conditions, especially at $(\mathcal{M}^+_\omega) (1) - (4)$, and $(\mathcal{M}^{++}) (3)$. Further research might tell whether there are still deeper connections behind this formal series.

More remarks on rules beyond rationality can be found in Section 3.2.2.3.

Summary

We can obtain all rules except $(RatM)$ and $(\approx)$ from (Opt), the monotony rules - $(iM), (eM\mathcal{I}), (eM\mathcal{F})$ -, and $(x * s)$ with increasing x.

Remark 3.2.2

There is, however, an important conceptual distinction to make here. Filters express "size" in an abstract way, in the context of non-monotonic logics, $\alpha \hspace{2pt}|\!\!\sim \beta$ iff the set of $\alpha \wedge \neg \beta$ is small in α. But here, we were interested in "small" changes in the reference set X (or α in our example). So we have two quite different uses of "size", one for non-monotonic logics, abstractly expressed by a filter, the other for coherence conditions. It is possible, but not necessary, to consider both essentially the same notions. But we should not forget that we have two conceptually different uses of size here.

3.2.2.2 A Partial Order View

Makinson (in personal communication) has suggested the following approach in the spririt of [FH96]:

Define a 2-place function f into a set with a linear order $\leq$, and with threshold elements s and l, $s \leq l$, and re-write $A \in \mathcal{I}(X)$ as $f(A, X) \leq s$, etc. To speak about medium size elements, we need another element m between s and l.

In this way, we can translate

3.2 Basic Definitions and Overview

- (Opt) to: $f(\emptyset, X) = s$, $f(X, X) = l$
- (im) to: if $A \subseteq X \subseteq Y$, then $f(X,Y) = s \Rightarrow f(A,Y) = s$ and $f(A,Y) = l \Rightarrow f(X,Y) = l$.
- (eMI) to: if $A \subseteq X \subseteq Y$, then $f(A,X) = s \Rightarrow f(A,Y) = s$
- (eMF) to: if $A \subseteq X \subseteq Y$, then $f(A,Y) = l \Rightarrow f(A,X) = l$.

In this translation, (eMI) and (eMF) are special cases of the principle

$(Order2)$ if $A \subseteq X \subseteq Y$, then $f(A,Y) \leq f(A,X)$

and (im) can be written as

$(Order1)$ if $A \subseteq X \subseteq Y$, then $f(A,Y) \leq f(X,Y)$,

both can be put together to

$(Order)$ if $A \subseteq X \subseteq Y$, then $f(A,Y) \leq min\{f(X,Y), f(A,X)\}$,

and also (Opt) be added to obtain finally

$(Order^+)$ if $A \subseteq X \subseteq Y$, then $s = f(\emptyset, X) \leq f(A,Y) \leq min\{f(X,Y), f(A,X)\} \leq f(X,X) = l$.

It might be a question of taste which language one prefers. Above principle $(Order2)$ has the advantage to unite the intuitively close (eMI) and (eMF), whereas our separate notation underlines that our intuition might be false, as there are examples which separate them.

3.2.2.3 Discussion of Other, Related, Rules

We discuss here briefly more rules found in the literature - and express our gratitude to the referee who pointed them out to us. One of them, (NR), fits into our "keeping proportions" line, and is integrated there, another one, (DR), is the unitary version of our $(Log \parallel)$, and put in the table "Logical rules" of Definition 1.2.10.

From [BMP97], we cite:

- Disjunctive rationality

  (DR) $\alpha \vee \beta \mathrel{|\!\sim} \gamma \Rightarrow \alpha \mathrel{|\!\sim} \gamma$ or $\beta \mathrel{|\!\sim} \gamma$

  holds in ranked structures and corresponds to our $(Log \parallel)$.

- Negation rationality

  (NR) $\alpha \mathrel{|\!\sim} \beta \Rightarrow \alpha \wedge \gamma \mathrel{|\!\sim} \beta$ or $\alpha \wedge \neg\gamma \mathrel{|\!\sim} \beta$

  corresponds to the $(\mathcal{M}^+)$ variant of $(\approx)$, $(\mathcal{M}^+ \cup disj)$.

The following four rules, also taken from [BMP97], all fall in a class we may call $(ULTRA)$. Looking at them from a size perspective, they say that some filters must be ultrafilters (i.e. there are no medium size sets, everything is either small or big). From the preferential perspective, it means that there is at most one minimal model.

- Determinacy preservation

  $(DP)\ \alpha\ \mid\!\sim \beta \Rightarrow \alpha \wedge \gamma \mid\!\sim \beta$ or $\alpha \wedge \gamma \mid\!\sim \neg\beta$

- Rational transitivity

  $(RT)\ \alpha \mid\!\sim \beta, \beta \mid\!\sim \gamma \Rightarrow \alpha \mid\!\sim \gamma$ or $\alpha \mid\!\sim \neg\gamma$

- Rational contraposition

  $(RC)\ \alpha \mid\!\sim \beta \Rightarrow \neg\beta \mid\!\sim \alpha$ or $\neg\beta \mid\!\sim \neg\alpha$

- Weak determinacy

  $(WD)\ True \mid\!\sim \neg\alpha \Rightarrow \alpha \mid\!\sim \beta$ or $\alpha \mid\!\sim \neg\beta$

We think that these rules may well be adapted to certain situations, but will not be able to solve more general problems like the blode Swedes problem. (Swedes are normally blonde and tall, but even not-blonde Swedes are tall. This is the problem of sub-ideal situations, where still as much as possible of the ideal is preserved.)

We cite from [HM07], see also [Haw96], [Haw07] the following rules, satisfied by probabilistic consequences relation (defined as validity above a fixed value threshold):

- Very cautious monotony

  $(VCM)\ \alpha \mid\!\sim \beta \wedge \gamma \Rightarrow \alpha \wedge \beta \mid\!\sim \gamma$

  which we can deduce from (I_ω) the same way as (CM_ω), using other standard rules.

- Weak OR

  $(WOR)\ \alpha \wedge \beta \mid\!\sim \gamma, \alpha \wedge \neg\beta \mid\!\sim \gamma \Rightarrow \alpha \mid\!\sim \gamma$

  corresponds to our $(disjOR)$.

- Weak AND

  $(WAND)\ \alpha \mid\!\sim \gamma, \alpha \wedge \neg\beta \mid\!\sim \beta \Rightarrow \alpha \mid\!\sim \beta \wedge \gamma$

  will not be discussed here, as the prerequisite $\alpha \wedge \neg\beta \mid\!\sim \beta$ entails that $\emptyset$ can be a big subset.

3.2.3 Coherent Systems

3.2.3.1 Definition and Basic Facts

Note that whenever we work with model sets, the rule

3.2 Basic Definitions and Overview

(LLE), left logical equivalence, $\vdash \alpha \leftrightarrow \alpha' \Rightarrow (\alpha \mid\sim \beta \Leftrightarrow \alpha' \mid\sim \beta)$ will hold. We will use this without further mentioning.

Definition 3.2.2

A coherent system of sizes, $\mathcal{CS}$, consists of a universe U, $\emptyset \notin \mathcal{Y} \subseteq \mathcal{P}(U)$, and for all $X \in \mathcal{Y}$ a system $\mathcal{I}(X) \subseteq \mathcal{P}(X)$ (dually $\mathcal{F}(X)$, i.e. $A \in \mathcal{F}(X) \Leftrightarrow X - A \in \mathcal{I}(X)$). $\mathcal{Y}$ may satisfy certain closure properties like closure under $\cup, \cap$, complementation, etc. We will mention this when needed, and not obvious.

We say that $\mathcal{CS}$ satisfies a certain property iff all $X, Y \in \mathcal{Y}$ satisfy this property.

$\mathcal{CS}$ is called basic or level 1 iff it satisfies (Opt), (iM), $(eM\mathcal{I})$, $(eM\mathcal{F})$, $(1 * s)$.

$\mathcal{CS}$ is level x iff it satisfies (Opt), (iM), $(eM\mathcal{I})$, $(eM\mathcal{F})$, $(x * s)$.

Fact 3.2.3

Note that, if for any Y $\mathcal{I}(Y)$ consists only of subsets of at most 1 element, then $(eM\mathcal{F})$ is trivially satisfied for Y and its subsets by (Opt).

□

Fact 3.2.4

Let a $\mathcal{CS}$ be given s.t. $\mathcal{Y} = \mathcal{P}(U)$. If $X \in \mathcal{Y}$ satisfies $(\mathcal{M}^{++})$, but not $(< \omega * s)$, then there is $Y \in \mathcal{Y}$ which does not satisfy $(2 * s)$.

Proof

Exercise, solution in the Appendix.

Fact 3.2.5

$(eM\mathcal{I})$ and $(eM\mathcal{F})$ are formally independent, though intuitively equivalent.

Proof

Let $U := \{x, y, z\}$, $X := \{x, z\}$, $\mathcal{Y} := \mathcal{P}(U) - \{\emptyset\}$

(1) Let $\mathcal{F}(U) := \{A \subseteq U : z \in A\}$, $\mathcal{F}(Y) = \{Y\}$ for all $Y \subset U$. (Opt), (iM) hold, $(eM\mathcal{I})$ holds trivially, so does $(< \omega * s)$, but $(eM\mathcal{F})$ fails for U and X.

(2) Let $\mathcal{F}(X) := \{\{z\}, X\}$, $\mathcal{F}(Y) := \{Y\}$ for all $Y \subseteq U, Y \neq X$. (Opt), (iM), $(< \omega * s)$ hold trivially, $(eM\mathcal{F})$ holds by Fact 3.2.3. $(eM\mathcal{I})$ fails, as $\{x\} \in \mathcal{I}(X)$, but $\{x\} \notin \mathcal{I}(U)$.

□

Fact 3.2.6

A level n system is strictly weaker than a level $n + 1$ system.

Proof

Consider $U := \{1, \ldots, n+1\}$, $\mathcal{Y} := \mathcal{P}(U) - \{\emptyset\}$. Let $\mathcal{I}(U) := \{\emptyset\} \cup \{\{x\} : x \in U\}$, $\mathcal{I}(X) := \{\emptyset\}$ for $X \neq U$. $(iM), (eM\mathcal{I}), (eM\mathcal{F})$ hold trivially. $(n * s)$ holds trivially for $X \neq U$, but also for U. $((n+1) * s)$ does not hold for U.

□

Remark 3.2.7

Note that our schemata allow us to generate infintely many new rules, here is an example:

Start with A, add $s_{1,1}$, $s_{1,2}$ two sets small in $A \cup s_{1,1}$ ($A \cup s_{1,2}$ respectively). Consider now $A \cup s_{1,1} \cup s_{1,2}$ and s_2 s.t. s_2 is small in $A \cup s_{1,1} \cup s_{1,2} \cup s_2$. Continue with $s_{3,1}$, $s_{3,2}$ small in $A \cup s_{1,1} \cup s_{1,2} \cup s_2 \cup s_{3,1}$ etc.

Without additional properties, this system creates a new rule, which is not equivalent to any usual rules.

□

3.2.3.2 Implications Between the Finite Versions

Fact 3.2.8

(1) $(I_n) + (eM\mathcal{I}) \Rightarrow (\mathcal{M}_n^+)$,

(2) $(I_n) + (eM\mathcal{I}) \Rightarrow (CM_n)$,

(3) $(I_n) + (eM\mathcal{I}) \Rightarrow (OR_n)$.

Proof

(1) Let $X_1 \subseteq \ldots \subseteq X_n$, so $X_n = X_1 \cup (X_2 - X_1) \cup \ldots \cup (X_n - X_{n-1})$. Let $X_i \in \mathcal{F}(X_{i+1})$, so $X_{i+1} - X_i \in \mathcal{I}(X_{i+1}) \subseteq \mathcal{I}(X_n)$ by $(eM\mathcal{I})$ for $1 \leq i \leq n-1$, so by (I_n) $X_1 \in \mathcal{M}^+(X_n)$.

(2) Suppose $\alpha \mathrel{\mid\!\sim} \beta_1, \ldots, \alpha \mathrel{\mid\!\sim} \beta_{n-1}$, but $\alpha \wedge \beta_1 \wedge \ldots \wedge \beta_{n-2} \mathrel{\mid\!\sim} \neg \beta_{n-1}$. Then $M(\alpha \wedge \neg \beta_1), \ldots, M(\alpha \wedge \neg \beta_{n-1}) \in \mathcal{I}(M(\alpha))$, and $M(\alpha \wedge \beta_1 \wedge \ldots \wedge \beta_{n-2} \wedge \beta_{n-1}) \in \mathcal{I}(M(\alpha \wedge \beta_1 \wedge \ldots \wedge \beta_{n-2})) \subseteq \mathcal{I}(M(\alpha))$ by $(eM\mathcal{I})$. But $M(\alpha) = M(\alpha \wedge \neg \beta_1) \cup \ldots \cup M(\alpha \wedge \neg \beta_{n-1}) \cup M(\alpha \wedge \beta_1 \wedge \ldots \wedge \beta_{n-2} \wedge \beta_{n-1})$ is now the union of n small subsets, contradiction.

(3) Exercise, solution in the Appendix.

□

In the following example, (OR_n), $(\mathcal{M}_n^+)$, (CM_n) hold, but $(\mathcal{I}_n)$ fails, so by Fact 3.2.8 $(\mathcal{I}_n)$ is strictly stronger than (OR_n), $(\mathcal{M}_n^+)$, (CM_n).

3.2 Basic Definitions and Overview

Example 3.2.1

Let $n \geq 3$.

Consider $X := \{1, \ldots, n\}$, $\mathcal{Y} := \mathcal{P}(X) - \{\emptyset\}$, $\mathcal{I}(X) := \{\emptyset\} \cup \{\{i\} : 1 \leq i \leq n\}$, and for all $Y \subset X$ $\mathcal{I}(Y) := \{\emptyset\}$.

(Opt), (iM), $(eM\mathcal{I})$, $(eM\mathcal{F})$ (by Fact 3.2.3), $(1 * s)$, $(2 * s)$ hold, (I_n) fails, of course.

(OR_n) holds:

Suppose $\alpha_1 \mathrel{|\!\sim} \beta, \ldots, \alpha_{n-1} \mathrel{|\!\sim} \beta, \alpha_1 \vee \ldots \vee \alpha_{n-1} \mathrel{|\!\sim} \neg\beta$.

Case 1: $\alpha_1 \vee \ldots \vee \alpha_{n-1} \vdash \neg\beta$, then for all i $\alpha_i \vdash \neg\beta$, so for no i $\alpha_i \mathrel{|\!\sim} \beta$ by $(1 * s)$ and thus (AND_1), contradiction.

Case 2: $\alpha_1 \vee \ldots \vee \alpha_{n-1} \nvdash \neg\beta$, then $M(\alpha_1 \vee \ldots \vee \alpha_{n-1}) = X$, and there is exactly 1 $k \in X$ s.t. $k \models \beta$. Fix this k. By prerequisite, $\alpha_i \mathrel{|\!\sim} \beta$. If $M(\alpha_i) = X$, $\alpha_i \vdash \beta$ cannot be, so there must be exactly 1 k' s.t. $k' \models \neg\beta$, but $card(X) \geq 3$, contradiction. So $M(\alpha_i) \subset X$, and $\alpha_i \vdash \beta$, so $M(\alpha_i) = \emptyset$ or $M(\alpha_i) = \{k\}$ for all i, so $M(\alpha_1 \vee \ldots \vee \alpha_{n-1}) \neq X$, contradiction.

$(\mathcal{M}_n^+)$ holds:

$(\mathcal{M}_n^+)$ is a consequence of $(\mathcal{M}_\omega^+)$, (3) so it suffices to show that the latter holds. Let $X_1 \in \mathcal{F}(X_2)$, $X_2 \in \mathcal{F}(X_3)$. Then $X_1 = X_2$ or $X_2 = X_3$, so the result is trivial.

(CM_n) holds:

Suppose $\alpha \mathrel{|\!\sim} \beta_1, \ldots, \alpha \mathrel{|\!\sim} \beta_{n-1}, \alpha \wedge \beta_1 \wedge \ldots \wedge \beta_{n-2} \mathrel{|\!\sim} \neg\beta_{n-1}$.

Case 1: For all i, $1 \leq i \leq n-2$, $\alpha \vdash \beta_i$, then $M(\alpha \wedge \beta_1 \wedge \ldots \wedge \beta_{n-2}) = M(\alpha)$, so $\alpha \mathrel{|\!\sim} \beta_{n-1}$ and $\alpha \mathrel{|\!\sim} \neg\beta_{n-1}$, contradiction.

Case 2: There is i, $1 \leq i \leq n-2$, $\alpha \nvdash \beta_i$, then $M(\alpha) = X$, $M(\alpha \wedge \beta_1 \wedge \ldots \wedge \beta_{n-2}) \subset M(\alpha)$, so $\alpha \wedge \beta_1 \wedge \ldots \wedge \beta_{n-2} \vdash \neg\beta_{n-1}$. $Card(M(\alpha \wedge \beta_1 \wedge \ldots \wedge \beta_{n-2})) \geq n - (n-2) = 2$, so $card(M(\neg\beta_{n-1})) \geq 2$, so $\alpha \mathrel{|\!\not\sim} \beta_{n-1}$, contradiction.

$\square$

3.2.3.3 Implications Between the ω Versions

Fact 3.2.9

$(CM_\omega) \Leftrightarrow (\mathcal{M}_\omega^+)$ (4)

Proof

"⇒": Suppose all sets are definable.
Let $A, B \in \mathcal{I}(X)$, $X = M(\alpha)$, $A = M(\alpha \wedge \neg\beta)$, $B = M(\alpha \wedge \neg\beta')$, so $\alpha \mathrel{\mid\!\sim} \beta$, $\alpha \mathrel{\mid\!\sim} \beta'$, so by (CM_ω) $\alpha \wedge \beta' \mathrel{\mid\!\sim} \beta$, so $A - B = M(\alpha \wedge \beta' \wedge \neg\beta) \in \mathcal{I}(M(\alpha \wedge \beta')) = \mathcal{I}(X - B)$.

"⇐": Let $\alpha \mathrel{\mid\!\sim} \beta$, $\alpha \mathrel{\mid\!\sim} \beta'$, so $M(\alpha \wedge \neg\beta) \in \mathcal{I}(M(\alpha))$, $M(\alpha \wedge \neg\beta') \in \mathcal{I}(M(\alpha))$, so by prerequisite $M(\alpha \wedge \neg\beta') - M(\alpha \wedge \neg\beta) = M(\alpha \wedge \beta \wedge \neg\beta') \in \mathcal{I}(M(\alpha) - M(\alpha \wedge \neg\beta)) = \mathcal{I}(M(\alpha \wedge \beta))$, so $\alpha \wedge \beta \mathrel{\mid\!\sim} \beta'$.

□

Fact 3.2.10

(1) $(I_\omega) + (eM\mathcal{I}) \Rightarrow (OR_\omega)$,

(2) $(I_\omega) + (eM\mathcal{I}) \Rightarrow (\mathcal{M}_\omega^+)$ (1),

(3) $(I_\omega) + (eM\mathcal{F}) \Rightarrow (\mathcal{M}_\omega^+)$ (2),

(4) $(I_\omega) + (eM\mathcal{I}) \Rightarrow (\mathcal{M}_\omega^+)$ (3),

(5) $(I_\omega) + (eM\mathcal{F}) \Rightarrow (\mathcal{M}_\omega^+)$ (4) (and thus, by Fact 3.2.9, (CM_ω)).

Proof

Exercise, solution in the Appendix.

We give three examples of independence of the various versions of $(\mathcal{M}_\omega^+)$.

Example 3.2.2

All numbers refer to the versions of $(\mathcal{M}_\omega^+)$.

For easier reading, we re-write for $A \subseteq X \subseteq Y$

$(\mathcal{M}_\omega^+)(1) : A \in \mathcal{F}(X), A \in \mathcal{I}(Y) \Rightarrow X \in \mathcal{I}(Y)$,

$(\mathcal{M}_\omega^+)(2) : X \in \mathcal{F}(Y), A \in \mathcal{I}(Y) \Rightarrow A \in \mathcal{I}(X)$.

Investigating all possibilities exhaustively seems quite tedious, and might best be done with the help of a computer. Fact 3.2.3 will be used repeatedly.

- (1), (2), (4) fail, (3) holds:

  Let $Y := \{a, b, c\}$, $\mathcal{Y} := \mathcal{P}(Y) - \{\emptyset\}$, $\mathcal{F}(Y) := \{\{a, c\}, \{b, c\}, Y\}$

  Let $X := \{a, b\}$, $\mathcal{F}(X) := \{\{a\}, X\}$, $A := \{a\}$, and $\mathcal{F}(Z) := \{Z\}$ for all $Z \neq X, Y$.

  (Opt), (iM), $(eM\mathcal{I})$, $(eM\mathcal{F})$ hold, (I_ω) fails, of course.

  (1) fails: $A \in \mathcal{F}(X)$, $A \in \mathcal{I}(Y)$, $X \notin \mathcal{I}(Y)$.

3.2 Basic Definitions and Overview

(2) fails: $\{a,c\} \in \mathcal{F}(Y)$, $\{a\} \in \mathcal{I}(Y)$, but $\{a\} \notin \mathcal{I}(\{a,c\})$.

(3) holds: If $X_1 \in \mathcal{F}(X_2)$, $X_2 \in \mathcal{F}(X_3)$, then $X_1 = X_2$ or $X_2 = X_3$, so (3) holds trivially (note that $X \notin \mathcal{F}(Y)$).

(4) fails: $\{a\}, \{b\} \in \mathcal{I}(Y)$, $\{a\} \notin \mathcal{I}(Y - \{b\}) = \mathcal{I}(\{a,c\}) = \{\emptyset\}$.

- (2), (3), (4) fail, (1) holds:

Let $Y := \{a,b,c\}$, $\mathcal{Y} := \mathcal{P}(Y) - \{\emptyset\}$, $\mathcal{F}(Y) := \{\{a,b\}, \{a,c\}, Y\}$

Let $X := \{a,b\}$, $\mathcal{F}(X) := \{\{a\}, X\}$, and $\mathcal{F}(Z) := \{Z\}$ for all $Z \neq X, Y$.

(Opt), (iM), $(eM\mathcal{I})$, $(eM\mathcal{F})$ hold, (I_ω) fails, of course.

(1) holds:

Let $X_1 \in \mathcal{F}(X_2)$, $X_1 \in \mathcal{I}(X_3)$, we have to show $X_2 \in \mathcal{I}(X_3)$. If $X_1 = X_2$, then this is trivial. Consider $X_1 \in \mathcal{F}(X_2)$. If $X_1 \neq X_2$, then X_1 has to be $\{a\}$ or $\{a,b\}$ or $\{a,c\}$. But none of these are in $\mathcal{I}(X_3)$ for any X_3, so the implication is trivially true.

(2) fails: $\{a,c\} \in \mathcal{F}(Y)$, $\{c\} \in \mathcal{I}(Y)$, $\{c\} \notin \mathcal{I}(\{a,c\})$.

(3) fails: $\{a\} \in \mathcal{F}(X)$, $X \in \mathcal{F}(Y)$, $\{a\} \notin \mathcal{F}(Y)$.

(4) fails: $\{b\}, \{c\} \in \mathcal{I}(Y)$, $\{c\} \notin \mathcal{I}(Y - \{b\}) = \mathcal{I}(\{a,c\}) = \{\emptyset\}$.

- (1), (2), (4) hold, (3) fails:

Let $Y := \{a,b,c\}$, $\mathcal{Y} := \mathcal{P}(Y) - \{\emptyset\}$, $\mathcal{F}(Y) := \{\{a,b\}, \{a,c\}, Y\}$

Let $\mathcal{F}(\{a,b\}) := \{\{a\}, \{a,b\}\}$, $\mathcal{F}(\{a,c\}) := \{\{a\}, \{a,c\}\}$, and $\mathcal{F}(Z) := \{Z\}$ for all other Z.

(Opt), (iM), $(eM\mathcal{I})$, $(eM\mathcal{F})$ hold, (I_ω) fails, of course.

(1) holds:

Let $X_1 \in \mathcal{F}(X_2)$, $X_1 \in \mathcal{I}(X_3)$, we have to show $X_2 \in \mathcal{I}(X_3)$. Consider $X_1 \in \mathcal{I}(X_3)$. If $X_1 = X_2$, this is trivial. If $\emptyset \neq X_1 \in \mathcal{I}(X_3)$, then $X_1 = \{b\}$ or $X_1 = \{c\}$, but then by $X_1 \in \mathcal{F}(X_2)$ X_2 has to be $\{b\}$, or $\{c\}$, so $X_1 = X_2$.

(2) holds: Let $X_1 \subseteq X_2 \subseteq X_3$, let $X_2 \in \mathcal{F}(X_3)$, $X_1 \in \mathcal{I}(X_3)$, we have to show $X_1 \in \mathcal{I}(X_2)$. If $X_1 = \emptyset$, this is trivial, likewise if $X_2 = X_3$. Otherwise $X_1 = \{b\}$ or $X_1 = \{c\}$, and $X_3 = Y$. If $X_1 = \{b\}$, then $X_2 = \{a,b\}$, and the condition holds, likewise if $X_1 = \{c\}$, then $X_2 = \{a,c\}$, and it holds again.

(3) fails: $\{a\} \in \mathcal{F}(\{a,c\})$, $\{a,c\} \in \mathcal{F}(Y)$, $\{a\} \notin \mathcal{F}(Y)$.

(4) holds:

If $A, B \in \mathcal{I}(X)$, and $A \neq B$, $A, B \neq \emptyset$, then $X = Y$ and e.g. $A = \{c\}$, $B = \{b\}$, and $\{c\} \in \mathcal{I}(Y - \{b\}) = \mathcal{I}(\{a,c\})$.

□

3.2.3.4 Rational Monotony

Fact 3.2.11

The three versions of $(\mathcal{M}^{++})$ are equivalent.

(We assume closure of the domain under set difference. For the third version of $(\mathcal{M}^{++})$, we use (iM).)

Proof

Exercise, solution in the Appendix.

Fact 3.2.12

We assume that all sets are definable by a formula.

$(RatM) \Leftrightarrow (\mathcal{M}^{++})$

Proof

We show equivalence of $(RatM)$ with version (1) of $(\mathcal{M}^{++})$.

"$\Rightarrow$": We have $A, B \subseteq X$, so we can write $X = M(\phi)$, $A = M(\phi \wedge \neg \psi)$, $B = M(\phi \wedge \neg \psi')$. $A \in \mathcal{I}(X)$, $B \notin \mathcal{F}(X)$, so $\phi \mathrel{|\!\sim} \psi$, $\phi \mathrel{|\!\not\sim} \neg \psi'$, so by $(RatM)$ $\phi \wedge \psi' \mathrel{|\!\sim} \psi$, so $A - B = M(\phi \wedge \neg \psi) - M(\phi \wedge \neg \psi') = M(\phi \wedge \psi' \wedge \neg \psi) \in \mathcal{I}(M(\phi \wedge \psi')) = \mathcal{I}(X - B)$.

"$\Leftarrow$": Let $\phi \mathrel{|\!\sim} \psi$, $\phi \mathrel{|\!\not\sim} \neg \psi'$, so $M(\phi \wedge \neg \psi) \in \mathcal{I}(M(\phi))$, $M(\phi \wedge \neg \psi') \notin \mathcal{F}(M(\phi))$, so by $(\mathcal{M}^{++})$ (1) $M(\phi \wedge \psi' \wedge \neg \psi) = M(\phi \wedge \neg \psi) - M(\phi \wedge \neg \psi') \in \mathcal{I}(M(\phi \wedge \psi'))$, so $\phi \wedge \psi' \mathrel{|\!\sim} \psi$.

$\Box$

3.2.3.5 Size and Principal Filter Logic

The connection with logical rules was shown in the table of Definition 1.2.10.

(1) to (7) of the following proposition (in different notation, as the more systematic connections were found only afterwards) was already published in [GS08c], we give it here in totality to complete the picture.

Proposition 3.2.13

(See Table 3.5, "Size and principal filter rules".) If $f(X)$ is the smallest A s.t. $A \in \mathcal{F}(X)$, then, given the property on the left, the one on the right follows.

Conversely, when we define $\mathcal{F}(X) := \{X' : f(X) \subseteq X' \subseteq X\}$, given the property on the right, the one on the left follows. For this direction, we assume that we can use the full powerset of some base set U - as is the case for the model sets of a finite

3.2 Basic Definitions and Overview

language. This is perhaps not too restrictive, as we mainly want to stress here the intuitive connections, without putting any weight on definability questions.

We assume (iM) to hold.

Note that there is no (μwCM), as the conditions $(\mu \ldots)$ imply that the filter is principal, and thus that (I_ω) holds - we cannot "see" (wCM) alone with principal filters, as (I_ω) will hold automatically in all filters, so a fortiori in all principal filters.

Proof

(1)

$(eM\mathcal{I}) \Rightarrow (\mu wOR)$:

$X - f(X)$ is small in X, so it is small in $X \cup Y$ by $(eM\mathcal{I})$, so $A := X \cup Y - (X - f(X)) \in \mathcal{F}(X \cup Y)$, but $A \subseteq f(X) \cup Y$, and $f(X \cup Y)$ is the smallest element of $\mathcal{F}(X \cup Y)$, so $f(X \cup Y) \subseteq A \subseteq f(X) \cup Y$.

$(\mu wOR) \Rightarrow (eM\mathcal{I})$:

Let $X \subseteq Y$, $X' := Y\text{-}X$. Let $A \in \mathcal{I}(X)$, so $X - A \in \mathcal{F}(X)$, so $f(X) \subseteq X\text{-}A$, so $f(X \cup X') \subseteq f(X) \cup X' \subseteq (X - A) \cup X'$ by prerequisite, so $(X \cup X') - ((X - A) \cup X') = A \in \mathcal{I}(X \cup X')$.

(2)

$(eM\mathcal{I}) + (I_\omega) \Rightarrow (\mu OR)$:

$X - f(X)$ is small in X, $Y - f(Y)$ is small in Y, so both are small in $X \cup Y$ by $(eM\mathcal{I})$, so $A := (X - f(X)) \cup (Y - f(Y))$ is small in $X \cup Y$ by (I_ω), but $X \cup Y - (f(X) \cup f(Y)) \subseteq A$, so $f(X) \cup f(Y) \in \mathcal{F}(X \cup Y)$, so, as $f(X \cup Y)$ is the smallest element of $\mathcal{F}(X \cup Y)$, $f(X \cup Y) \subseteq f(X) \cup f(Y)$.

$(\mu OR) \Rightarrow (eM\mathcal{I}) + (I_\omega)$:

Let again $X \subseteq Y$, $X' := Y\text{-}X$. Let $A \in \mathcal{I}(X)$, so $X - A \in \mathcal{F}(X)$, so $f(X) \subseteq X\text{-}A$. $f(X') \subseteq X'$, so $f(X \cup X') \subseteq f(X) \cup f(X') \subseteq (X - A) \cup X'$ by prerequisite, so $(X \cup X') - ((X - A) \cup X') = A \in \mathcal{I}(X \cup X')$.

(I_ω) holds by definition.

(3)

$(eM\mathcal{I}) + (I_\omega) \Rightarrow (\mu PR)$:

Let $X \subseteq Y$. $Y - f(Y)$ is the largest element of $\mathcal{I}(Y)$, $X - f(X) \in \mathcal{I}(X) \subseteq \mathcal{I}(Y)$ by $(eM\mathcal{I})$, so $(X - f(X)) \cup (Y - f(Y)) \in \mathcal{I}(Y)$ by (I_ω), so by "largest" $X - f(X) \subseteq Y - f(Y)$, so $f(Y) \cap X \subseteq f(X)$.

$(\mu PR) \Rightarrow (eM\mathcal{I}) + (I_\omega)$

Let again $X \subseteq Y$, $X' := Y\text{-}X$. Let $A \in \mathcal{I}(X)$, so $X - A \in \mathcal{F}(X)$, so $f(X) \subseteq X\text{-}A$, so by prerequisite $f(Y) \cap X \subseteq X\text{-}A$, so $f(Y) \subseteq X' \cup (X\text{-}A)$, so $(X \cup X') - (X' \cup (X - A)) = A \in \mathcal{I}(Y)$.

Again, (I_ω) holds by definition.

(4)

$(I \cup disj) \Rightarrow (\mu disjOR)$:

If $X \cap Y = \emptyset$, then (1) $A \in \mathcal{I}(X), B \in \mathcal{I}(Y) \Rightarrow A \cup B \in \mathcal{I}(X \cup Y)$ and (2) $A \in \mathcal{F}(X), B \in \mathcal{F}(Y) \Rightarrow A \cup B \in \mathcal{F}(X \cup Y)$ are equivalent. (By $X \cap Y = \emptyset$, $(X - A) \cup (Y - B) = (X \cup Y) - (A \cup B)$.) So $f(X) \in \mathcal{F}(X), f(Y) \in \mathcal{F}(Y)$ $\Rightarrow$ (by prerequisite) $f(X) \cup f(Y) \in \mathcal{F}(X \cup Y)$. $f(X \cup Y)$ is the smallest element of $\mathcal{F}(X \cup Y)$, so $f(X \cup Y) \subseteq f(X) \cup f(Y)$.

$(\mu disjOR) \Rightarrow (I \cup disj)$:

Let $X \subseteq Y$, $X' := Y\text{-}X$. Let $A \in \mathcal{I}(X)$, $A' \in \mathcal{I}(X')$, so $X - A \in \mathcal{F}(X)$, $X' - A' \in \mathcal{F}(X')$, so $f(X) \subseteq X\text{-}A$, $f(X') \subseteq X' - A'$, so $f(X \cup X') \subseteq f(X) \cup f(X') \subseteq (X - A) \cup (X' - A')$ by prerequisite, so $(X \cup X') - ((X - A) \cup (X' - A')) = A \cup A' \in \mathcal{I}(X \cup X')$.

(5)

$(\mathcal{M}_\omega^+) \Rightarrow (\mu CM)$:

$f(X) \subseteq Y \subseteq X \Rightarrow X - Y \in \mathcal{I}(X)$, $X - f(X) \in \mathcal{I}(X) \Rightarrow$ (by $(\mathcal{M}_\omega^+)$, (4)) $A := (X - f(X)) - (X - Y) \in \mathcal{I}(Y) \Rightarrow Y - A = f(X) - (X - Y) \in \mathcal{F}(Y)$ $\Rightarrow f(Y) \subseteq f(X) - (X - Y) \subseteq f(X)$.

$(\mu CM) \Rightarrow (\mathcal{M}_\omega^+)$

Let $X - A \in \mathcal{I}(X)$, so $A \in \mathcal{F}(X)$, let $B \in \mathcal{I}(X)$, so $f(X) \subseteq X - B \subseteq X$, so by prerequisite $f(X - B) \subseteq f(X)$. As $A \in \mathcal{F}(X)$, $f(X) \subseteq A$, so $f(X - B) \subseteq f(X) \subseteq A \cap (X - B) = A\text{-}B$, and $A - B \in \mathcal{F}(X\text{-}B)$, so $(X - A) - B = X - (A \cup B) = (X - B) - (A - B) \in \mathcal{I}(X\text{-}B)$, so $(\mathcal{M}_\omega^+)$, (4) holds.

(6)

$(\mathcal{M}^{++}) \Rightarrow (\mu RatM)$:

Let $X \subseteq Y$, $X \cap f(Y) \neq \emptyset$. If $Y - X \in \mathcal{F}(Y)$, then $A := (Y - X) \cap f(Y) \in \mathcal{F}(Y)$, but by $X \cap f(Y) \neq \emptyset$ $A \subset f(Y)$, contradicting "smallest" of $f(Y)$. So $Y - X \notin \mathcal{F}(Y)$, and by $(\mathcal{M}^{++})$ $X - f(Y) = (Y - f(Y)) - (Y - X) \in \mathcal{I}(X)$, so $X \cap f(Y) \in \mathcal{F}(X)$, so $f(X) \subseteq f(Y) \cap X$.

$(\mu RatM) \Rightarrow (\mathcal{M}^{++})$

Let $A \in \mathcal{F}(Y), B \notin \mathcal{F}(Y)$. $B \notin \mathcal{F}(Y) \Rightarrow Y - B \notin \mathcal{I}(Y) \Rightarrow (Y - B) \cap f(Y) \neq \emptyset$. Set $X := Y\text{-}B$, so $X \cap f(Y) \neq \emptyset$, $X \subseteq Y$, so $f(X) \subseteq f(Y) \cap X$ by prerequisite. $f(Y) \subseteq A \Rightarrow f(X) \subseteq f(Y) \cap X = f(Y) - B \subseteq A\text{-}B$.

(7)

Trivial in both directions.

(8.1)

Let $f(X) \subseteq Y \subseteq X$. $Y - f(Y) \in \mathcal{I}(Y) \subseteq \mathcal{I}(X)$ by (eMI). $f(X) \subseteq Y$ $\Rightarrow X - Y \subseteq X - f(X) \in \mathcal{I}(X)$, so by (iM) $X - Y \in \mathcal{I}(X)$. Thus by

3.2 Basic Definitions and Overview

(I_ω) $X - f(Y) = (X - Y) \cup (Y - f(Y)) \in \mathcal{I}(X)$, so $f(Y) \in \mathcal{F}(X)$, so $f(X) \subseteq f(Y)$ by definition.

(8.2)

(μCUT) is too special to allow to deduce $(eM\mathcal{I})$. Consider $U := \{a, b, c\}$, $X := \{a, b\}$, $\mathcal{F}(X) = \{X, \{a\}\}$, $\mathcal{F}(Z) = \{Z\}$ for all other $X \neq Z \subseteq U$. Then $(eM\mathcal{I})$ fails, as $\{b\} \in \mathcal{I}(X)$, but $\{b\} \notin \mathcal{I}(U)$. (iM) and $(eM\mathcal{F})$ hold. We have to check $f(A) \subseteq B \subseteq A \Rightarrow f(A) \subseteq f(B)$. The only case where it might fail is $A = X$, $B = \{a\}$, but it holds there, too.

(9.1)

Obviously $(\mu CM) + (\mu CUT) \Leftrightarrow (\mu CUM)$, so the result follows from (5.1) and (8.1).

(9.2)

Consider the same example as in (8.2). $f(A) \subseteq B \subseteq A \Rightarrow f(A) = f(B)$ holds there, too, by the same argument as above.

(10.1)

Let $f(X) \subseteq Y$, $f(Y) \subseteq X$. So $f(X), f(Y) \subseteq X \cap Y$, and $X - (X \cap Y) \in \mathcal{I}(X)$, $Y - (X \cap Y) \in \mathcal{I}(Y)$ by (iM). Thus $f(X), f(Y) \in \mathcal{F}(X \cap Y)$ by $(eM\mathcal{F})$ and $f(X) \cap f(Y) \in \mathcal{F}(X \cap Y)$ by (I_ω). So $X \cap Y - (f(X) \cap f(Y)) \in \mathcal{I}(X \cap Y)$, so $X \cap Y - (f(X) \cap f(Y)) \in \mathcal{I}(X), \mathcal{I}(Y)$ by $(eM\mathcal{I})$, so $(X - (X \cap Y)) \cup (X \cap Y - f(X) \cap f(Y)) = X - f(X) \cap f(Y) \in \mathcal{I}(X)$ by (I_ω), so $f(X) \cap f(Y) \in \mathcal{F}(X)$, likewise $f(X) \cap f(Y) \in \mathcal{F}(Y)$, so $f(X) \subseteq f(X) \cap f(Y)$, $f(Y) \subseteq f(X) \cap f(Y)$, and $f(X) = f(Y)$.

(10.2)

Consider again the same example as in (8.2), we have to show that $f(A) \subseteq B$, $f(B) \subseteq A \Rightarrow f(A) = f(B)$. The only interesting case is when one of A, B is X, but not both. Let e.g., $A = X$. We then have $f(X) = \{a\}$, $f(B) = B \subseteq X$, and $f(X) = \{a\} \subseteq B$, so $B = \{a\}$, and the condition holds.

□

3.2.4 Multiplicative Properties

(For an overview, see Table 3.4.)

We are mainly interested in non-monotonic logic. In this domain, independence is strongly connected to multiplication of abstract size, and an important part of the present text treats this connection and its repercussions.

We have at least two scenarios for multiplication; one is decribed in Figure 3.1, the second in Figure 3.2. In the first scenario, we have nested sets, in the second, we have set products. In the first scenario, we consider subsets which behave as the big set does; in the second scenario we consider subspaces, and decompose the behaviour of the big space into behaviour of the subspaces. In both cases, this results naturally in multiplication of abstract sizes. When we look at the corresponding relation properties, they are quite different (rankedness vs. some kind of modularity). But this is perhaps to be expected, as the two scenarios are quite different.

Other scenarios which might be interesting to consider in our framework are as follows:

- When we have more than two truth values, say three, and two is considered a big subset, and we have n propositional variables, and m of them are considered many, then 2^m might give a "big" subset of the total of 3^n situations.
- Similarly, when we fix one variable, consider two cases of the possible three, and multiply this with a "big" set of models.
- We may also consider the utility or cost of a situation, and work with a "big" utility, and "many" situations, etc.
- Note that, in the case of distances, subspaces add distances, and do not multiply them: $d(xy, x'y') = d(x, x') + d(y, y')$.

These questions are left for further research.

3.2.4.1 Multiplication of Size for Subsets

Here we have nested sets, $A \subseteq X \subseteq Y$, A is a certain proportion of X, and X of Y, resulting in a multiplication of relative size or proportions. This is a classical subject of non-monotonic logic; see the last section, taken from [GS09a]; it is partly repeated here to stress the common points with the other scenario.

Properties

Figure 3.1 is to be read as follows: The whole set Y is split into X and $Y - X$, X is split into A and $X - A$. X is a small/medium/big part of Y, and A is a small/medium/big part of X. The question is, is A a small/medium/big part of Y?

Note that the relation of A to X is conceptually different from that of X to Y, as we change the base set by going from X to Y, but not when going from A to X. Thus, in particular, when we read the diagram as expressing multiplication, commutativity is not necessarily true.

We looked at this scenario in [GS09a], but from an additive point of view, using various basic properties like (iM), $(eM\mathcal{I})$, $(eM\mathcal{F})$; see Section 3.2.2. Here, we use just multiplication — except sometimes for motivation.

3.2 Basic Definitions and Overview

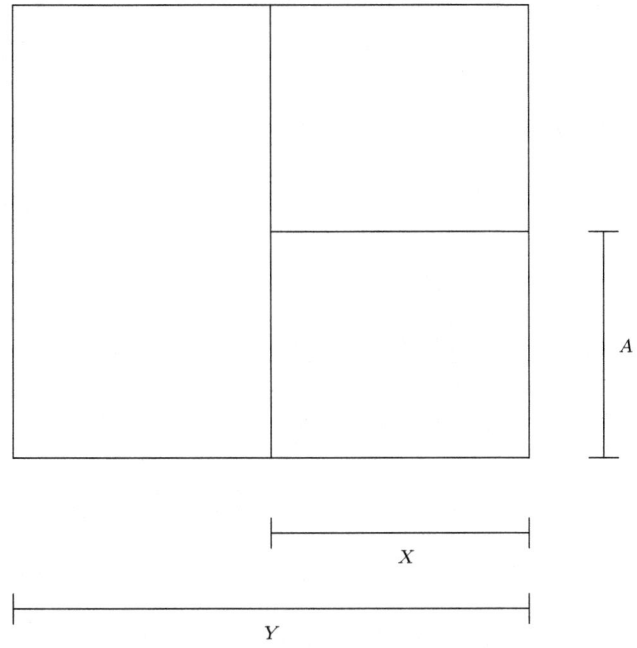

Fig. 3.1 Scenario 1

We examine different rules:

If $Y = X$ or $X = A$, there is nothing to show, so 1 is the neutral element of multiplication.

If $X \in \mathcal{I}(Y)$ or $A \in \mathcal{I}(X)$, then we should have $A \in \mathcal{I}(Y)$. (Use for motivation (iM) or $(eM\mathcal{I})$ respectively.)

So it remains to look at the following cases, with the "natural" answers given already:

(1) $X \in \mathcal{F}(Y), A \in \mathcal{F}(X) \Rightarrow A \in \mathcal{F}(Y)$,
(2) $X \in \mathcal{M}^+(Y), A \in \mathcal{F}(X) \Rightarrow A \in \mathcal{M}^+(Y)$,
(3) $X \in \mathcal{F}(Y), A \in \mathcal{M}^+(X) \Rightarrow A \in \mathcal{M}^+(Y)$,
(4) $X \in \mathcal{M}^+(Y), A \in \mathcal{M}^+(X) \Rightarrow A \in \mathcal{M}^+(Y)$.

But (1) is case (3) of $(\mathcal{M}_\omega^+)$ in [GS09a]; see Table "Rules on size" in Section 3.2.2.

(2) is case (1) of $(\mathcal{M}_\omega^+)$ there,

(3) is case (2) of $(\mathcal{M}_\omega^+)$ there, and finally,

(4) is $(\mathcal{M}^{++})$ there.

So the first three correspond to various expressions of (AND_ω), (OR_ω), (CM_ω), the last one to $(RatM)$.

But we can read them also the other way round, e.g.:

(1) corresponds to $\alpha \mathrel{|\!\sim} \beta,\ \alpha \wedge \beta \mathrel{|\!\sim} \gamma \Rightarrow \alpha \mathrel{|\!\sim} \gamma$,

(2) corresponds to $\alpha \mathrel{|\!\not\sim} \neg\beta,\ \alpha \wedge \beta \mathrel{|\!\sim} \gamma \Rightarrow \alpha \mathrel{|\!\not\sim} \neg(\beta \wedge \gamma)$,

(3) corresponds to $\alpha \mathrel{|\!\sim} \beta,\ \alpha \wedge \beta \mathrel{|\!\not\sim} \neg\gamma \Rightarrow \alpha \mathrel{|\!\not\sim} \neg(\beta \wedge \gamma)$.

All these rules might be seen as too idealistic, so as we did in [GS09a], we can consider milder versions: We might for instance consider a rule which says that $big * \ldots * big$, n times, is not small. Consider for instance the case $n = 2$. So we would conclude that A is not small in Y. In terms of logic, we then have $\alpha \mathrel{|\!\sim} \beta$, $\alpha \wedge \beta \mathrel{|\!\sim} \gamma \Rightarrow \alpha \mathrel{|\!\not\sim} (\neg\beta \vee \neg\gamma)$. We can obtain the same logical property from $3 * small \neq all$.

3.2.4.2 Multiplication of Size for Subspaces

Our main interest here is multiplication for subspaces, which we discuss now. The reason for this interest is that it is an abstract view on model sets defined for sublanguages and their composition. Consequently, it offers an abstract view on properties, assuring interpolation for non-monotonic logics. Here we will, however, only look at these laws in general, which are interesting also beyond interpolation. In particular, they also describe what may happen when we go from one language L to a bigger language L', and to the consequence relations in both languages. A priori, they may be totally different, but it often seems reasonable to postulate certain coherence properties between those consequence relations. For instance, if $\alpha \mathrel{|\!\sim} \beta$ in L, then we will often implicitly assume that also $\alpha \mathrel{|\!\sim} \beta$ in L' — but this is not trivial. This corresponds to the following coherence property for sizes: If $\Sigma \subseteq \Pi L$ is a big subset, then $\Sigma \times \Pi(L' - L) \subseteq \Pi L'$ is also a big subset. This is the kind of property we investigate here.

Properties

In this scenario, Σ_i are sets of sequences, (see Figure 3.2) corresponding, intuitively, to a set of models in language $\mathcal{L}_i$; Σ_i will be the set of α_i-models and the subsets Γ_i are to be seen as the "best" models where β_i will hold. The languages are supposed to be disjoint sublanguages of a common language $\mathcal{L}$. As the Σ_i have symmetrical roles, there is no intuitive reason for multiplication not to be commutative.

We can interpret the situation twofold:

First, we work separately in sublanguages $\mathcal{L}_1$ and $\mathcal{L}_2$, and, say, α_i and β_i are both defined in $\mathcal{L}_i$, we look at $\alpha_i \mathrel{|\!\sim} \beta_i$ in the sublanguage $\mathcal{L}_i$ or we consider both α_i

3.2 Basic Definitions and Overview

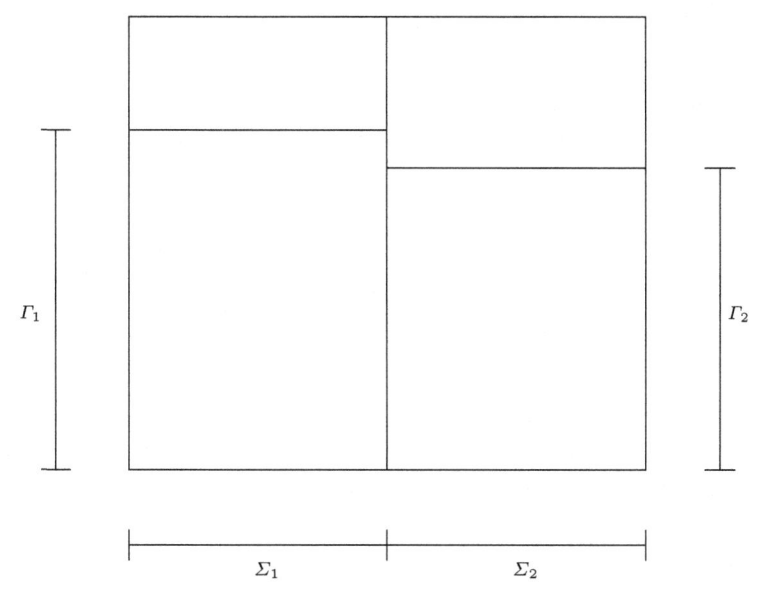

Fig. 3.2 Scenario 2

and β_i in the big language $\mathcal{L}$ and look at $\alpha_i \mathrel{|\!\sim} \beta_i$ in $\mathcal{L}$. These two ways are a priori completely different. Speaking in preferential terms, it is not at all clear why the orderings on the submodels should have anything to do with the orderings on the whole models. It seems a very desirable property, but we have to postulate it, which we do now (an overview is given in Table 3.4). We first give informally a list of such rules, mainly to show the connection with the first scenario. Later (see Definition 3.2.3), we will introduce formally some rules for which we show a connection with interpolation. Here "$(big * big \Rightarrow big)$" stands for "if both factors are big, so will be the product"; this will be abbreviated by "$b * b \Rightarrow b$" in Table 3.4.

We have

$(big * 1 \Rightarrow big)$ Let $\Gamma_1 \subseteq \Sigma_1$; if $\Gamma_1 \in \mathcal{F}(\Sigma_1)$, then $\Gamma_1 \times \Sigma_2 \in \mathcal{F}(\Sigma_1 \times \Sigma_2)$ (and we have the dual rule for Σ_2 and Γ_2).

This property preserves proportions, so it seems intuitively quite uncontested whenever we admit coherence over products. (Recall that there was nothing to show in the first scenario.)

When we reconsider the above case, and suppose $\alpha \mathrel{|\!\sim} \beta$ is in the sublanguage, so $M(\beta) \in \mathcal{F}(M(\alpha))$ holds in the sublanguage, so by $(big*1 \Rightarrow big)$, $M(\beta) \in \mathcal{F}(M(\alpha))$ in the big language $\mathcal{L}$.

We obtain the dual rule for small (and likewise, medium size) sets:

$(small*1 \Rightarrow small)$ Let $\Gamma_1 \subseteq \Sigma_1$; if $\Gamma_1 \in \mathcal{I}(\Sigma_1)$, then $\Gamma_1 \times \Sigma_2 \in \mathcal{I}(\Sigma_1 \times \Sigma_2)$ (and we have the dual rule for Σ_2 and Γ_2),

establishing $All = 1$ as the neutral element for multiplication.

We look now at other plausible rules:

$(small * x \Rightarrow small)$ $\Gamma_1 \in \mathcal{I}(\Sigma_1), \Gamma_2 \subseteq \Sigma_2 \Rightarrow \Gamma_1 \times \Gamma_2 \in \mathcal{I}(\Sigma_1 \times \Sigma_2)$.
$(big * big \Rightarrow big)$ $\Gamma_1 \in \mathcal{F}(\Sigma_1), \Gamma_2 \in \mathcal{F}(\Sigma_2) \Rightarrow \Gamma_1 \times \Gamma_2 \in \mathcal{F}(\Sigma_1 \times \Sigma_2)$.
$(big * medium \Rightarrow medium)$ $\Gamma_1 \in \mathcal{F}(\Sigma_1), \Gamma_2 \in \mathcal{M}^+(\Sigma_2) \Rightarrow \Gamma_1 \times \Gamma_2 \in \mathcal{M}^+(\Sigma_1 \times \Sigma_2)$.
$(medium * medium \Rightarrow medium)$ $\Gamma_1 \in \mathcal{M}^+(\Sigma_1), \Gamma_2 \in \mathcal{M}^+(\Sigma_2) \Rightarrow \Gamma_1 \times \Gamma_2 \in \mathcal{M}^+(\Sigma_1 \times \Sigma_2)$.

When we accept all the above rules, we can invert $(big*big \Rightarrow big)$, as a big product must be composed of big components. Likewise, at least one component of a small product has to be small; see Proposition 3.2.14.

We see that these properties give a lot of modularity. We can calculate the consequences of α and α' separately — provided α, α' use disjoint alphabets — and put the results together afterwards. Such properties are particularly interesting for classification purposes, where subclasses are defined with disjoint alphabets.

Recall that we work here with a notion of "big" and "small" subsets, which may be thought of as defined by a filter (ideal), though we usually will not need the full strength of a filter (ideal). But assume as usual that $A \subseteq B \subseteq C$ and $A \subseteq C$ is big together imply $B \subseteq C$ is big, and that $C \subseteq C$ is big, and define $A \subseteq B$ as small iff $(B - A) \subseteq B$ is big; call all subsets which are neither big nor small medium size. For an extensive discussion, see [GS09a].

Notation 3.2.1

For the intuition, we work with (classical) models, thus with sequences 0/1, for some language $\mathcal{L}$, and sublanguages.

We denote the set of propositional variables I, and those of sublanguages J, J', J'' etc. Those sets may, however, be any (nonempty) sets. The J, J', J'' will always be pairwise disjoint, and often $I = J \cup J'$ will be the case, context will tell. ΠI etc. will be the full set of sequences over I, $X \subseteq \Pi I$ etc. some subset. Single sequences will be denoted σ, τ, etc. We have $\Pi I = \Pi J \times \Pi J'$, if $I = J \cup J'$, etc. This will all be evident. We frequently use the restriction operator $\upharpoonright$, $\sigma \upharpoonright J$ will be the restriction of a full sequence σ over I to J, likewise $X \upharpoonright J$ for a set of such σ.

Definition 3.2.3

$\mu(X) \subseteq X$ will be the smallest big subset of X, as in the definition of a (minimal) preferential structure. We suppose this is defined on all sets considered.

3.2 Basic Definitions and Overview

Let $J' \cup J'' = I$, so $\Pi I = \Pi J' \times \Pi J''$. We consider subsets X, etc. of ΠI.

We define first the following two sets of three finite product rules about size and μ, $(\mu * 1) - (\mu * 3)$ and $(S * 1) - (S * 3)$. Both sets will be shown to be equivalent in Proposition 3.2.14.

$(\mu * 1)$: $\mu(X \times X') = \mu(X) \times \mu(X')$.

Note that μ is defined here on different sets. So, if $I = J \cup J'$, $J \cap J' = \emptyset$, and $X \subseteq \Pi J$, $X' \subseteq \Pi J'$, so $X \times X' \subseteq \Pi I$, then we can write more precisely, as we will do in more complicated situations.

$(\mu * 1)$: $\mu_I(X \times X') = \mu_J(X) \times \mu_{J'}(X')$.

$(\mu * 2)$ $\mu(X) \subseteq Y \Rightarrow \mu(X \upharpoonright J') \subseteq Y \upharpoonright J'$.

$(\mu * 3)$ $\mu(\Pi J' \times X'') \upharpoonright J'' \subseteq \mu(X) \upharpoonright J''$.

$(s * s)$ Let $Y_i \subseteq X_i$; then $Y_1 \times Y_2 \subseteq X_1 \times X_2$ is small iff $Y_1 \subseteq X_1$ is small or $Y_1 \subseteq X_1$ is small.

A generalization to more than two factors is obvious.

$(S * 1)$ $A \subseteq X' \times X''$ is big iff there is $Y = Y' \times Y'' \subseteq A$ such that $Y' \subseteq X'$ and $Y'' \subseteq X''$ are big.

$(S * 2)$ $Y \subseteq X$ is big $\Rightarrow Y \upharpoonright J' \subseteq X \upharpoonright J'$ is big, where X is not necessarily a product.

$(S * 3)$ $A \subseteq X$ is big $\Rightarrow$ there is a $B \subseteq \Pi J' \times X''$ big such that $B \upharpoonright J'' \subseteq A \upharpoonright J''$; again, X is not necessarily a product.

One can also consider weakenings, e.g.,

$(S * 1')$ $Y' \times X'' \subseteq X' \times X''$ is big iff $Y' \subseteq X'$ is big.

A variant of $(\mu * 1)$ is $(\mu * 4)$:

$(\mu * 4)$: $\mu(X \times X') = \mu(X \times \Pi J') \upharpoonright J \times \mu(\Pi J \times X') \upharpoonright J'$

Here, μ is always μ_I, so it is not necessary to distinguish different μ's.

The difference between $(\mu * 1)$ and $(\mu * 4)$ is important. For the first, we consider big subsets on different products, for the latter, we work within the same product. Thus, the first is a stronger coherence condition.

Example 3.2.3 shows that $(\mu * 2)$ seems too strong when compared to probability-defined size.

Proposition 3.2.14

(1) Let $(S * 1)$ hold. Then

$Y' \times Y'' \subseteq X' \times X''$ is small iff $Y' \subseteq X'$ or $Y'' \subseteq X''$ is small.

(2) If the filters over A are principal filters, generated by $\mu(A)$, i.e. $B \subseteq A$ is big iff $\mu(A) \subseteq B \subseteq A$ for some $\mu(A) \subseteq A$, then

$(S * i)$ is equivalent to $(\mu * i)$, $i = 1, 2, 3$.

(3) Let the notion of size satisfy (Opt), (iM), and $(< \omega * s)$; see the tables "Rules on size" in Section 3.2.2. Then $(\mu * 1)$ and $(s * s)$ are equivalent.

Proof

(1)

Exercise, solution in the Appendix.

(2.1)

"$\Rightarrow$"

"$\subseteq$": $\mu(X') \subseteq X'$ and $\mu(X'') \subseteq X''$ are big, so by $(S * 1)$ $\mu(X') \times \mu(X'') \subseteq X' \times X''$ is big, so $\mu(X' \times X'') \subseteq \mu(X') \times \mu(X'')$.

"$\supseteq$": $\mu(X' \times X'') \subseteq X' \times X''$ is big $\Rightarrow$ by $(S * 1)$ there is $Y' \times Y'' \subseteq \mu(X' \times X'')$ and $Y' \subseteq X', Y'' \subseteq X''$ big $\Rightarrow \mu(X') \subseteq Y', \mu(X'') \subseteq Y'' \Rightarrow \mu(X') \times \mu(X'') \subseteq \mu(X' \times X'')$.

"$\Leftarrow$"

Let $Y' \subseteq X'$ be big, $Y'' \subseteq X''$ be big, $Y' \times Y'' \subseteq \Delta$; then $\mu(X') \subseteq Y'$, $\mu(X'') \subseteq Y''$, so by $(\mu * 1)$ $\mu(X) = \mu(X') \times \mu(X'') \subseteq Y' \times Y'' \subseteq \Delta$, so Δ is big.

Let $\Delta \subseteq X$ be big; then by $(\mu * 1)$ $\mu(X') \times \mu(X'') = \mu(X) \subseteq \Delta$.

(2.2)

"$\Rightarrow$"

$\mu(X) \subseteq Y \Rightarrow Y \subseteq X$ big $\Rightarrow$ by $(S * 2)$ $Y \restriction J' \subseteq X \restriction J'$ big $\Rightarrow \mu(X \restriction J') \subseteq Y \restriction J'$.

"$\Leftarrow$"

$Y \subseteq X$ big $\Rightarrow \mu(X) \subseteq Y \Rightarrow$ by $(\mu * 2)$ $\mu(X \restriction J') \subseteq Y \restriction J' \Rightarrow Y \restriction J' \subseteq X \restriction J'$ big.

(2.3)

"$\Rightarrow$"

$\mu(X) \subseteq X$ big $\Rightarrow \exists B \subseteq \Pi J' \times X''$ big such that $B \restriction J'' \subseteq \mu(X) \restriction J''$ by $(S * 3)$; thus in particular $\mu(\Pi J' \times X'') \restriction J'' \subseteq \mu(X) \restriction J''$.

"$\Leftarrow$"

$A \subseteq X$ big $\Rightarrow \mu(X) \subseteq A$. $\mu(\Pi J' \times X'') \subseteq \Pi J' \times X''$ is big, and by $(\mu * 3)$ $\mu(\Pi J' \times X'') \restriction J'' \subseteq \mu(X) \restriction J'' \subseteq A \restriction J''$.

(3)

"⇒": (1) Let $Y' \subseteq X'$ be small; we show that $Y' \times Y'' \subseteq X' \times X''$ is small. So $X' - Y' \subseteq X'$ is big, so by (Opt) and $(\mu * 1)$ $(X' - Y') \times X'' \subseteq X' \times X''$ is big, so $Y' \times X'' = (X' \times X'') - ((X' - Y') \times X'') \subseteq X' \times X''$ is small, so by (iM) $Y' \times Y'' \subseteq X' \times X''$ is small.

(2) Suppose $Y' \subseteq X'$ and $Y'' \subseteq X''$ are not small; we show that $Y' \times Y'' \subseteq X' \times X''$ is not small. So $X' - Y' \subseteq X'$ and $X'' - Y'' \subseteq X''$ are not big. We show that $Z := ((X' \times X'') - (Y' \times Y'')) \subseteq X' \times X''$ is not big. $Z = (X' \times (X'' - Y'')) \cup ((X' - Y') \times X'')$.

Suppose $X_1 \times X_2 \subseteq Z$; then $X_1 \subseteq X' - Y'$ or $X_2 \subseteq X'' - Y''$. Proof: Let $X_1 \not\subseteq X' - Y'$ and $X_2 \not\subseteq X'' - Y''$, but $X_1 \times X_2 \subseteq Z$. Let $\sigma' \in X_1 - (X' - Y')$, $\sigma'' \in X_2 - (X'' - Y'')$; consider $\sigma'\sigma''$. $\sigma'\sigma'' \notin (X' - Y') \times X''$, as $\sigma' \notin X' - Y'$, $\sigma'\sigma'' \notin X' \times (X'' \times Y'')$, as $\sigma'' \notin X'' - Y''$, so $\sigma'\sigma'' \notin Z$.

By the prerequisite, $X' - Y' \subseteq X'$ is not big, $X'' - Y'' \subseteq X''$ is not big, so by (iM) no X_1 with $X_1 \subseteq X' - Y'$ is big, no X_2 with $X_2 \subseteq X'' - Y''$ is big, so by $(\mu * 1)$ or $(S * 1)$ $Z \subseteq X' \times X''$ is not big, so $Y' \times Y'' \subseteq X' \times X''$ is not small.

"⇐": (1) Suppose $Y' \subseteq X'$ is big; $Y'' \subseteq X''$ is big, we have to show $Y' \times Y'' \subseteq X' \times X''$ is big. $X' - Y' \subseteq X'$ is small, $X'' - Y'' \subseteq X''$ is small, so by $(s * s)$ $(X' - Y') \times X'' \subseteq X' \times X''$ is small and $X' \times (X'' - Y'') \subseteq X' \times X''$ is small, so by $(<\omega * s)$ $(X' \times X'') - (Y' \times Y'') = ((X' - Y') \times X'') \cup (X' \times (X'' - Y'')) \subseteq X' \times X''$ is small, so $Y' \times Y'' \subseteq X' \times X''$ is big.

(2) Suppose $Y' \times Y'' \subseteq X' \times X''$ is big; we have to show $Y' \subseteq X'$ is big, and $Y'' \subseteq X''$ is big. By the prerequisite, $(X' \times X'') - (Y' \times Y'') = ((X' - Y') \times X'') \cup (X' \times (X'' - Y'')) \subseteq X' \times X''$ is small, so by (iM) $X' \times (X'' - Y'') \subseteq X' \times X''$ is small, so by (Opt) and $(s * s)$, $X'' - Y'' \subseteq X''$ is small, so $Y'' \subseteq X''$ is big, and likewise $Y' \subseteq X'$ is big.

□

Discussion

We compare these rules to probability defined size.

Let "big" be defined by "more than 50%". If $\Pi J'$ and $\Pi J''$ have three elements each, then subsets of $\Pi J'$ or $\Pi J''$ of $card \geq 2$ are big. But taking the product may give $4/9 < 1/2$. So the product rule "$big * big = big$" will not hold there. One direction will hold, of course.

Next, we discuss the prerequisite $X = X' \times X''$. Consider the following example:

Example 3.2.3

Take a language of five propositional variables, with $J' := \{a, b, c\}$, $J'' := \{d, e\}$. Consider the model set $X := \{\pm a \pm b \pm cde, -a - b - c - d \pm e\}$, i.e., of eight models of de and two models of $-d$. The models of de are 8/10 of all elements of X, so it is reasonable to call them a big subset of X. But its projection on J'' is only 1/3 of X''.

So we have a potential *decrease* when going to the coordinates.

This shows that weakening the prerequisite about X as done in $(S * 2)$ is not innocent.

Remark 3.2.15

When we set small sets to 0, big sets to 1, we have the following Boolean rules for filters:

(1) $0 + 0 = 0$

(2) $1 + x = 1$

(3) $-0 = 1, -1 = 0$

(4) $0 * x = 0$

(5) $1 * 1 = 1$.

There are no such rules for medium size sets, as the union of two medium size sets may be big, but also stay medium.

Such multiplication rules capture the behaviour of Reiter defaults and of defeasible inheritance.

We summarize in this section properties related to multiplicative laws in Table 3.4.

$pr(b) = b$ means the projection of a big set on one of its coordinates is big again.

Note that $A \times B \subseteq X \times Y$ big $\Rightarrow A \subseteq X$ big is intuitively better justified than the other direction, as the proportion might increase in the latter, and decrease in the former. See the table "Rules on size", Section 3.2.2, "increasing proportions".

3.2.4.3 Conditions for Abstract Multiplication and Generating Relations

Remark 3.2.16

This is trivial, but helpful:

When checking $(\mu*4)$, and $\Sigma = \{\sigma\}$, a singleton, $\Gamma \neq \emptyset$, and $\mu(X) = \emptyset \rightarrow X = \emptyset$, then $\mu(\Sigma \times \Pi J') \upharpoonright J = \mu(\Sigma \times \Gamma) \upharpoonright J = \{\sigma\}$. This simplifies the condition and its verification considerably. Likewise for $\Gamma = \{\gamma\}$, of course.

3.2 Basic Definitions and Overview

Example 3.2.4

$(\mu * 1)$ does not imply any of the following:

(1) $\mu_{J'}(X \upharpoonright J') \subseteq \mu_I(X) \upharpoonright J'$, even if the generating relations are smooth.

(2) transitivity or smoothness of the generating relation (even if each element has size 2).

(3) $\mu(A) \subseteq B, \mu(B) \subseteq A \Rightarrow \mu(A) = \mu(B)$, so a consequence of smoothness of the generating relation may fail.

Proof

(1) Let $I = J \cup J'$, in that order, i.e. J is the first, J' the second coordinate. Consider $\{00, 01, 10, 11\}$, with the order $00 \prec 01 \prec 10 \prec 11$, closed under transitivity, thus, it is smooth. This is the order in I. In J and J' set $0 \prec 1$.

This order satisfies $(\mu * 1)$, as we show now.

The possible products for this set are:

(1) $1 \times \{0, 1\}$ and $0 \times \{0, 1\}$,

(2) $\{1, 0\} \times 0$ and $\{1, 0\} \times 1$,

(3) $\{1, 0\} \times \{0, 1\}$.

We have for (1): $\mu_I(1 \times \{0, 1\}) = \{10\} = \mu_J\{1\} \times \mu_{J'}\{0, 1\} = 10$, likewise for $0 \times \{0, 1\}$.

We have for (2): $\mu_I(\{1, 0\} \times 0) = 00 = \mu_J\{1, 0\} * \mu_{J'}(0) = 00$, likewise for $\{1, 0\} \times 1$.

We have for (3): $\mu_I(\{1, 0\} \times \{0, 1\}) = 00 = \mu_J(1, 0) \times \mu_{J'}(0, 1) = 00$.

Set now $X = \{10, 01\}$, then $\mu_I(X) = \{01\}, \mu_I(X) \upharpoonright J' = \{1\}$, but $X \upharpoonright J' = \{0, 1\}$, so $\mu_{J'}(X \upharpoonright J') = \{0\}$. Thus, $\mu_{J'}(X \upharpoonright J') \not\subseteq \mu_I(X) \upharpoonright J'$.

(2) Exercise, solution in the Appendix.

(3) Consider the example for (2), and $A := \{00, 10, 11\}, B := \{00, 11\}$. Then $\mu(A) = \{00\}, \mu(B) = B$, so $\mu(A) \subseteq B, \mu(B) \subseteq A$, but $\mu(A) \neq \mu(B)$.

□

Fact 3.2.17

Let $I = J \cup J' \cup J''$, $\Delta := \Sigma \upharpoonright J'$ for some Σ. Then

$(\mu * 4)$ entails

$$\mu[\Pi J \times \Delta \times \Pi J''] = \Big(\mu(\Pi I) \upharpoonright J\Big) \times \Big(\mu[\Pi J \times \Delta \times \Pi J''] \upharpoonright J'\Big) \times \Big(\mu(\Pi I) \upharpoonright J''\Big).$$

Proof

Consider

(1) $\mu[\Pi J \times \Delta \times \Pi J'']$.

By applying $(\mu * 4)$ to the first $\times$ in (1),

$\mu[\Pi J \times \Delta \times \Pi J''] =$

(2) $\Big(\mu(\Pi I) \upharpoonright J\Big) \times \Big(\mu[\Pi J \times \Delta \times \Pi J''] \upharpoonright J' \cup J''\Big)$.

By applying $(\mu * 4)$ to the second $\times$ in (1),

$\mu[\Pi J \times \Delta \times \Pi J''] =$

$\Big(\mu[\Pi J \times \Delta \times \Pi J''] \upharpoonright J \cup J'\Big) \times \Big(\mu(\Pi I) \upharpoonright J''\Big)$,

Thus, we have for the second part of (2):

$\mu[\Pi J \times \Delta \times \Pi J''] \upharpoonright J' \cup J'' =$

$\Big(\Big(\mu[\Pi J \times \Delta \times \Pi J''] \upharpoonright J \cup J'\Big) \times \Big(\mu(\Pi I) \upharpoonright J''\Big)\Big) \upharpoonright J' \cup J'' =$

$\Big(\mu[\Pi J \times \Delta \times \Pi J''] \upharpoonright J'\Big) \times \Big(\mu(\Pi I) \upharpoonright J''\Big)$,

the second equality holds by Fact 1.2.5.

So we finally have $\mu[\Pi J \times \Delta \times \Pi J''] = \Big(\mu(\Pi I) \upharpoonright J\Big) \times \Big(\mu[\Pi J \times \Delta \times \Pi J''] \upharpoonright J'\Big) \times \Big(\mu(\Pi I) \upharpoonright J''\Big)$.

□

3.2.5 Modular Relations and Multiplication of Size

Remark 3.2.18

There is a multitude of possible modifications in this context, which may all be interesting for special situations, e.g. resulting in restricted, but not full, interpolation.

(1) We can restrict the J's for which the conditions hold.
(2) We can restrict the X's and Y's for which the conditions hold.
(3) We can examine validity for μ generated by relations with various properties: smooth vs. not necessarily smooth, transitive vs. not necessarily transitive, etc.

We define the following conditions for a relation, and will relate them to the $(\mu * i)$ properties:

3.2 Basic Definitions and Overview

Definition 3.2.4

Let again $X \subseteq \Pi J, Y \subseteq \Pi J'$. Let $\sigma, \tau \in \Pi J, \sigma', \tau' \in \Pi J'$, and let relations $\prec$ be defined on $\Pi I, \Pi J, \Pi J$. (We will not distinguish them as $\prec_I$ etc., as the context will disambiguate.) As usual, $\preceq$ will denote $\prec$ or $=$, and $x \prec y :\Leftrightarrow x \preceq y$ and $x \neq y$.

Analogously, σ_i will be an element of ΠJ_i, and we suppose relations $\prec$ are defined on all ΠJ_i, and on arbitrary ΠJ for (R1) below.

$(GH1)$: $\sigma \preceq \tau \wedge \sigma' \preceq \tau' \wedge (\sigma \prec \tau \, \sigma' \prec \tau') \Rightarrow \sigma\sigma' \prec \tau\tau'$,

$(GH2)$: $\sigma\sigma' \prec \tau\tau' \Rightarrow \sigma \prec \tau \, \sigma' \prec \tau'$.

$(GH2)$ means that some compensation is possible, e.g., $\tau' \prec \sigma'$ might be the case, but $\sigma'' \prec \tau''$ wins in the end, so $\sigma'\sigma'' \prec \tau'\tau''$.

(GH) (= general Hamming) will stand for $(GH1)$ and $(GH2)$ together.

$(GH3)$: $\sigma'\sigma'' \preceq \tau'\tau'' \Leftrightarrow \sigma' \preceq \tau'$ and $\sigma'' \preceq \tau''$.

Thus:

$(GH3')$ $\sigma'\sigma'' \prec \tau'\tau''$ iff $\sigma'\sigma'' \preceq \tau'\tau''$ and $(\sigma' \prec \tau'$ or $\sigma'' \prec \tau'')$

(R1) $\sigma \prec \sigma'$ iff $\forall i(\sigma_i = \sigma'_i$ or $\sigma_i \prec \sigma'_i)$ and $\exists i(\sigma_i \prec \sigma'_i)$.

$(R1')$ $\sigma \preceq \sigma'$ iff $\forall i(\sigma_i = \sigma'_i$ or $\sigma_i \prec \sigma'_i)$.

($<$-Abs): $\sigma\tau \prec \sigma\tau'$ iff $\sigma'\tau \prec \sigma'\tau'$ (for all σ, σ' etc.)

This is an absoluteness condition for $\prec$.

Remark 3.2.19

There is a multitude of possible definitions and relations between those definitions, e.g. from longer to shorter sequences and inversely.

This is left to the interested reader to explore.

For instance, we might weaken ($<$-Abs), to $\sigma\tau \prec \sigma\tau' \Rightarrow \sigma'\tau' \not\prec \sigma'\tau$ for all σ, τ etc. We can introduce lexicographic relations, context dependent relations, like $fly \prec \neg fly$ for birds, but $\neg fly \prec fly$ for penguins, determine the order between longer sequences by an (abstract or probabilistic) majority rule from the smaller sequences, etc., ad libitum.

Example 3.2.5

Consider a binary system of length $I, \Pi I$. For $\sigma, \sigma' \in \Pi I$, set

$\sigma \prec \sigma'$ iff σ is the bitwise complement of σ', and $\sigma(0) = 0$.

Thus, only complementary sequences are comparable, and the comparison is decided by the first element.

The prerequisite of (<-Abs) is never satisfied (except for the empty sequence), so (<-Abs) holds. But, e.g., (R1) will not hold, as comparison is only by the first element. (GH1) and (GH2) will hold.

Example 3.2.6

The following are examples of GH relations:

Define on all components X_i a relation $\prec_i$.

(1) The set variant Hamming relation:

Let the relation $\prec$ be defined on $\Pi\{X_i : i \in I\}$ by $\sigma \prec \tau$ iff for all j $\sigma_j \preceq_j \tau_j$, and there is at least one i such that $\sigma_i \prec_i \tau_i$.

(2) The counting variant Hamming relation:

Let the relation $\prec$ be defined on $\Pi\{X_i : i \in I\}$ by $\sigma \prec \tau$ iff the number of i such that $\sigma_i \prec_i \tau_i$ is bigger than the number of i such that $\tau_i \prec_i \sigma_i$.

(3) The weighed counting Hamming relation:

Like the counting relation, but we give different (numerical) importance to different i. E.g., $\sigma_1 \prec \tau_1$ may count 1, $\sigma_2 \prec \tau_2$ may count 2, etc.

□

Example 3.2.7

The circumscription relation satisfies $(GH3)$ with $\neg p \leq p$ and $\bigwedge \pm q_i \leq \bigwedge \pm q'_i$ iff $\forall i (\pm q_i \leq \pm q'_i)$.

Remark 3.2.20

(1) The independence makes sense because the concept of models, and thus the usual interpolation for classical logic, rely on the independence of the assignments.

(2) This corresponds to social choice for many independent dimensions.

(3) We can also consider such factorisation as an approximation: we can do part of the reasoning independently.

The following proposition summarizes various properties for the different Hamming relations:

Proposition 3.2.21

Let $\Pi I = \Pi J' \times \Pi J''$, $X \subseteq \Pi I$.

(1) Let $\preceq$ be a smooth relation satisfying $(GH3)$. Then $(\mu * 2)$ holds, and thus $(S * 2)$ by Proposition 3.2.14, (2).

3.2 Basic Definitions and Overview

(2) Let again $X' := X \upharpoonright J'$, $X'' := X \upharpoonright J''$. Let $\preceq$ be a smooth relation satisfying $(GH3)$. Then

$$\mu(X') \times \mu(X'') \subseteq X \Rightarrow \mu(X) = \mu(X') \times \mu(X'').$$

(Here $X = X' \times X''$ will not necessarily hold.)

(3) Let again $X' := X \upharpoonright J'$, $X'' := X \upharpoonright J''$. Let $\preceq$ be a relation satisfying $(GH3)$, and $X = X' \times X''$. Then $(\mu * 1)$ holds, and thus, by Proposition 3.2.14, (2), $(S * 1)$ holds.

(4) Let $\preceq$ be a smooth relation satisfying $(GH3)$; then $(\mu * 3)$ holds, and thus by Proposition 3.2.14, (2), $(S * 3)$ holds.

(5) $(\mu*1)$ and $(\mu*2)$ and the usual axioms for smooth relations characterize smooth relations satisfying $(GH3)$.

(6) Let $\sigma \prec \tau \Leftrightarrow \tau \notin \mu(\{\sigma, \tau\})$ and $\prec$ be smooth. Then μ satisfies $(\mu * 1)$ (or, by Proposition 3.2.14, equivalently $(s * s)$) iff $\prec$ is a GH relation.

(7) Let $Y' \subseteq X'$, $Y'' \subseteq X''$, $Y' \times Y'' \subseteq X' \times X''$ be small; let $(GH2)$ hold; then $Y' \subseteq X'$ is small or $Y'' \subseteq X''$ is small.

(8) Let $Y' \subseteq X'$ be small, $Y'' \subseteq X''$; let $(GH1)$ hold; then $Y' \times Y'' \subseteq X' \times X''$ is small.

Proof

(1) Suppose $\mu(X) \subseteq Y$ and $\sigma' \in X \upharpoonright J' - Y \upharpoonright J'$; we show $\sigma' \notin \mu(X \upharpoonright J')$.

Let $\sigma = \sigma'\sigma'' \in X$; then $\sigma \notin Y$, so $\sigma \notin \mu(X)$. So here is $\rho \prec \sigma$, $\rho \in \mu(X) \subseteq Y$ by smoothness. Let $\rho = \rho'\rho''$. We have $\rho' \preceq \sigma'$ by $(GH3)$. $\rho' = \sigma'$ cannot be, as $\rho' \in Y \upharpoonright J'$, and $\sigma' \notin Y \upharpoonright J'$. So $\rho' \prec \sigma'$, and $\sigma' \notin \mu(X \upharpoonright J')$.

(2) "$\supseteq$": Let $\sigma' \in \mu(X')$, $\sigma'' \in \mu(X'')$. By the prerequisite, $\sigma'\sigma'' \in X$. Suppose $\tau \prec \sigma'\sigma''$; then $\tau' \prec \sigma'$ or $\tau'' \prec \sigma''$, a contradiction.

"$\subseteq$": Let $\sigma \in \mu(X)$; suppose $\sigma' \notin \mu(X')$ or $\sigma'' \notin \mu(X'')$. So there are $\tau' \preceq \sigma'$, $\tau'' \preceq \sigma''$ with $\tau' \in \mu(X')$, $\tau'' \in \mu(X'')$ by smoothness. Moreoever, $\tau' \prec \sigma'$ or $\tau'' \prec \sigma''$. By the prerequisite, $\tau'\tau'' \in X$ and $\tau'\tau'' \prec \sigma$, so $\sigma \notin \mu(X)$.

(3) "$\supseteq$": As in (2), the prerequisite holds trivially.

"$\subseteq$": As in (2), but we do not need $\tau' \in \mu(X')$, $\tau'' \in \mu(X'')$, as $\tau'\tau''$ will be in X trivially. So smoothness is not needed.

(4) Exercise, solution in the Appendix.

(5) If $\preceq$ is smooth and satisfies $(GH3)$, then $(\mu * 1)$ and $(\mu * 2)$ hold by (1) and (3). For the converse define as usual $\sigma \prec \tau :\Leftrightarrow \tau \notin \mu(\{\sigma, \tau\})$. Let $\sigma = \sigma'\sigma''$, $\tau = \tau'\tau''$.

We have to show

$\sigma \prec \tau$ iff $\sigma' \preceq \tau'$ and $\sigma'' \preceq \tau''$ and $(\sigma' \prec \tau'$ or $\sigma'' \prec \tau'')$.

"⇐":

Suppose $\sigma' \prec \tau'$ and $\sigma'' \preceq \tau''$. Then $\mu(\{\sigma', \tau'\}) = \{\sigma'\}$, and $\mu(\{\sigma'', \tau''\}) = \{\sigma''\}$ (either $\sigma'' \prec \tau''$ or $\sigma'' = \tau''$, so in both cases $\mu(\{\sigma'', \tau''\}) = \{\sigma''\}$). As $\tau' \notin \mu(\{\sigma', \tau'\})$, $\tau \notin \mu(\{\sigma', \tau'\} \times \{\sigma'', \tau''\}) =$ (by $(\mu * 1)$) $\mu(\{\sigma', \tau'\}) \times \mu(\{\sigma'', \tau''\}) = \{\sigma'\} \times \{\sigma''\} = \{\sigma\}$, so by smoothness $\sigma \prec \tau$.

"⇒":

Conversely, if $\sigma \prec \tau$, so $Y := \{\sigma\} = \mu(X)$ for $X := \{\sigma, \tau\}$, so by $(\mu * 2)$ $\mu(X \upharpoonright J') = \mu(\{\sigma', \tau'\}) \subseteq Y \upharpoonright J' = \{\sigma'\}$, so $\sigma' \preceq \tau'$; analogously $\mu(X \upharpoonright J'') = \mu(\{\sigma'', \tau''\}) \subseteq Y \upharpoonright J'' = \{\sigma''\}$, so $\sigma'' \preceq \tau''$, but both cannot be equal.

(6) (6.1) $(\mu * 1)$ entails the GH relation conditions.

$(GH1)$: Suppose $\sigma' \prec \tau'$ and $\sigma'' \preceq \tau''$. Then $\tau' \notin \mu(\{\sigma', \tau'\}) = \{\sigma'\}$, and $\mu(\{\sigma'', \tau''\}) = \{\sigma''\}$ (either $\sigma'' \prec \tau''$ or $\sigma'' = \tau''$, so in both cases $\mu(\{\sigma'', \tau''\}) = \{\sigma''\}$). As $\tau' \notin \mu(\{\sigma', \tau'\})$, $\tau'\tau'' \notin \mu(\{\sigma', \tau'\} \times \{\sigma'', \tau''\}) =_{(\mu*1)} \mu(\{\sigma', \tau'\}) \times \mu(\{\sigma'', \tau''\}) = \{\sigma'\} \times \{\sigma''\} = \{\sigma'\sigma''\}$, so by smoothness $\sigma'\sigma'' \prec \tau'\tau''$.

$(GH2)$: Let $X := \{\sigma', \tau'\}$, $Y := \{\sigma'', \tau''\}$, so $X \times Y = \{\sigma'\sigma'', \sigma'\tau'', \tau'\sigma'', \tau'\tau''\}$. Suppose $\sigma'\sigma'' \prec \tau'\tau''$, so $\tau'\tau'' \notin \mu(X \times Y) =_{(\mu*1)} \mu(X) \times \mu(Y)$. If $\sigma' \not\prec \tau'$, then $\tau' \in \mu(X)$; likewise if $\sigma'' \not\prec \tau''$, then $\tau'' \in \mu(Y)$, so $\tau'\tau'' \in \mu(X \times Y)$, a contradiction.

(6.2) The GH relation conditions generate $(\mu * 1)$.

$\mu(X \times Y) \subseteq \mu(X) \times \mu(Y)$: Let $\tau' \in X$, $\tau'' \in Y$, $\tau'\tau'' \notin \mu(X) \times \mu(Y)$; then $\tau' \notin \mu(X)$ or $\tau'' \notin \mu(Y)$. Suppose $\tau' \notin \mu(X)$; let $\sigma' \in X$, $\sigma' \prec \tau'$; so by condition $(GH1)$ $\sigma'\tau'' \prec \tau'\tau''$, so $\tau'\tau'' \notin \mu(X \times Y)$.

$\mu(X) \times \mu(Y) \subseteq \mu(X \times Y)$: Let $\tau' \in X$, $\tau'' \in Y$, $\tau'\tau'' \notin \mu(X \times Y)$; so there is $\sigma'\sigma'' \prec \tau'\tau''$, $\sigma' \in X$, $\sigma'' \in Y$, so by $(GH2)$ either $\sigma' \prec \tau'$ or $\sigma'' \prec \tau''$, so $\tau' \notin \mu(X)$ or $\tau'' \notin \mu(Y)$, so $\tau'\tau'' \notin \mu(X) \times \mu(Y)$.

(7) Suppose $Y' \subseteq X'$ is not small; so there is $\gamma' \in Y'$ and no $\sigma' \in X'$ with $\sigma' \prec \gamma'$. Fix this γ'. Consider $\{\gamma'\} \times Y''$. As $Y' \times Y'' \subseteq X' \times X''$ is small, there is for each $\gamma'\gamma''$, $\gamma'' \in Y''$ some $\sigma'\sigma'' \in X' \times X''$, $\sigma'\sigma'' \prec \gamma'\gamma''$. By $(GH2)$ $\sigma' \prec \gamma'$ or $\sigma'' \prec \gamma''$, but $\sigma' \prec \gamma'$ was excluded, so for all $\gamma'' \in Y''$ there is $\sigma'' \in X''$ with $\sigma'' \prec \gamma''$, so $Y'' \subseteq X''$ is small.

(8) Let $\gamma' \in Y'$; so there is $\sigma' \in X'$ and $\sigma' \prec \gamma'$. By $(GH1)$, for any $\gamma'' \in Y''$, $\sigma'\gamma'' \prec \gamma'\gamma''$, so no $\gamma'\gamma'' \in Y' \times Y''$ is minimal.

□

Example 3.2.8

Even for smooth relations satisfying $(GH3)$, the converse of $(\mu * 2)$ is not necessarily true:

3.2 Basic Definitions and Overview

Let $\sigma' \prec \tau', \tau'' \prec \sigma'', X := \{\sigma, \tau\}$; then $\mu(X) = X$, but $\mu(X') = \{\sigma'\}, \mu(X'') = \{\tau''\}$, so $\mu(X) \neq \mu(X') \times \mu(X'')$.

We need the additional assumption that $\mu(X') \times \mu(X'') \subseteq X$; see Proposition 3.2.21 (2).

Note

Note that $(\mu * 1)$ results in a strong independence result in the second scenario: Let $\sigma\rho' \prec \tau\rho'$; then $\sigma\rho'' \prec \tau\rho''$ for all ρ''. Thus, whether $\{\rho''\}$ is small or medium size (i.e., $\rho'' \in \mu(X')$), the behaviour of $X \times \{\rho''\}$ is the same. This we do not have in the first scenario, as small sets may behave very differently from medium size sets. (But, still, their internal structure is the same; only the minimal elements change.) When $(\mu * 2)$ holds, if $\sigma\sigma' \prec \tau\tau'$ and $\sigma \neq \tau$, then $\sigma \prec \tau$, i.e., we need not have $\sigma' = \tau'$.

Fact 3.2.22

If (R1) holds for $\prec$, then so does $<$-Abs.

(This follows also from Fact 3.2.23 and Fact 3.2.24.)

Proof

Trivial.

$\square$

Fact 3.2.23

$(\mu * 4)$ implies $(< Abs)$.

Proof

Let σ be defined on J, τ, τ' on J'.

$\sigma\tau \prec \sigma\tau'$ iff $\sigma\tau' \notin \mu(\{\sigma\} \times \{\tau, \tau'\}) = \mu(\{\sigma\} \times \Pi J') \upharpoonright J \times \mu(\Pi J \times \{\tau, \tau'\}) \upharpoonright J'$
$= \{\sigma\} \times \mu(\Pi J \times \{\tau, \tau'\}) \upharpoonright J'$, but the second part of this product is independent of σ.

$\square$

Fact 3.2.24

If (R1) holds for $\prec$, then so does $(\mu * 4)$.

Proof

Let J, J' be as in the definition of $(\mu * 4)$.

Let $\sigma\gamma \in (X \times Y) - \mu(X \times Y)$, then there ist $\sigma'\gamma' \in X \times Y$, $\sigma'\gamma' \prec \sigma\gamma$. Let $K := \{i : (\sigma'\gamma')_i \prec (\sigma\gamma)_i\} \neq \emptyset$. Suppose $K \cap J \neq \emptyset$. Then by (R1) for all $\gamma'' \in \Pi J'$ $\sigma'\gamma'' \prec \sigma\gamma''$, so there is no $\rho \in \mu(X \times \Pi J')$ s.t. $\rho \upharpoonright J = \sigma$, so $\sigma\gamma \notin (\mu(X \times \Pi J') \upharpoonright J) \times (\mu(\Pi J \times Y) \upharpoonright J')$. The case for $K \cap J' \neq \emptyset$ is analogous.

Conversely:

Let $\prec$ be given on the $p_i/\neg p_i$. Define $\delta : I := J \cup J' \to \{0, 1\}$ as follows: if $p_i \prec \neg p_i$, then $\delta_i := 1$, if $p_i \succ \neg p_i$, then $\delta_i := 0$, otherwise δ_i is arbitrary, say 1. Then δ is minimal in ΠI. Let $\sigma\gamma \in X \times Y$, but $\sigma\gamma \notin (\mu(X \times \Pi J') \upharpoonright J) \times (\mu(\Pi J \times Y) \upharpoonright J')$. Then $\sigma \notin \mu(X \times \Pi J') \upharpoonright J$ or $\gamma \notin \mu(\Pi J \times Y) \upharpoonright J'$. Suppose $\sigma \notin \mu(X \times \Pi J') \upharpoonright J$. Consider $\sigma(\delta \upharpoonright J')$, the concatenation of σ with $\delta \upharpoonright J'$. As $\sigma \notin \mu(X \times \Pi J') \upharpoonright J$, there must be $\tau\tau' \in X \times \Pi J'$, $\tau\tau' \prec \sigma(\delta \upharpoonright J')$, $\tau \neq \sigma$, and by choice of δ, $\tau \prec \sigma$ (by abuse of language, $\prec$ in J), so $\tau\gamma \prec \sigma\gamma$, thus $\sigma\gamma \notin \mu(X \times Y)$.

The case $\gamma \notin \mu(\Pi J \times Y) \upharpoonright J'$ is analogous.

$\square$

Fact 3.2.25

(R1) implies $(\mu * 1)$

Proof

Exercise, solution in the Appendix.

Fact 3.2.26

(1) If (R1) holds, and all individual relations are transitive, then so is the global relation.

(2) In the finite and loop-free, 2-value case, if (R1) holds, then smoothness holds, $(\mu * 1)$ holds, and equivalence interpolation holds, too, by Proposition 6.5.11.

Proof

(1) This is trivial by $\rho \prec \sigma \prec \tau \to \rho_i \prec \sigma_i \prec \tau_i \to \rho \prec \tau$.

(2) By Fact 3.2.25, (R1) implies $(\mu * 1)$, so by smoothness (which follows from transitivity, finiteness, and absence of loops), and Proposition 6.5.11, there is a common interpolant for both directions.

$\square$

3.2 Basic Definitions and Overview

3.2.5.1 Hamming Distances

This material is strongly related to the above. For this reason, we treat it here, and not with Theory Revision, one of the subjects where it may be applied.

Definition 3.2.5

Given $x, y \in X$, a set of sequences over an index set I, the Hamming distance comes in two flavours:

$d_s(x, y) := \{i \in I : x(i) \neq y(i)\}$, the set variant,

$d_c(x, y) := card(d_s(x, y))$, the counting variant.

We define $d_s(x, y) \leq d_s(x', y')$ iff $d_s(x, y) \subseteq d_s(x', y')$;

thus, s-distances are not always comparable. Consequently, readers should be aware that d_s-values are *not* always comparable, even though $<$ and $\leq$ may suggest a linear order. We use these symbols to be in line with other distances.

There are straightforward generalizations of the counting variant:

We can also give different importance to different i in the counting variant, so, e.g., $d_c(\langle x, x' \rangle, \langle y, y' \rangle)$ might be 1 if $x \neq y$ and $x' = y'$, but 2 if $x = y$ and $x' \neq y'$.

If the $x \in X$ may have more than two different values, then a varying individual distance may also reflect in the distances in X. So, (for any distance d) if $d(x(i), x'(i)) < d(x(i), x''(i))$, then (the rest being equal) we may have $d(x, x') < d(x, x'')$.

Fact 3.2.27

(1) If the $x \in X$ have only two values, say TRUE and FALSE, then $d_s(x, y) = \{i \in I : x(i) = TRUE\} \triangle \{i \in I : y(i) = TRUE\}$, where $\triangle$ is the symmetric set difference.

(2) d_c has the normal addition, set union takes the role of addition for d_s, $\emptyset$ takes the role of 0 for d_s; both are distances in the following sense:

(2.1) $d(x, y) = 0$ iff $x = y$,

(2.2) $d(x, y) = d(y, x)$,

(2.3) the triangle inequality holds for the set variant in the form $d_s(x, z) \subseteq d_s(x, y) \cup d_s(y, z)$.

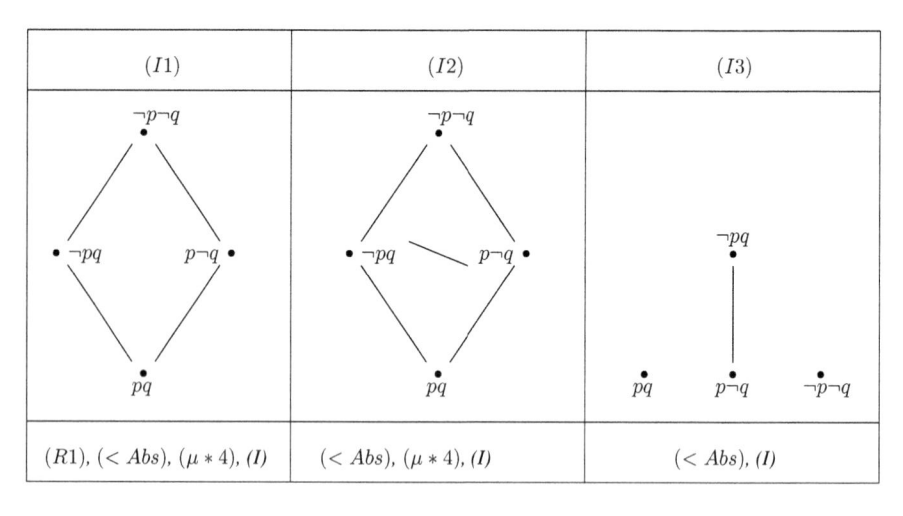

(I1)	(I2)	(I3)
¬p¬q at top, ¬pq left, p¬q right, pq bottom (diamond)	¬p¬q at top, ¬pq left, p¬q right (disconnected), pq bottom	¬pq — pq (vertical), and isolated p¬q, ¬p¬q
$(R1), (< Abs), (\mu * 4), (I)$	$(< Abs), (\mu * 4), (I)$	$(< Abs), (I)$

Fig. 3.3 Example 3.2.9 (1)

Proof

(2.3) If $i \notin d_s(x,y) \cup d_s(y,z)$, then $x(i) = y(i) = z(i)$, so $x(i) = z(i)$ and $i \notin d_s(x,z)$.

The others are trivial.

□

Definition 3.2.6

Let d be an abstract distance on some product space $X \times Y$ and its components. (We require of distances only that they be comparable, that $d(x,y) = 0$ iff $x = y$, and that $d(x,y) \geq 0$.)

d is called a generalized Hamming distance (GHD) iff it satisfies the following two properties:

$(GHD1)$ $d(\sigma, \tau) \leq d(\alpha, \beta)$ and $d(\sigma', \tau') \leq d(\alpha', \beta')$ and $(d(\sigma, \tau) < d(\alpha, \beta)$ or $d(\sigma', \tau') < d(\alpha', \beta')) \Rightarrow d(\sigma\sigma', \tau\tau') < d(\alpha\alpha', \beta\beta')$,

$(GHD2)$ $d(\sigma\sigma', \tau\tau') < d(\alpha\alpha', \beta\beta') \Rightarrow d(\sigma, \tau) < d(\alpha, \beta)$ or $d(\sigma', \tau') < d(\alpha', \beta')$.

(Compare this definition to Definition 3.2.4.)

3.2 Basic Definitions and Overview

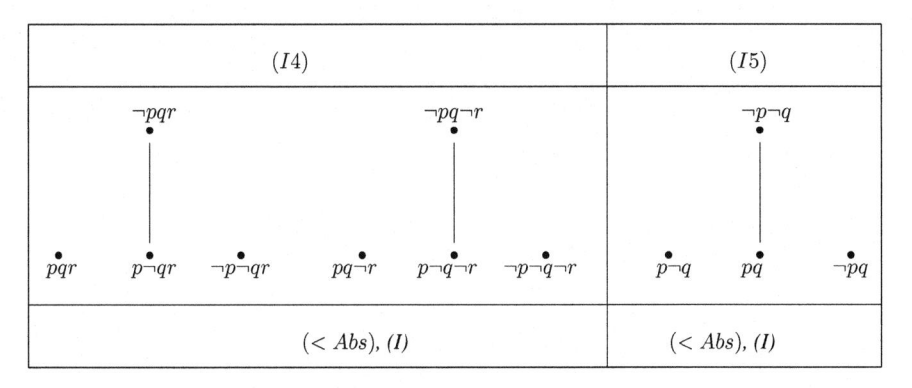

(I4)	(I5)
¬pqr — pqr, p¬qr, ¬p¬qr ; ¬pq¬r — pq¬r, p¬q¬r, ¬p¬q¬r	¬p¬q — p¬q, pq, ¬pq
(< Abs), (I)	(< Abs), (I)

Fig. 3.4 Example 3.2.9 (2)

3.2.5.2 Some Examples

Example 3.2.9

Consider the following examples. The relations should always be closed under transitivity. All work with the language $\{p, q\}$, except (I4), which works with $\{p, q, r\}$. We only give the relation between the models.

(I1) $pq \prec \neg pq \prec \neg p \neg q, pq \prec p \neg q \prec \neg p \neg q$

(I2) $pq \prec \neg pq \prec \neg p \neg q, pq \prec p \neg q \prec \neg p \neg q, p \neg q \prec \neg pq$

(I3) $p \neg q \prec \neg pq$

(I4) $p \neg qr \prec \neg pqr, p \neg q \neg r \prec \neg pq \neg r$

(I5) $pq \prec \neg p \neg q$

(I6) $pq \prec p \neg q \prec \neg pq, p \neg q \prec \neg p \neg q$

(I7) $pq \prec p \neg q, \neg p \neg q \prec \neg pq$

See Figure 3.3, Figure 3.4, Figure 3.5. The lower end of the lines indicates the smaller element, the upper end the bigger one. The lower boxes indicate the valid properties, (I) stands for interpolation. A summary of the properties is also found in Table 3.1.

Fact 3.2.28

$(\mu * 4)$ holds for $(I1) - (I2)$, and fails for $(I3) - (I7)$.

Proof

Exercise, solution in the Appendix.

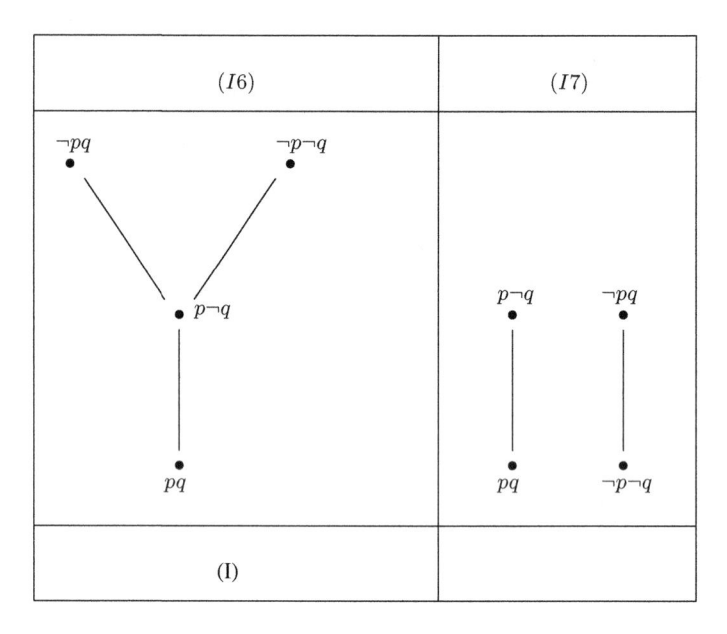

Fig. 3.5 Example 3.2.9 (3)

Fact 3.2.29

(R1) holds for (I1), and fails for $(I2) - (I7)$.

Proof

(I1) The order can be defined by $p \prec \neg p, q \prec \neg q$.

(I2) If $p\neg q \prec \neg pq$ were defined componentwise, then $\neg q \prec q$, contradicting $pq \prec p\neg q$.

(I3) If (R1) were to hold, then by $p\neg q \prec \neg pq$, and as $p \neq \neg p$, $q \neq \neg q$, both $p \prec \neg p$ and $\neg q \prec q$ have to hold, but then e.g. $pq \prec \neg pq$ has to hold, too, contradiction.

$(I4) - (I5)$: similar to (I3).

$(I6) - (I7)$ From $pq \prec p\neg q$ conclude $\neg pq \prec \neg p\neg q$.

□

Fact 3.2.30

$(< Abs)$ holds for $(I1) - (I5)$, and fails for $(I6) - (I7)$.

3.2 Basic Definitions and Overview

Proof

If there is no $\sigma\tau \prec \sigma\tau'$, then $(< Abs)$ holds trivially.

(I3), (I5): $(< Abs)$ holds thus trivially.

(I1) We have $p \prec \neg p$, $q \prec \neg q$, irrespective of $q/\neg q$, $p/\neg p$, so it holds.

(I2) Same as (I1), the condition does not apply to the new pair $p\neg q \prec \neg pq$.

(I4) The only comparisons are $(p\neg q)r \prec (\neg pq)r$, and $(p\neg q)\neg r \prec (\neg pq)\neg r$ with constant parts $\sigma = r$, $\sigma' = \neg r$, $\tau = p\neg q$, $\tau' = \neg pq$. For these, the condition holds, and τ, τ' do not occur with any other σ''.

$(I6) - (I7)$: We have $pq \prec p\neg q$, but not $\neg pq \prec \neg p\neg q$.

□

Remark 3.2.31

Note that we have limited (R1) for (I4), when we treat $\{p, q\}$ together as a group. This is important for "real life", where we may treat parts of the language separately.

Table 3.1 Properties of (I1)-(I7)

Diagram	$(R1)$	$(< Abs)$	$(\mu * 4)$	Interpolation
$(I1)$	+	+	+	+
$(I2)$	-	+	+	+
$(I3)$	-	+	-	+
$(I4)$	-	+	-	+
$(I5)$	-	+	-	+
$(I6)$	-	-	-	+
$(I7)$	-	-	-	-

3.2.6 Tables for Abstract Size

The following Tables 3.2 and 3.3 are discussed in Section 3.2.2.1. For Table 3.5 see Section 3.2.4; for Table 3.5 see Proposition 3.2.13.

Table 3.2 Rules on size – Part I (discussion in Section 3.2.2.1)

	"Ideal"	"Filter"	$\mathcal{M}^+$	∇
Optimal proportion				
(Opt)	$\emptyset \in \mathcal{I}(X)$	$X \in \mathcal{F}(X)$		$\forall x \alpha \to \nabla x \alpha$
Monotony (Improving proportions). (iM): internal monotony. $(eM\mathcal{I})$: external monotony for ideals, $(eM\mathcal{F})$: external monotony for filters				
(iM)	$A \subseteq B \in \mathcal{I}(X)$ $\Rightarrow$ $A \in \mathcal{I}(X)$	$A \in \mathcal{F}(X),$ $A \subseteq B \subseteq X$ $\Rightarrow B \in \mathcal{F}(X)$		$\nabla x \alpha \wedge \forall x (\alpha \to \alpha')$ $\to \nabla x \alpha'$
$(eM\mathcal{I})$	$X \subseteq Y \Rightarrow$ $\mathcal{I}(X) \subseteq \mathcal{I}(Y)$			$\nabla x (\alpha : \beta) \wedge$ $\forall x (\alpha' \to \beta) \to$ $\nabla x (\alpha \vee \alpha' : \beta)$
$(eM\mathcal{F})$		$X \subseteq Y \Rightarrow$ $\mathcal{F}(Y) \cap \mathcal{P}(X) \subseteq$ $\mathcal{F}(X)$		$\nabla x (\alpha : \beta) \wedge$ $\forall x (\beta \wedge \alpha \to \alpha') \to$ $\nabla x (\alpha \wedge \alpha' : \beta)$
Keeping proportions				
$(\approx)$	$(\mathcal{I} \cup disj)$ $A \in \mathcal{I}(X),$ $B \in \mathcal{I}(Y),$ $X \cap Y = \emptyset \Rightarrow$ $A \cup B \in \mathcal{I}(X \cup Y)$	$(\mathcal{F} \cup disj)$ $A \in \mathcal{F}(X),$ $B \in \mathcal{F}(Y),$ $X \cap Y = \emptyset \Rightarrow$ $A \cup B \in \mathcal{F}(X \cup Y)$	$(\mathcal{M}^+ \cup disj)$ $A \in \mathcal{M}^+(X),$ $B \in \mathcal{M}^+(Y),$ $X \cap Y = \emptyset \Rightarrow$ $A \cup B \in \mathcal{M}^+(X \cup Y)$	$\nabla x (\alpha : \beta) \wedge$ $\nabla x (\alpha' : \beta) \wedge$ $\neg \exists x (\alpha \wedge \alpha') \to$ $\nabla x (\alpha \vee \alpha' : \beta)$

3.2 Basic Definitions and Overview

Robustness of proportions: $n * small \neq All$

$(1 * s)$	$(\mathcal{I}_1)$ $X \notin \mathcal{I}(X)$	$(\mathcal{F}_1)$ $\emptyset \notin \mathcal{F}(X)$		(∇_1) $\nabla x \alpha \to \exists x \alpha$
$(2 * s)$	$(\mathcal{I}_2)$ $A, B \in \mathcal{I}(X) \Rightarrow$ $A \cup B \neq X$	$(\mathcal{F}_2)$ $A, B \in \mathcal{F}(X) \Rightarrow$ $A \cap B \neq \emptyset$		(∇_2) $\nabla x \alpha \wedge \nabla x \beta$ $\to \exists x (\alpha \wedge \beta)$
$(n * s)$ $(n \geq 3)$	$(\mathcal{I}_n)$ $A_1, \ldots, A_n \in \mathcal{I}(X)$ $\Rightarrow$ $A_1 \cup \ldots \cup A_n \neq X$	$(\mathcal{F}_n)$ $A_1, \ldots, A_n \in \mathcal{F}(X)$ $\Rightarrow$ $A_1 \cap \ldots \cap A_n \neq \emptyset$	$(\mathcal{M}_n^+)$ $X_1 \in \mathcal{F}(X_2), \ldots,$ $X_{n-1} \in \mathcal{F}(X_n) \Rightarrow$ $X_1 \in \mathcal{M}^+(X_n)$	(∇_n) $\nabla x \alpha_1 \wedge \ldots \wedge \nabla x \alpha_n$ $\to$ $\exists x (\alpha_1 \wedge \ldots \wedge \alpha_n)$
$(< \omega * s)$	$(\mathcal{I}_\omega)$ $A, B \in \mathcal{I}(X) \Rightarrow$ $A \cup B \in \mathcal{I}(X)$	$(\mathcal{F}_\omega)$ $A, B \in \mathcal{F}(X) \Rightarrow$ $A \cap B \in \mathcal{F}(X)$	$(\mathcal{M}_\omega^+)$ (1) $A \in \mathcal{F}(X), X \in \mathcal{M}^+(Y)$ $\Rightarrow A \in \mathcal{M}^+(Y)$ (2) $A \in \mathcal{M}^+(X), X \in \mathcal{F}(Y)$ $\Rightarrow A \in \mathcal{M}^+(Y)$ (3) $A \in \mathcal{F}(X), X \in \mathcal{F}(Y)$ $\Rightarrow A \in \mathcal{F}(Y)$ (4) $A, B \in \mathcal{I}(X) \Rightarrow$ $A - B \in \mathcal{I}(X\text{-}B)$	(∇_ω) $\nabla x \alpha \wedge \nabla x \beta \to$ $\nabla x (\alpha \wedge \beta)$

Robustness of $\mathcal{M}^+$

| $(\mathcal{M}^{++})$ | | | $(\mathcal{M}^{++})$
(1)
$A \in \mathcal{I}(X), B \notin \mathcal{F}(X)$
$\Rightarrow A - B \in \mathcal{I}(X - B)$
(2)
$A \in \mathcal{F}(X), B \notin \mathcal{F}(X)$
$\Rightarrow A - B \notin \mathcal{F}(X - B)$
(3)
$A \in \mathcal{M}^+(X),$
$X \in \mathcal{M}^+(Y)$
$\Rightarrow A \in \mathcal{M}^+(Y)$ | |

Table 3.3 Rules on size – Part II (discussion in Section 3.2.2.1)

	various rules	AND	OR	Caut./Rat. Mon.
Optimal proportion				
(Opt)	$\alpha \vdash \beta \Rightarrow \alpha \mathrel{\vert\!\sim} \beta$			
Monotony (Improving proportions)				
(iM)	(RW) $\alpha \mathrel{\vert\!\sim} \beta, \beta \vdash \beta' \Rightarrow$ $\alpha \mathrel{\vert\!\sim} \beta'$			
$(eM\mathcal{I})$	(PR') $\alpha \mathrel{\vert\!\sim} \beta, \alpha \vdash \alpha',$ $\alpha' \wedge \neg \alpha \vdash \beta \Rightarrow$ $\alpha' \mathrel{\vert\!\sim} \beta$ (μPR) $X \subseteq Y \Rightarrow$ $\mu(Y) \cap X \subseteq \mu(X)$		(wOR) $\alpha \mathrel{\vert\!\sim} \beta, \alpha' \vdash \beta \Rightarrow$ $\alpha \vee \alpha' \mathrel{\vert\!\sim} \beta$ (μwOR) $\mu(X \cup Y) \subseteq \mu(X) \cup Y$	
$(eM\mathcal{F})$				(wCM) $\alpha \mathrel{\vert\!\sim} \beta, \alpha' \vdash \alpha,$ $\alpha \wedge \beta \vdash \alpha' \Rightarrow$ $\alpha' \mathrel{\vert\!\sim} \beta$
Keeping proportions				
$(\approx)$	(NR) $\alpha \mathrel{\vert\!\sim} \beta \Rightarrow$ $\alpha \wedge \gamma \mathrel{\vert\!\sim} \beta$ or $\alpha \wedge \neg \gamma \mathrel{\vert\!\sim} \beta$		$(disjOR)$ $\alpha \mathrel{\vert\!\sim} \beta, \alpha' \mathrel{\vert\!\sim} \beta'$ $\alpha \vdash \neg \alpha', \Rightarrow$ $\alpha \vee \alpha' \mathrel{\vert\!\sim} \beta \vee \beta'$ $(\mu disjOR)$ $X \cap Y = \emptyset \Rightarrow$ $\mu(X \cup Y) \subseteq \mu(X) \cup \mu(Y)$	

3.2 Basic Definitions and Overview

Robustness of proportions: $n * small \neq All$

$(1 * s)$	(CP) $\alpha \mathrel{\|\!\sim} \bot \Rightarrow \alpha \vdash \bot$		
$(2 * s)$		(AND_1) $\alpha \mathrel{\|\!\sim} \beta \Rightarrow \alpha \not\mathrel{\|\!\sim} \neg\beta$	
		(AND_2) $\alpha \mathrel{\|\!\sim} \beta, \alpha \mathrel{\|\!\sim} \beta' \Rightarrow \alpha \not\mathrel{\|\!\sim} \neg\beta \vee \neg\beta'$	(OR_2) $\alpha \mathrel{\|\!\sim} \beta \Rightarrow \alpha \not\mathrel{\|\!\sim} \neg\beta$	(CM_2) $\alpha \mathrel{\|\!\sim} \beta \Rightarrow \alpha \not\mathrel{\|\!\sim} \neg\beta$
$(n * s)$ $(n \geq 3)$		(AND_n) $\alpha \mathrel{\|\!\sim} \beta_1, \ldots, \alpha \mathrel{\|\!\sim} \beta_n \Rightarrow \alpha \not\mathrel{\|\!\sim} \neg\beta_1 \vee \cdots \vee \neg\beta_n$	(OR_n) $\alpha_1 \mathrel{\|\!\sim} \beta, \ldots, \alpha_{n-1} \mathrel{\|\!\sim} \beta \Rightarrow \alpha_1 \vee \cdots \vee \alpha_{n-1} \not\mathrel{\|\!\sim} \neg\beta$	(CM_n) $\alpha \mathrel{\|\!\sim} \beta_1, \ldots, \alpha \mathrel{\|\!\sim} \beta_{n-1} \Rightarrow \alpha \wedge \beta_1 \wedge \cdots \wedge \beta_{n-2} \not\mathrel{\|\!\sim} \neg\beta_{n-1}$
$(< \omega * s)$		(AND_ω) $\alpha \mathrel{\|\!\sim} \beta, \alpha \mathrel{\|\!\sim} \beta' \Rightarrow \alpha \mathrel{\|\!\sim} \beta \wedge \beta'$	(OR_ω) $\alpha \mathrel{\|\!\sim} \beta, \alpha' \mathrel{\|\!\sim} \beta \Rightarrow \alpha \vee \alpha' \mathrel{\|\!\sim} \beta$ (μOR) $\mu(X \cup Y) \subseteq \mu(X) \cup \mu(Y)$	(CM_ω) $\alpha \mathrel{\|\!\sim} \beta, \alpha \mathrel{\|\!\sim} \beta' \Rightarrow \alpha \wedge \beta \mathrel{\|\!\sim} \beta'$ (μCM) $\mu(X) \subseteq Y \subseteq X \Rightarrow \mu(Y) \subseteq \mu(X)$

Robustness of $\mathcal{M}^+$

$(\mathcal{M}^{++})$	$(RatM)$ $\alpha \mathrel{\|\!\sim} \beta, \alpha \not\mathrel{\|\!\sim} \neg\beta' \Rightarrow \alpha \wedge \beta' \mathrel{\|\!\sim} \beta$ $(\mu RatM)$ $X \subseteq Y$, $X \cap \mu(Y) \neq \emptyset \Rightarrow \mu(X) \subseteq \mu(Y) \cap X$

Table 3.4 Multiplication laws (see Section 3.2.4)

Multiplication law	Scenario 1 (see Figure 3.1)			Scenario 2 (∗ symmetrical, only 1 side shown) (see Figure 3.2)		Interpolation		
	Corresponding algebraic addition property	Logical property	Relation property	Algebraic property ($\Gamma_i \subseteq \Sigma_i$)	Logical property (α, β in L_1, α', β' in L_2 $L = L_1 \cup L_2$ (disjoint))	Multiplic. law	Relation property	Interpolation
Non-monotonic logic								
$x \ast 1 \Rightarrow x$	trivial			$\Gamma_1 \in \mathcal{F}(\Sigma_1) \Rightarrow$ $\Gamma_1 \times \Sigma_2 \in \mathcal{F}(\Sigma_1 \times \Sigma_2)$	$\alpha \mathrel{\|\!\sim}_{L_i} \beta \Rightarrow \alpha \mathrel{\|\!\sim}_L \beta$			
$1 \ast x \Rightarrow x$	trivial							
$x \ast s \Rightarrow s$ (iM)	$A \subseteq B \in \mathcal{I}(X) \Rightarrow A \in \mathcal{I}(X)$	$\alpha \mathrel{\|\!\sim} \neg\beta \Rightarrow$ $\alpha \mathrel{\|\!\sim} \neg\beta \vee \gamma$	-	$\Gamma_1 \in I(\Sigma_1) \Rightarrow$ $\Gamma_1 \times \Gamma_2 \in I(\Sigma_1 \times \Sigma_2)$	$\alpha \mathrel{\|\!\sim}_{L_1} \beta, \beta' \vdash_{L_2} \alpha' \Rightarrow$ $\alpha \wedge \alpha' \mathrel{\|\!\sim} (\beta \wedge \alpha') \vee (\alpha \wedge \beta')$			
$s \ast x \Rightarrow s$ (eMI)	$X \subseteq Y \Rightarrow \mathcal{I}(X) \subseteq \mathcal{I}(Y)$, $X \subseteq Y \Rightarrow$ $\mathcal{F}(Y) \cap \mathcal{P}(X) \subseteq \mathcal{F}(X)$	$\alpha \wedge \beta \mathrel{\|\!\sim} \neg\gamma \Rightarrow$ $\alpha \mathrel{\|\!\sim} \neg\beta \vee \neg\gamma$	-					
$b \ast b \Rightarrow b$ $(\mu \ast 1)$	$(<\omega \ast s), (\mathcal{M}_\omega^+) (3)$ $A \in \mathcal{F}(X), X \in \mathcal{F}(Y) \Rightarrow$ $A \in \mathcal{F}(Y)$	$\alpha \mathrel{\|\!\sim} \beta, \alpha \wedge \beta \mathrel{\|\!\sim} \gamma$ $\Rightarrow \alpha \mathrel{\|\!\sim} \gamma$	- (Filter)	$\Gamma_1 \in \mathcal{F}(\Sigma_1), \Gamma_2 \in \mathcal{F}(\Sigma_2)$ $\Rightarrow \Gamma_1 \times \Gamma_2 \in \mathcal{F}(\Sigma_1 \times \Sigma_2)$	$\alpha \mathrel{\|\!\sim}_{L_1} \beta, \alpha' \mathrel{\|\!\sim}_{L_2} \beta' \Rightarrow$ $\alpha \wedge \alpha' \mathrel{\|\!\sim}_L \beta \wedge \beta'$	$b \ast b \Leftrightarrow b$ $(\mu \ast 1)$	(GH)	$\mathrel{\|\!\sim} \circ \mathrel{\|\!\sim}$
$b \ast m \Rightarrow m$	$(<\omega \ast s), (\mathcal{M}_\omega^+) (2)$ $A \in \mathcal{M}^+(X), X \in \mathcal{F}(Y) \Rightarrow$ $A \in \mathcal{M}^+(Y)$	$\alpha \mathrel{\|\!\sim} \beta, \alpha \wedge \beta \mathrel{\not\|\!\sim} \neg\gamma$ $\Rightarrow \alpha \mathrel{\not\|\!\sim} \neg\beta \vee \neg\gamma$	- (Filter)	$\Gamma_1 \in \mathcal{F}(\Sigma_1)$, $\Gamma_2 \in \mathcal{M}^+(\Sigma_2) \Rightarrow$ $\Gamma_1 \times \Gamma_2 \in \mathcal{M}^+(\Sigma_1 \times \Sigma_2)$	$\alpha \mathrel{\not\|\!\sim}_{L_1} \neg\beta, \alpha' \mathrel{\|\!\sim}_{L_2} \beta' \Rightarrow$ $\alpha \wedge \alpha' \mathrel{\not\|\!\sim}_L \neg(\beta \wedge \beta')$			

3.2 Basic Definitions and Overview

$m*b \Rightarrow m$	$(<\omega*s), (\mathcal{M}_\omega^+)(1)$ $A \in \mathcal{F}(X), X \in \mathcal{M}^+(Y) \Rightarrow$ $A \in \mathcal{M}^+(Y)$	$\alpha \not\vdash \neg\beta, \alpha \wedge \beta \vdash \neg\gamma$ $\Rightarrow \alpha \not\vdash \neg\beta \vee \neg\gamma$	- (Filter)			
$m*m \Rightarrow m$	$(\mathcal{M}^{++})$ $A \in \mathcal{M}^+(X), X \in \mathcal{M}^+(Y)$ $\Rightarrow A \in \mathcal{M}^+(Y)$	Rational Monotony	ranked	$\Gamma_1 \in \mathcal{M}^+(\Sigma_1),$ $\Gamma_2 \in \mathcal{M}^+(\Sigma_2) \Rightarrow$ $\Gamma_1 \times \Gamma_2 \in \mathcal{M}^+(\Sigma_1 \times \Sigma_2)$	$\alpha \not\vdash_{L_1} \neg\beta,\ \alpha' \not\vdash_{L_2} \neg\beta'$ $\Rightarrow$ $\alpha \wedge \alpha' \not\vdash_L \neg(\beta \wedge \beta')$	
$b*b \Leftrightarrow b,$ $pr(b) = b$ $(\mu*2)$					$\alpha \vdash \beta \Rightarrow \alpha \upharpoonright L_1 \vdash \beta \upharpoonright L_1$ and $\alpha \vdash_{L_1} \beta,\ \alpha' \vdash_{L_2} \beta' \Rightarrow$ $\alpha \wedge \alpha' \vdash_L \beta \wedge \beta'$	$(\mu*1)$ $+$ $(\mu*2)$ (GH3) $\vdash \circ \vdash$
J' small					$\alpha \wedge \alpha' \vdash \beta \wedge \beta' \Leftrightarrow$ $\alpha \vdash \beta, \alpha' \vdash \beta'$	$forget(J')$ -

Theory revision

543.44763pt			$(|*):$ $(\Sigma_1 \times \Sigma_1') | (\Sigma_2 \times \Sigma_2') =$ $(\Sigma_1 | \Sigma_2) \times (\Sigma_1' | \Sigma_2')$	$(\phi \wedge \phi') * (\psi \wedge \psi') =$ $(\phi * \psi) \wedge (\phi' * \psi')$	(GHD) $(\phi \wedge \phi') * (\psi \wedge \psi')$ $\vdash \rho$ $\Rightarrow \phi' * \psi' \vdash \rho$ ϕ, ψ in $J,$ ϕ', ψ', ρ in $L - J$

Table 3.5 Size and principal filter rules (see Proposition 3.2.13)

(1)	$(eM\mathcal{I})$	$\Leftrightarrow$	(μwOR)
(2)	$(eM\mathcal{I}) + (I_\omega)$	$\Leftrightarrow$	(μOR)
(3)	$(eM\mathcal{I}) + (I_\omega)$	$\Leftrightarrow$	(μPR)
(4)	$(I \cup disj)$	$\Leftrightarrow$	$(\mu disjOR)$
(5)	$(\mathcal{M}_\omega^+)(4)$	$\Leftrightarrow$	(μCM)
(6)	$(\mathcal{M}^{++})$	$\Leftrightarrow$	$(\mu RatM)$
(7)	(I_ω)	$\Leftrightarrow$	(μAND)
(8.1)	$(eM\mathcal{I}) + (I_\omega)$	$\Rightarrow$	(μCUT)
(8.2)		$\not\Leftarrow$	
(9.1)	$(eM\mathcal{I}) + (I_\omega) + (\mathcal{M}_\omega^+)(4)$	$\Rightarrow$	(μCUM)
(9.2)		$\not\Leftarrow$	
(10.1)	$(eM\mathcal{I}) + (I_\omega) + (eM\mathcal{F})$	$\Rightarrow$	$(\mu \subseteq \supseteq)$
(10.2)		$\not\Leftarrow$	

3.3 Defaults as Generalized Quantifiers

3.3.1 Introduction

We treat here defaults as generalized quantifiers in a first order setting:

Recall the definition of a "weak filter" or $\mathcal{N}$-system, see Definition 3.2.1.

We use weak filters on the semantical side, and add the following axioms on the syntactical side to a FOL axiomatisation:

1. $\nabla x \phi(x) \land \forall x(\phi(x) \to \psi(x)) \to \nabla x \psi(x)$,
2. $\nabla x \phi(x) \to \neg \nabla x \neg \phi(x)$,
3. $\forall x \phi(x) \to \nabla x \phi(x)$ and $\nabla x \phi(x) \to \exists x \phi(x)$.

A model is now a pair, consisting of a classical FOL model M, and a weak filter over its universe. Both sides are connected by the following definition, where $\mathcal{N}(M)$ is the weak filter on the universe of the classical model M:

3.3 Defaults as Generalized Quantifiers

$\langle M, \mathcal{N}(M) \rangle \models \nabla x \phi(x)$ iff there is an $A \in \mathcal{N}(M)$ such that $\forall a \in A$ ($\langle M, \mathcal{N}(M) \rangle \models \phi[a]$).

The extension to defaults with prerequisites by restricted quantifiers is straightforward.

This approach was developped within IBM's LILOG project (IBM Germany, Stuttgart), which used a KL-ONE style order sorted language, see [BHR90].

For more background, see [Sch04], [Sch95-1], and [Sch97-2].

3.3.1.1 In More Detail

We augment proof theory and semantics of classical FOL to deal with the (open normal, in the sense of [Rei80]) default "normally, $\phi(x)$ holds". We say that $\phi(x)$ holds normally (in a model) iff there is a "large" subset of the universe of that model in which $\phi(x)$ "really" holds. In other words, "normally" is read as a generalized quantifier. In that sense, our semantics is local, as we work in one model only, and give a direct interpretation of defaults, in contrast to the various types of preferential models (see e.g. [Bou90b], [Del87], [Del88], [KLM90], [Lif85], [Lif86], [LM92], [McC80], [McC86], [Sho87b]) which are global, working with several classical models. We consider as basic semantic structures pairs $\mathcal{M} = \langle M, \mathcal{N}(M) \rangle$, where M is a model of classical FOL, and $\mathcal{N}(M)$ is a system of large subsets of (the universe of) M. Validity of $\nabla x \phi(x)$ (= $\phi(x)$ holds normally) in $\mathcal{M}$ is defined in Definition 3.3.2. Suitable axioms for the quantifier ∇ are given in Definition 3.3.3, soundness and completeness of the axioms wrt. the semantics are shown in Lemma 3.3.6 and Theorem 3.3.7 there. The system permits full nestedness and boolean combinations of defaults. In particular, a notion of consistency of defaults results. The rest of Section 3.3.2 is devoted to extensions of the basic idea: (Open normal) defaults with prerequisites are seen as restricted generalized quantifiers, defaults of various strengths of "normality" are introduced, soundness and completeness shown. In conclusion, we examine various strengthenings of the axioms, which correspond to widespread use of defaults and other systems of defeasible reasoning discussed in the literature.

The common use of defaults seems to presuppose strong assumptions on the structure of the universe - and seems thus to have common aspects with learning, and even philosophy of science. E.g. applying both defaults "normally $\phi(x)$" and "normally $\psi(x)$" seems to presuppose that not only "normally $\phi(x)$" and "normally $\psi(x)$", but also "normally $\phi(x) \wedge \psi(x)$". Even if our base system is too weak to permit such reasoning, we can easily adapt it by introducing a corresponding axiom schema, and strengthening the system of "important" subsets to a filter. Problems of this kind are discussed in the Introduction, Section 1.3.11.

Section 3.3.2 contains the central definitions and results of Section 3.3. $\mathcal{N}$-systems formalize the notion of a "large" or "important" subset, we introduce the generalized quantifier ∇ - read e.g. "for almost all". $\mathcal{N}$-systems then provide us with a semantics: A ∇-structure (or -model) is a classical first order structure with an $\mathcal{N}$-system over its universe, and a formula $\nabla x \phi(x)$ is defined to hold in an ∇-structure iff there is a large subset A of the universe, i.e. some A in the $\mathcal{N}$-system over the universe, s.th. for all elements $a \in A$ $\phi(a)$ holds. An axiomatisation is given, soundness and completeness for our semantics is shown. (For simplicity, we first treat the case of normal open defaults without prerequisites - corresponding to $\nabla x \phi(x)$ in our notation - the extensions to those with prerequisites is straightforward: $\nabla x \phi(x) : \psi(x)$ is interpreted as a generalized quantifier restricted to $\{x : \phi(x)\}$.

The weakness - choosing $\mathcal{N}$-systems weaker than filters, and making no but trivial connections between the $\mathcal{N}$-systems over different subsets of the universe for the relativized ∇-quantifier - is deliberate: We make as little "philosophical" commitments as possible, and permit easy and straightforward strengthenings into many different directions. In particular, the axiom systems P and R of [KLM90] and [LM92] can be incorporated easily into our system. Several such extensions are discussed in Section 3.3.3.

In Section 3.3.4, we use the notion of consistency of defaults given in Section 3.3.2 to choose a reasonable consistent subset from a possibly inconsistent set of default information. We exploit the reliability relation given by the (partial) order of specificity in the sorted language of the LILOG project. The basic idea is to consider minimal inconsistent subsets, and to eliminate at least one suitably chosen (according to specificity, and determined by the function f there) element from each. Several approaches are discussed, one of which has essentially been implemented.

3.3.2 Semantics and Proof Theory

3.3.2.1 Overview of This Section

An $\mathcal{N}$-system over M is supposed to capture the notion of a "large" or "important" subset, or one containing the "important" or "typical" elements of a base set M. Let me emphasize again that we have made the properties of an $\mathcal{N}$-system deliberately extremely weak, weaker than a filter. Thus, our approach is generic on the sense that, if need be, we can strengthen the notion of an $\mathcal{N}$-system, add the corresponding axioms to the proof theory and still have a sound and complete system. Such modifications would be harder to achieve if, instead, we had to take away some undesirable strong properties.

The third property of weak filters or $\mathcal{N}$-systems is the most interesting one. A filter would require the intersection of two large subsets to be large itself, we content ourselves with non-empty intersection. This permits e.g. a simple probabilistic interpretation by "more than half". It is property (3) which permits - in the intended in-

3.3 Defaults as Generalized Quantifiers

terpretation - to deduce that "normally $\phi(x)$" and "normally $\neg\phi(x)$" together can't be: there is no element which satisfies both $\phi(x)$ and $\neg\phi(x)$.

The further development proceeds in two stages. We first treat normal open defaults without prerequisites, and then extend the discussion to normal open defaults with prerequisites (and $\mathcal{N}$-families, see below). All essential techniques and ideas are present in the case without prerequisite, and the reader so inclined may just leaf through the later subsections.

We introduce the new (unbounded) quantifier ∇ into the language (and its dual $\heartsuit$ for proof theoretical purposes only). The crucial definition linking language and semantics is Definition 3.3.2, where we define an $\mathcal{N}$-model and validity of a ∇-formula in an additional inductive step: An $\mathcal{N}$-model is a classical first order structure, with an $\mathcal{N}$-system of large subsets over its universe M. $\nabla x\phi(x)$ is defined to hold in the $\mathcal{N}$-model, iff there is a large subset $A \subseteq M$ such that for all $a \in A$ $\phi(a)$ "really" holds. As ϕ was not necessarily a classical formula, we can treat nested ∇'s. Moreover, the new quantifier is fully embedded into the classical setting, so we can form boolean combinations, quantify classically over defaults in the sense of e.g. $\exists x \nabla y \phi(x,y)$ etc., and give these formulas a precise meaning, preserving the constructive spirit of FOL. This seems to me one advantage over "global" Kripke style semantics where we have to "look elsewhere" for the interpretation of normality. Here, we can say "look, you see its holds in this universe on a large subset, so it normally holds here".

A corresponding axiomatisation follows in Definition 3.3.3: The first axiom says that implication preserves normality: If $\phi(x)$ normally holds, and $\phi(x)$ implies $\psi(x)$, then $\psi(x)$ normally holds. This corresponds to the second property of $\mathcal{N}$-systems. The second axiom says that normally ϕ and normally $\neg\phi$ can't be (we have discussed this above), and the third that the ∇-quantifier lies between the two classical ones, corresponding to the fact that M is a large subset of itself. (The second half of the axiom could be derived by $\neg\exists x\phi(x) \rightarrow \forall x\neg\phi(x) \rightarrow \nabla x\neg\phi(x) \rightarrow \neg\nabla x\phi(x)$ - we prefer to state it explicitely.) The last two axioms are auxiliary.

Lemma 3.3.4 states some basic and trivial consequences of the axioms. We then give a normal form for ∇-formulas to facilitate the completeness proof.

The central result of Section 3.3 is given in Lemma 3.3.6: A consistent ∇-theory has a model. The idea is to consider first order consequences of pairs of ∇-formulas: Assume T to be deductively closed under our axiomatisation. Let $\nabla x\psi(x), \nabla y\psi'(y) \in T$ with, for simplicity, ψ, ψ' classical formulas. By Lemma 3.3.4, a) $\nabla x\psi(x) \wedge \nabla y\psi'(y) \rightarrow \exists x(\psi \wedge \psi')(x)$. We now take a classical structure M satisfying all those first order consequences, say $M \models \psi \wedge \psi'(c_{\psi \wedge \psi'})$. For fixed $\nabla x\psi(x) \in T$, let $X_{\nabla x\psi(x)} := \{c_{\psi \wedge \psi'} : \nabla x\psi'(x) \in T\}$. $X_{\nabla x\psi(x)}$ will be one of the large subsets of the system to be constructed. It remains to show that the intersections of those sets are non-empty, and that the defined structure really is a model of T. But this is not difficult.

The soundness and completeness theorem is a direct consequence of this Lemma.

The extension to normal open defaults with prerequisites is straightforward: $\nabla x\phi(x) : \psi(x)$ ("if $\phi(x)$, then normally $\psi(x)$") is interpreted as a generalized quantifier relativized to $\{x : \phi(x)\}$, i.e. we consider an $\mathcal{N}$-system not over the whole universe, but only over the subset where $\phi(x)$ holds. Again, we keep our system deliberately very weak, in the sense that we do not demand any connections between the $\mathcal{N}$-systems over the different $\{x : \phi(x)\}$ and $\{x : \phi'(x)\}$ - besides the trivial ones when e.g. $\{x : \phi(x)\} = \{x : \phi'(x)\}$. We even do not postulate $A \in \mathcal{N}(B) \land A \subseteq B' \subseteq B \to A \in \mathcal{N}(B')$, which a purely quantitative reading would justify: A large subset of B is a fortiori large in B', when $A \subseteq B' \subseteq B$. Again, we want to leave open all possibly intended developments and strengthenings.

The next extension concerns different degrees of normality. This is again straightforward and the reader is referred directly to this subsection.

3.3.2.2 Semantics

To facilitate proofs and enable normal forms, we introduce a complementary quantifier, $\heartsuit$, too, with the meaning $\heartsuit x\phi(x) :\leftrightarrow \neg\nabla x\neg\phi(x)$. The intuitive reading of $\heartsuit x\phi(x)$ is thus roughly: "for at least a few x, $\phi(x)$ holds".

Remark: Our semantics covers the two extremes:

- fix one element a of the universe U, then $\{A \subseteq U: a \in A\}$ will be a $\mathcal{N}$-system

- let some probability measure be given on U, then $\{A \subseteq U: p(A) > 0.5\}$ will be a $\mathcal{N}$-system.

(Note, however, that the former can also be expressed by a suitable point measure on U.) We can thus cover both the "prototypical" and the "average" case.

Remark 3.3.1

a) $\mathcal{N}$ stands for normal. We formalize "ϕ is normally valid in c" by $\exists a \in \mathcal{N}(c).\forall x \in a.\phi(x)$.

b) There is nothing to prevent e.g. $\mathcal{N}(c) = \{a \subseteq c : x \in a\}$ for some fixed $x \in c$. This might seem pathological. Two comments: First, compare to topo- logy. Suitable choice of topology will make e.g. the function $d : \Re \to \Re$,

$$d(x) := \begin{cases} 0 & \text{iff } x \text{ is rational} \\ 1 & \text{otherwise} \end{cases}$$

continous, certainly a pathological case too. Second, this x might be a very prototypical case, and thus have intuitive meaning.

3.3 Defaults as Generalized Quantifiers

Lemma 3.3.2

Let $\mathcal{L} \subseteq \mathcal{P}(X)$ be such that $A, B \in \mathcal{L} \rightarrow A \cap B \neq \emptyset$. Then $\mathcal{N}(X) := \{A \subseteq X : \exists B \in \mathcal{L}.B \subseteq A\} \cup \{X\}$ is a $\mathcal{N}$-system over X.

□

Example 3.3.1

Let α be any ordinal, and $U := \{f : \alpha \rightarrow 2 = \{0,1\}\}$. For $i < \alpha$ let $X_i := \{f \in U : f(i) = 1\}$, $X_i'' := \{f \in U : f(i) = 0\} = U - X_i$. For $j < \alpha$ let $I_j \subseteq \alpha$, $I_j'' \subseteq \alpha - \{j\}$ such that $I_j \cap I_j'' = \emptyset$, and let $\mathcal{L}_j := \{X_j \cap X_i : i \in I_j\} \cup \{X_j \cap X_i'' : i \in I_j''\}$. Then $\mathcal{L}_j$ satisfies the prerequisites of Lemma 3.3.2 for X_j and all j. Consequently, $\mathcal{L}_j$ so defined generates a $\mathcal{N}$-system over X_j for all $j < \alpha$.

Proof

$\mathcal{L}_j \subseteq \mathcal{P}(X_j)$ is trivial. As $I_j \cap I_j'' = \emptyset$ and $j \notin I_j''$, the function

$$f(i) := \begin{cases} 1 & \text{if } i = j \text{ or } i \in I_j \\ 0 & \text{otherwise} \end{cases}$$

is well-defined and in all $X_j \cap X_i$, $X_j \cap X_i''$.

□

Definition 3.3.1

We augment the language of first order logic by the new quantifiers: If ϕ and ψ are formulas, then so are $\nabla x\phi(x)$, $\heartsuit x\phi(x)$, $\nabla x\phi(x) : \psi(x)$, $\heartsuit x\phi(x) : \psi(x)$ for any variable x. We call any formula of $\mathcal{L}$, possibly containing ∇ or $\heartsuit$ a $\nabla - \mathcal{L}$-formula.

Definition 3.3.2

($\mathcal{N}$-Model) Let $\mathcal{L}$ be a first order language, and M be a $\mathcal{L}$-structure. Let $\mathcal{N}(M)$ be a $\mathcal{N}$-system over M. Define $\langle M, \mathcal{N}(M)\rangle \models \phi$ for any $\nabla - \mathcal{L}$-formula inductively as usual, with two additional induction steps:

- $\langle M, \mathcal{N}(M)\rangle \models \nabla x\phi(x)$ iff there is $A \in \mathcal{N}(M)$ s.th. $\forall a \in A \,(\langle M, \mathcal{N}(M)\rangle \models \phi[a])$
- $\langle M, \mathcal{N}(M)\rangle \models \heartsuit x\phi(x)$ iff $\{a \in M: <M, \mathcal{N}(M)> \models \neg\phi[a]\} \notin \mathcal{N}(M)$.

Lemma 3.3.3

$\langle M, \mathcal{N}(M)\rangle \models \heartsuit x\phi(x)$ iff $\forall A \in \mathcal{N}(M) \exists a \in A(\langle M, \mathcal{N}(M)\rangle \models \phi[a])$.

□

3.3.2.3 Proof Theory

Definition 3.3.3

Let any axiomatization of predicate calculus be given. Augment this with the axiom schemata

1. $\nabla x \phi(x) \wedge \forall x (\phi(x) \to \psi(x)) \to \nabla x \psi(x)$
2. $\nabla x \phi(x) \to \neg \nabla x \neg \phi(x)$
3. $\forall x \phi(x) \to \nabla x \phi(x) \to \exists x \phi(x)$
4. $\heartsuit x \phi(x) :\leftrightarrow \neg \nabla x \neg \phi(x)$
5. $\nabla x \phi(x) \leftrightarrow \nabla y \phi(y)$ if x does not occurr free in $\phi(y)$ and y does not occurr free in $\phi(x)$

(for all ϕ, ψ).

We also denote the corresponding notion of derivability by $\vdash_\nabla$.

Lemma 3.3.4

The following formulae are derivable:

a. $\nabla x \phi(x) \wedge \nabla x \psi(x) \to \exists x (\phi \wedge \psi)(x)$
b. $\nabla x \phi(x) \wedge \neg \nabla x \psi(x) \to \exists x (\phi \wedge \neg \psi)(x)$
c. $\neg \nabla x \neg \phi(x) \to \exists x \phi(x)$
d. $\heartsuit x \phi(x) \to \exists x \phi(x)$
e. $\nabla x \phi(x) \wedge \heartsuit x \psi(x) \to \exists x (\phi \wedge \psi)(x)$
f. $\forall x (\phi(x) \leftrightarrow \psi(x)) \to (\nabla x \phi(x) \leftrightarrow \nabla x \psi(x)) \wedge (\heartsuit x \phi(x) \leftrightarrow \heartsuit x \psi(x))$
g. $\nabla x \phi(x) \to \heartsuit x \phi(x)$

It is usually not derivable: $\heartsuit x \phi(x) \wedge \heartsuit x \psi(x) \to \exists x (\phi \wedge \psi)(x)$. (To see this, use Theorem 3.3.7 below and argue semantically.)

□

3.3.2.4 Soundness and Completeness

To prepare the proof of completeness, we introduce ∇-normal forms (∇-NF).

Definition 3.3.4

ϕ is in ∇-normal form (∇-NF) iff

3.3 Defaults as Generalized Quantifiers

1. ϕ contains only $\neg, \wedge$, as propositional operators
2. only atomic FOL formulas are in the scope of $\neg$.

Lemma 3.3.5

For every ϕ there is ϕ' in ∇-NF s.th. $\vdash_\nabla \phi \leftrightarrow \phi'$.

Proof

By induction on the depth of $\nabla + \heartsuit$ - nesting.

Lemma 3.3.6

Let T be a $\nabla - \mathcal{L}$-theory. Then T is consistent under the axioms of Definition 3.3.3 iff T has a model as defined in Definition 3.3.2.

Proof

The consistency of T when it has a model is trivial.

Let T be a $\vdash_\nabla -consistent$ $\nabla - \mathcal{L}$-theory. We have to show that it has a model. Throughout the proof, let "$\vdash_\nabla -consistent$" be abbreviated by "consistent". We give a constructive proof, to make the reader comfortable with the new logic. By the above, assume wlog. that all $\phi \in T$ are in ∇–NF.

We first construct a consistent $T' \supseteq T$.

We add $c_\alpha : \alpha < \kappa$ new constants to $\mathcal{L}$, where κ is the size of $\mathcal{L}$, and inductively construct $T' = \bigcup \{T_\gamma : \gamma < \beta\}$ (T_γ ascending, β large enough) with $T_0 := T$, by adding new formulas to T, preserving consistency. (For simplicity, we omit the exact enumeration process - it does not matter anyway.) Let $\phi \in T_\gamma$, depending on the topmost operator, we add 0, 1, or several new formulas. It should be noted that all added formulas are in ∇-NF too.

Case 1: $\phi = \neg \psi$: We do nothing, by ∇-NF, ψ is a classical atomic formula

Case 2: $\phi = \psi \wedge \psi'$: We add ψ, ψ', obviously preserving consistency.

Case 3: $\phi = \psi \vee \psi'$: Both $T_\gamma + \psi$ and $T_\gamma + \psi'$ can't be inconsistent, as $\phi \in T_\gamma$, so add one (or both) which preserves consistency.

Case 4: $\phi = \forall x \psi(x)$: Add all $\psi(c_\alpha)$, $\alpha < \kappa$

Case 5: $\phi = \exists x \psi(x)$: Add some $\psi(c_\alpha)$ which preserves consistency

Case 6: $\phi = \nabla x \psi(x)$: Add $\exists x \psi(x)$, and for each $\nabla y \psi'(y) \in T_\gamma$ $\exists x(\psi \wedge \psi')(x)$ and for each $\heartsuit y \psi'(y) \in T_\gamma$ $\exists x(\psi \wedge \psi')(x)$. (after suitable renaming, preserving consistency by Lemma 3.3.4)

Case 7: $\phi = \heartsuit x \psi(x)$: Add $\exists x \psi(x)$, and for each $\nabla y \psi'(y) \in T_\gamma$ $\exists x(\psi \wedge \psi')(x)$ (after suitable renaming, preserving consistency by Lemma 3.3.4)

In case 6 and 7, we mark all new $\exists x \psi(x)$ / $\exists x(\psi \wedge \psi')(x)$ as children of $\phi = \nabla x \psi(x)$ / $\phi = \nabla x \psi(x)$ and $\phi = \nabla x \psi'(x)$ etc.

Let $T' := \bigcup\{T_\gamma : \gamma < \beta\}$, β large enough, and $T'' \subseteq T'$ be the set of FOL-formulas of T'. By $FOL - completeness$, T' has a model M with universe U, where each $u \in U$ is denoted by some c_α.

Next, we define $\mathcal{N}(U)$.

Case 1: T' contains no $\nabla x \psi(x)$: Set $\mathcal{N}(U) := \{U\}$.

Case 2: Otherwise. Let $\nabla x \psi(x)$ be in T', and its children be $\exists x \psi(x)$, $\exists x(\psi \wedge \psi_i)(x)$, $i \in I$ (with $\nabla y \psi_i(y) / \heartsuit y \psi_i(y) \in T'$), so there are $\psi(c_\alpha)$, $(\psi \wedge \psi_i)(c_{\alpha_i}) \in T'$. Let $X_{\nabla x \psi(x)} := \{c_\alpha\} \cup \{c_{\alpha_i} : i \in I\}$ (we identify the c_α with their interpretation), and set $\mathcal{N}(U) := \{V \subseteq U : X_{\nabla x \psi(x)} \subseteq V$ for some $\nabla x \psi(x) \in T'\}$ Obviously, for $\nabla x \psi(x), \nabla x \psi'(x) \in T'$, $X_{\nabla x \psi(x)} \cap X_{\nabla x \psi'(x)} \neq \emptyset$, as they have the common child $\exists x(\psi \wedge \psi')(x)$, so $\mathcal{N}(U)$ is a $\mathcal{N}$-system.

It remains to show that T holds in $\mathcal{M} := \langle M, \mathcal{N}(U) \rangle$. We show by induction on the complexity of ϕ that all $\phi \in T'$ hold in $\mathcal{M}$.

□

Theorem 3.3.7

The axioms given in Definition 3.3.3 are sound and complete for the semantics of Definition 3.3.2, "they capture the $\mathcal{N}$-semantics of ∇".

Proof

Let $T \not\models \phi$. Then there is a model M, s.th. $M \models T \wedge \neg \phi$. Thus, $Con(T \wedge \neg \phi)$, so $T \not\vdash \phi$. The other direction is analogous.

□

3.3.2.5 Extension to Normal Defaults with Prerequisites

Definition 3.3.5

Call $\mathcal{N}^+(M) = \langle \mathcal{N}(N) : N \subseteq M \rangle$ a $\mathcal{N}^+-$ system over M iff for each $N \subseteq M$ $\mathcal{N}(N)$ is a $\mathcal{N}$-system over N. (It suffices to consider the definable subsets of M.)

Definition 3.3.6

Extend the logic of first order predicate calculus by adding the axiom schemata

(1) a. $\nabla x \phi(x) \leftrightarrow \nabla x(x = x) : \phi(x)$

 b. $\forall x(\sigma(x) \leftrightarrow \tau(x)) \wedge \nabla x \sigma(x) : \phi(x) \rightarrow \nabla x \tau(x) : \phi(x)$

(2) $\nabla x \phi(x) : \psi(x) \wedge \forall x(\phi(x) \wedge \psi(x) \rightarrow \vartheta(x)) \rightarrow \nabla x \phi(x) : \vartheta(x)$

3.3 Defaults as Generalized Quantifiers

(3) $\exists x\phi(x) \wedge \nabla x\phi(x) : \psi(x) \rightarrow \neg \nabla x\phi(x) : \neg\psi(x)$

(4) $\forall x(\phi(x) \rightarrow \psi(x)) \rightarrow \nabla x\phi(x) : \psi(x) \rightarrow [\exists x\phi(x) \rightarrow \exists x(\phi(x) \wedge \psi(x))]$

(5) $\heartsuit x\phi(x) : \psi(x) \leftrightarrow \neg \nabla x\phi(x) : \neg\psi(x)$

(6) $\nabla x\phi(x) : \psi(x) \leftrightarrow \nabla y\phi(y) : \psi(y)$ (under the usual caveat for substitution.)

(for all $\phi, \psi, \vartheta, \sigma, \tau$).

Lemma 3.3.8

The following are derivable:

a) the axioms of Definition 3.3.3, and the formulae of Lemma 3.3.4 (via the above Definition 3.3.6 1.a. and the corresponding relativized versions).

b) the relativized versions of Lemma 3.3.4, where the existential statements have to be weakened by an existential assumption as in Definition 3.3.6, 4.

□

Definition 3.3.7

Let $\mathcal{L}$ be a first order language, and M a $\mathcal{L}$-structure. Let $\mathcal{N}^+(M)$ be a $\mathcal{N}^+$– system over M.

Define $\langle M, \mathcal{N}^+(M) \rangle \models \phi$ for any formula inductively as usual, with the additional induction steps:

(1) $\langle M, \mathcal{N}^+(M) \rangle \models \nabla x\phi(x)$ iff there is $A \in \mathcal{N}(M)$ s.th. $\forall a \in A$ ($\langle M, \mathcal{N}^+(M) \rangle \models \phi[a]$).

(2) $\langle M, \mathcal{N}^+(M) \rangle \models \heartsuit x\phi(x)$ iff $\{a \in M : <M, \mathcal{N}^+(M)> \models \neg\phi[a]\} \notin \mathcal{N}(M)$.

(3) $\langle M, \mathcal{N}^+(M) \rangle \models \nabla x\phi(x) : \psi(x)$ iff there is $A \in \mathcal{N}(\{x : \langle M, \mathcal{N}^+(M) \rangle \models \phi(x)\})$ s.th. $\forall a \in A$ ($\langle M, \mathcal{N}^+(M) \rangle \models \psi[a]$)

(4) $\langle M, \mathcal{N}^+(M) \rangle \models \heartsuit x\phi(x) : \psi(x)$ iff $\{a \in M : \langle M, \mathcal{N}^+(M) \rangle \models \phi[a] \wedge \neg\psi[a]\} \notin \mathcal{N}(\{x : \langle M, \mathcal{N}^+(M) \rangle \models \phi(x)\})$.

Theorem 3.3.9

The axioms of Definition 3.3.6 capture $\mathcal{N}^+$– semantics of ∇.

3.3.2.6 Extension to $\mathcal{N}$-Families

So far, we can describe normal cases and the classical situation. We would like to generalize now to $\mathcal{N}$-families, sequences of $\mathcal{N}$-systems of increasing strength, i.e. "approaching" hard = classical information. For a motivation, the reader is referred to a semantics for defeasible inheritance, discussed in [Sch90].

Definition 3.3.8

(1) Let γ be any ordinal. Call $< \mathcal{N}_i(c) : i\langle\gamma\rangle, \mathcal{N}_i(c) \subseteq \mathcal{P}(c)$ a $\mathcal{N}$-family over c iff

    a. $c \in \mathcal{N}_i(c)$ for all i

    b. $a \in \mathcal{N}_i(c), a \subseteq b \subseteq c \to b \in \mathcal{N}_i(c)$ for all $i < \gamma$

    c. $< \mathcal{N}_i(c) : i\langle\gamma\rangle$ is decreasing, i.e. $a \in \mathcal{N}_i(c), j < i \to a \in \mathcal{N}_j(c)$

    d. $a \in \mathcal{N}_i(c), b \in \mathcal{N}_j(c) \to a \cap b \neq \emptyset$ for all i, j, if $c \neq \emptyset$.

(2) Call $\mathcal{N}^+(M) = \langle \mathcal{N}(N) : N \subseteq M \rangle$ a $\mathcal{N}^+$-family over M iff for each $N \subseteq M$ $\mathcal{N}^+(N)$ is a $\mathcal{N}$-family over N.

Remark 3.3.10

We have now formalized γ degrees of normality. Condition d. says, that all $\mathcal{N}_i(c)$ ($\mathcal{N}_i(c)$) are $\mathcal{N}$ ($\mathcal{N}^+$)− systems over c.

Definition 3.3.9

We introduce $\gamma * 2$ quantifiers $\nabla^i, \heartsuit^i, i < \gamma$ into first order predicate calculus, and the following axioms: Let any axiomatization of predicate calculus be given. Augment this with the axiom schemata

(1) For $\mathcal{N}$-families

    1. $\nabla^i x \phi(x) \wedge \forall x(\phi(x) \to \psi(x)) \to \nabla^i x \psi(x)$

    2. $\nabla^i x \phi(x) \to \neg \nabla^i x \neg \phi(x)$

    3. $\forall x \phi(x) \to \nabla^i x \phi(x) \to \exists x \phi(x)$

    4. $\nabla^i x \phi(x) \to \nabla^j x \phi(x)$ for $j < i$

    5. $\heartsuit^i x \phi(x) \leftrightarrow \neg \nabla^i x \neg \phi(x)$

    6. $\nabla^i x \phi(x) \leftrightarrow \nabla^i y \phi(y)$ for safe substitution

    (for all $\phi, \psi, i, j < \gamma$).

(2) For $\mathcal{N}^+$-families:

    1. a. $\nabla^i x \phi(x) \leftrightarrow \nabla^i x (x = x) : \phi(x)$

       b. $\forall x(\sigma(x) \leftrightarrow \tau(x)) \wedge \nabla^i x \sigma(x) : \phi(x) \to \nabla^i x \tau(x) : \phi(x)$

    2. $\nabla^i x \phi(x) : \psi(x) \wedge \forall x(\phi(x) \wedge \psi(x) \to \vartheta(x)) \to \nabla^i x \phi(x) : \vartheta(x)$

    3. $\exists x \phi(x) \wedge \nabla^i x \phi(x) : \psi(x) \to \neg \nabla^i x \phi(x) : \neg \psi(x)$

    4. $\forall x(\phi(x) \to \psi(x)) \to \nabla^i x \phi(x) : \psi(x) \to [\exists x \phi(x) \to \exists x(\phi(x) \wedge \psi(x))]$

    5. $\nabla^i x \phi(x) : \psi(x) \to \nabla^j x \phi(x) : \psi(x)$ for $j < i$

    6. $\heartsuit^i x \phi(x) : \psi(x) \leftrightarrow \neg \nabla^i x \phi(x) : \neg \psi(x)$

7. $\nabla^i x \phi(x) : \psi(x) \leftrightarrow \nabla^i y \phi(y) : \psi(y)$ for safe substitution

(for all $\phi, \psi, \vartheta, \sigma, \tau, i, j < \gamma$).

Definition 3.3.10

Let $\mathcal{L}$ be a first order language, and M a $\mathcal{L}$-structure. Let $\mathcal{N}(M)$ ($\mathcal{N}^+(M)$) be a $\mathcal{N}$ ($\mathcal{N}^+(M)$)-family over M. The definitions of $\langle M, \mathcal{N}(M) \rangle \models \phi$ etc. are straightforward, like $\langle M, \mathcal{N}(M) \rangle \models \nabla^i x \phi(x)$ iff there is $A \in \mathcal{N}_i(M)$ s.th. $\forall a \in A$ ($\langle M, \mathcal{N}(M) \rangle \models \phi[a]$) etc.

Theorem 3.3.11

(1) The axioms of Definition 3.3.9, 1) capture $\mathcal{N} - family$-semantics of ∇^i.

(2) The axioms of Definition 3.3.9, 2) capture $\mathcal{N}^+ - family$-semantics of ∇^i.

Proof

Analogous to the proofs of Lemma 3.3.6 and Theorem 3.3.9.

□

Remark:

To improve performance of a system, we might have several layers of information. The very general (and simple) information might be formulated with ∇^0-quantifiers, the more specific one with ∇^1- quantifiers etc., and the most specific (and hard) information classically, and all levels kept apart in the axiomatisation. Thus, when pressed for a fast answer, the system might "jump to a conclusion" by using only the ∇^0- information, without checking consistency (i.e. without using $\nabla^i- (i > 0)$ and hard information).

3.3.3 Strengthening the Axioms

3.3.3.1 Overview of This Section

In this section, we discuss various strengthenings of the axioms. They consist in adding properties to the single $\mathcal{N}$-systems (corresponding to part of "iterability" - see below), in particular making $\mathcal{N}$-systems filters, or coherence properties between the $\mathcal{N}$-systems over different subsets of the universe for the relativized case. They are motivated partly by the common use of defaults, partly by discussion elsewhere in the literature. In particular, we show how to interpret the axiom systems P and R of [KLM90] and [LM92] in our first order setting, several of those axioms trivially hold in our system, "And" corresponds again to the filter property, and others translate into more subtle interdependencies among the $\mathcal{N}$-systems of different subsets.

The ease with which we can incorporate these strengthenings is, of course, due to the fact that our system was chosen so weak at the outset, while making the language very expressive.

3.3.3.2 The Details

In the introduction to Section 3.3, we have discussed problems of homogenousness and iterability. What do they look like in Normal Case language? But, this is simple, iterability will be discussed presenty, and homogenousness with respect to $\sigma(x)$ amounts to $\nabla x \phi(x) \to \nabla x \sigma(x) : \phi(x)$. We can add these as single axioms, or as axiom schemata for all formulae σ, and all defaults $\nabla \phi$, $\nabla \psi$, when justified. So, we are allowed to apply the conjunction of two defaults only if we have deduced $\nabla x (\phi \wedge \psi)(x)$. And, if we know something which is more than a mere tautology about a, let's say $T \vdash \sigma(a)$, then we are allowed to apply the default $\nabla \phi$ only if we can deduce $T \vdash \nabla x \sigma(x) : \phi(x)$.

We can now give a formal definition of iterability within our extended language. If a default theory is closed under iterability, and the defaults are of a simple form, then the construction of a model is especially easy, this is the content of the subsequent lemma.

Definition 3.3.11

Iterability for defaults without prerequisites) We call the following axiom schema Iterability for defaults without prerequisites: $\nabla x \phi(x) \wedge \nabla x \psi(x) \to \nabla x (\phi \wedge \psi)(x)$

Lemma 3.3.12

Let D be a finite set of defaults of the form $\nabla x \phi_i(x)$, $i < n$, where ϕ_i is a classical formula, and let T' be a set of classical formulae, and suppose that D is closed under iterability. Then $D \cup T'$ is consistent iff $T' \cup \{\exists x (\phi_1 \wedge \ldots \wedge \phi_{n-1})(x)\}$ is consistent.

Proof

Trivial.

□

In the presence of defaults with prerequisites, iterability takes a slightly extended form. In addition, "normal use" of defaults sanctions still another axiom schema, which we call chaining. Again, we have a simplified model construction for simple defaults closed under iterability.

Definition 3.3.12

(Iterability for defaults with prerequisites) We call the following axiom schemata Iterability for defaults with prerequisites:

3.3 Defaults as Generalized Quantifiers

$\nabla x \phi(x) \wedge \nabla x \sigma(x) : \psi(x) \to \nabla x \sigma(x) : (\phi \wedge \psi)(x)$

$\nabla x \sigma(x) : \psi(x) \wedge \nabla x \sigma'(x) : \psi'(x) \to \nabla x (\sigma \wedge \sigma')(x) : (\psi \wedge \psi')(x)$

Definition 3.3.13

(Chaining)

We call the following axiom schemata Chaining:

$\nabla x \phi(x) \wedge \nabla x \phi(x) : \psi(x) \to \nabla x \psi(x)$

$\nabla x \sigma(x) : \psi(x) \wedge \nabla x \psi(x) : \psi'(x) \to \nabla x \sigma(x) : \psi'(x)$

Lemma 3.3.13

Let D be a finite set of defaults of the form $\nabla x \phi_i(x)$, $i < n$, let P be a finite set of defaults of the form $\nabla x \sigma_i(x) : \psi_i(x)$, $i < m$, where all ϕ_i, σ_i, ψ_i are classical formulae, and let T' be a set of classical formulae, and suppose that D and P are closed under iterability.

Then $D \cup P \cup T'$ is consistent iff the following set of (classical) formulae is consistent:

$C := T' \cup \{\exists x (\bigwedge \{\phi_i : i < n\})(x)\} \cup \{ \exists x (\bigwedge \{\sigma_i : i \in p\})(x) \to \exists x (\bigwedge \{\sigma_i : i \in p\} \wedge \bigwedge \{\psi_i : i \in p\} \wedge \bigwedge \{\phi_i : i < n\})(x) : p \subseteq m \}$

Definition 3.3.14

(Smoothness)

Call $\mathcal{N}^+(M)$ a smooth $\mathcal{N}^+$− system over M, iff it is a $\mathcal{N}^+$− system over M, and for each $N, N' \subseteq M, N \subseteq N', A \in \mathcal{N}(N'), A \subseteq N$ implies $A \in \mathcal{N}(N)$. This will be captured by adding the axiom schema $\nabla x \phi(x) : \psi(x) \wedge \forall x (\sigma(x) \to \phi(x)) \wedge \forall x (\phi(x) \wedge \psi(x) \to \sigma(x)) \to \nabla x \sigma(x) : \psi(x)$.

3.3.3.3 An Alternative Semantics for a Predicate Logic Version of P and R

We conclude by giving predicate logic versions of the systems P and R (see [KLM90] and [LM92]) an alternative semantics by translating $\phi(x) \mathrel{\mid\!\sim} \psi(x)$ into $\nabla \phi(x) : \psi(x)$.

For the convenience of the reader, we repeat these axiom systems in their propositional form:

We see that Right Weakening, Reflexivity, and Left Logical Equivalence are already built into our system.

The others have to be introduced by additional axiom schemata, like $\nabla x \alpha(x) : \beta(x) \to (\nabla x \alpha(x) : \neg \gamma(x) \vee \nabla x (\alpha \wedge \gamma)(x) : \beta(x))$ for Rational Monotony. The translation is obvious, so we discuss the semantical counterparts.

"And" says that $\mathcal{N}$-systems are closed under finite intersections, i.e. ω-complete filters. (This is part of the above iterability.)

"Or", Cautious and Rational Monotony concern the relation of the $\mathcal{N}$−systems of the subsets:

Or: $A \in \mathcal{N}(X), B \in \mathcal{N}(Y) \to A \cup B \in \mathcal{N}(X \cup Y)$

Cautious Monotony: $A, B \in \mathcal{N}(X) \to A \cap B \in \mathcal{N}(A)$

Rational Monotony: $A \in \mathcal{N}(X), B \subseteq X \to B \in \mathcal{N}(X)$ or $A - B \in \mathcal{N}(X\text{-B})$

3.3.4 Sceptical Revision of Partially Ordered Defaults

3.3.4.1 Overview of This Section

Recollect from Section 3.3.1 that the LILOG setting consisted of an order sorted language, which determines the strength of default information through the specificity of the sort to which the default is applied. This partial order on the default information will now be exploited to select, in case of conflict, a reasonable consistent subset of the information. Recall also that "conflict" is made precise as inconsistency in our generalized quantifier logic introduced in Section 3.3.2.

We present and discuss four alternative approaches to theory revision, and decide for the third, which seems to present the intuitively best results. More precisely, this solution is a generic one, as it still leaves much liberty of choice for the "parameters" f and $\prec \bullet$ (see below).

As usual in theory Revison, we try to retain as much information as possible - under some restrictions. In particular, when two formulas of equal strength are in conflict, we will be fair and exclude both. On the other hand, if ϕ and ψ are in conflict, and so are ψ and σ, we will sometimes like to preserve ϕ and σ, eliminating just ψ, and thus preserve a maximum of information. This is basically the problem we investigate here: Given such ϕ, ψ, σ and the partial order on formulas, what are the intuitions that might guide us in the selection of a subset $\{\phi, \psi, \sigma\}$, and how can we formalize our choice? In our proposed solution, not only the relation between $\{\phi, \psi\}$ and $\{\psi, \sigma\}$ - e.g., is ϕ stronger than σ? - will enter the picture, but also the certainty (to be called definiteness of choice below) with which we would eliminate e.g. ψ from $\{\phi, \psi\}$.

For more background, see again [Sch04].

3.3.4.2 Introduction

We will treat here a special case of theory revision based on axiom systems. Consider a situation of a (possibly inconsistent) partially ordered set of defaults. Our task will be to choose a consistent subset of those defaults, - where consistency is determined by the above discussed logic - using the given partial order, but being fair otherwise. Consequently, we will proceed sceptically, i.e. not necessarily choose a maximal consistent subset, as two contradictory defaults of the same or incomparable quality should both be excluded - as fairness dictates. The reader may as well assume that all formulae considered are classical ones, and the notion of consistency is the usual one. As a matter of fact, logic will play only a marginal role, being restricted to the notion of consistency.

First, we will introduce some basic definitions, and subsequently discuss several approaches to making a good and fair choice.

The reader will find many ideas from nonmonotonic inheritance theories applied to and generalized in the following.

3.3.4.3 Basic Definitions and Approaches

Let Σ be a set of sorts S, partially ordered by $\leq$, and Δ a set of defaults, where each $\delta \in \Delta$ belongs to some sort s, written $s \models \delta$, thus Δ inherits the order $\leq$ from the sorts. To avoid inessential complications, we assume Σ and Δ to be finite. "$s \models \delta$" is supposed to read something like "normally, all elements of the sort s, have the property $\delta(x)$". (In addition, at sorts s we might have classical information, but we shall always assume that the total classical information is consistent, and also that the defeasible information written directly to some s is always consistent too (even when taking the classical information into account). This will simplify matters, but is not essential.) Thus, for some $x \in s$, $s < t < u$, $t \models \phi$, $u \models \psi$, it is natural to consider $\phi(x)$ to be the stronger information than $\psi(x)$, because it is the more specific information: In case of conflict, ϕ should win over ψ. (Thus, quality decreases with increase by $\leq$!) If, however, t and u are incomparable, fairness dictates that a conflict between ϕ and ψ should result in disbelief of both ϕ and ψ. We thus consider $f := \mathcal{P}(\Delta) - \{\emptyset\} \to \mathcal{P}(\Delta)$ with $f(\alpha) \subsetneq \alpha$, where f is supposed to choose the "best" elements of α - if there are none, $f(\alpha)$ will be empty. Let further $\overline{f}(\alpha) := \alpha - f(\alpha)$.

Any such f gives rise naturally to a notion of quality of the choice: Suppose $card(\alpha) > 1$. If $card(\overline{f}(\alpha)) = 1$, then $f(\alpha)$ is a very definite choice, if $f(\alpha) = \emptyset$, i.e. $card(\overline{f}(\alpha)) = card(\alpha)$, $f(\alpha)$ is very indefinite. We can thus define the definiteness of the choice $f(\alpha)$ by

$$d(f(\alpha)) := \begin{cases} 1 & \text{iff } card(\alpha) = 1 \\ \frac{card(f(\alpha))}{card(\alpha)-1} & \text{otherwise} \end{cases}$$

(Thus, in the first case, $d(f(\alpha)) = 1$, in the second one $d(f(\alpha)) = 0$.) Speaking in terms of defeasible inheritance, $d(f(\alpha)) = 1$ corresponds to preclusion, $d(f(\alpha)) = 0$ to contradiction. We shall not use the notion of definiteness until the third approach, when we need a relation $\prec \bullet$ on minimal inconsistent subsets.

Examples: Let $g(\alpha)$ be the greatest element of α - i.e. for all $x \in \alpha$, $x \neq g(\alpha)$, $x < g(\alpha)$ - if it exists.

$$f_1(\alpha) := \begin{cases} \alpha - \{g(\alpha)\}, & \text{iff } g(\alpha) \text{ is defined} \\ \emptyset & \text{otherwise} \end{cases}$$

$$f_2(\alpha) := \begin{cases} \{\, x \in \alpha : x \text{ is no non-trivial maximum, i.e.} & \text{iff there are } x, y \in \alpha \\ \exists\, y \in \alpha (x < y) \text{ or } \neg \exists y \in \alpha(y < x)\,\} & \text{such that } x < y \\ \emptyset & \text{otherwise} \end{cases}$$

$f_3(\alpha) := \{\, x \in \alpha : x \text{ is a non-maximal element, i.e. } \exists y \in \alpha(x < y)\,\}$ By $f(\alpha) \subset \alpha$, $f(\alpha)$ will be consistent, if α is minimal inconsistent.

This suggests the probably simplest

Approach 1

Iterate some fixed f, starting on Δ, until consistency is reached. By finiteness of Δ and antitony of f, this will always work. This approach has some effects which may not always be desirable: Let $s < t < u$, $s \models \phi$, $t \models \neg \phi$, $u \models \psi$, where $\vdash \psi \to \phi$. ψ being the weakest information, it should always be eliminated first by f. On the other hand, one might argue that $\{\phi, \psi\}$ is a good choice, as ϕ should be accepted by being the best information, thus $\neg \phi$ should be out, leaving the way open to ψ.

Approach 2

We shall leave momentarily the above introduced function f, and even the order $\leq$ on sorts and defaults, and discuss a quite different way. We now consider arguments, which we identify with subsets of Δ, choosing the best ones, and hoping that the union of the best arguments is consistent. This meets with some problems. An order $\prec$ on arguments should respect the following properties: Let α, β etc be subsets of Δ. (Again, strength will decrease with increasing $\prec$.)

1. $\prec$ should be transitive, i.e. $\alpha \prec \beta \prec \gamma \to \alpha \prec \gamma$

2. $\alpha \prec \beta \subseteq \gamma \to \alpha \prec \gamma$. Reason: simply adding some information to an argument should not make it stronger. Adding the truth $2 + 2 = 4$ to an argument should

3.3 Defaults as Generalized Quantifiers

not give it more power. It is rather that the weakest part should determine its force.

3. $\forall i \in I(\alpha_i \prec \beta) \to \bigcup\{\alpha_i i \in I\} \prec \beta$: just putting arguments together should not violate a common upper bound. This seems to be well in accord with many natural definitions based purely on an order $\leq$ as given above - though not on all. (Consider e.g. $\alpha \prec \beta :\leftrightarrow \exists x \in \beta \forall y \in \alpha(y < x)$.)

4. The inverse seems to be very doubtful: $\beta \prec \bigcup\{\alpha_i : i \in I\} \to \exists i \in I(\beta \prec \alpha_i)$. This is already violated by $\alpha \prec \beta :\leftrightarrow \forall x \in \alpha \exists y \in \beta(x < y)$.

Yet rule 4 seems to suggest itself in our present approach, when trying to prove consistency: Let α, β, γ be pairwise consistent, but $\neg Con(\alpha \cup \beta \cup \gamma)$. If $\alpha \prec \beta$ and $\beta \prec \gamma$, then by 3., $\alpha \cup \gamma \prec \gamma$, and γ will be omitted in the inductive procedure, resulting in a consistent choice. If, however, $\alpha \perp \gamma$ (incom- parable), and $\beta \perp \gamma$, then maybe $\gamma \prec \alpha \cup \beta$ (unless 4. holds), so $\alpha \cup \beta$ will not be chosen, but maybe each of α, β, γ, resulting in an inconsistent theory.

Approach 3

We now work more closely again with the introduced functions f. First, a combinatorial result.

Definition 3.3.15

Let D be a finite set, $M' \subseteq \mathcal{P}(D)$, $\prec \bullet$ an acyclic binary relation on M', for $A \in M'$ let $\emptyset \neq X_A \subseteq A$ be defined.

Define inductively for $i \in \omega$:

$M'_i := \{ A \in M' : A \prec \bullet\text{-minimal in } M' - \bigcup\{M'_j : j < i\} \}$,
$M_i := \{ A \in M'_i : \forall j < i \forall B \in M_j (B \prec \bullet A \to A \cap X_B = \emptyset) \}$,
$X_i := \bigcup\{X_A : A \in M_i\}$ and
$M := \bigcup\{M_i : i \in \omega\}$, $X := \bigcup\{X_i : i \in \omega\}$, $D' := D - X$.

Let, in the sequel, the situation of the definition be given.

Lemma 3.3.14

For all $A \in M'$, $A \cap X \neq \emptyset$.

Remark 3.3.15

1) It suffices to choose $\prec \bullet$ well-founded in the above Definition, D and M' can then be infinite, too. We do not need the more general result, however.

2) Evidently, $\prec \bullet$ may be chosen as the empty relation.

3) Let $A \in M'$. Then $A \in M \leftrightarrow \forall B \prec \bullet A(B \in M \to A \cap X_B = \emptyset)$.

Proof by induction on i: Let $A \in M'_i$ be minimal such that the result fails. "$\to$": If $A \in M_i$ and $\exists B \prec \bullet A(B \in M, A \cap X_B \neq \emptyset)$, then $B \in M_j$ for some $j < i$, and $A \notin M_i$ by construction. "$\leftarrow$": Let $A \in M'_i - M_i$ s.th. $\forall B \prec \bullet A(B \in M \to A \cap X_B = \emptyset)$. Thus, for all $j < i, B \in M_j, B \prec \bullet A$ $A \cap X_B = \emptyset$, and $A \in M_i$ by construction again.

$\square$

Corollary 3.3.16

Let D be a finite set of formulae, $M' \subseteq \mathcal{P}(D)$ the set of minimal inconsistent sets of formulae from D. If X is defined as above, D' will be consistent.

Thus, letting f be as above, defining $X_A := A - f(A)$, any acyclic $\prec \bullet$ on minimal inconsistent subsets of Δ will give a consistent subset $\Delta' \subseteq \Delta$. Choosing $\prec \bullet$ non-empty will lead to some inconsistencies "being invisible", since some elements responsible are eliminated already. The effect can be seen easily when looking back at the example discussed in the first approach: Letting $\{\phi, \neg\phi\} \prec \bullet \{\neg\phi, \psi\}, \neg\phi$ might be eliminated (by suitable f and X) as $\{\phi, \neg\phi\}$ is $\prec \bullet$-minimal, leaving $\{\neg\phi, \psi\}$ out of consideration. The discussion of $\prec \bullet$ will be resumed in a moment.

Next, we look at the compatibility of this approach with an order on arguments (subsets of Δ).

It is natural to define for $\alpha, \beta \subseteq \Delta$: $\alpha \prec \beta :\leftrightarrow \alpha \subseteq f(\alpha \cup \beta)$ - leaving all logic aside for the moment. We have discussed the Postulates 1.-3. for orders on arguments above, and examine now which of the above introduced f_i satisfy some or all of these postulates. f_1 will violate 2., as the addition of any element, such that the largest element does not exist any more, will show. Consider f_2, and let $\alpha := \{x\}$, $\beta := \{y, z\}$, ordered by $y < z$ (as defaults) only. Then $f(\alpha \cup \beta) = \{x, y\}$, thus $\alpha \prec \beta$ (as arguments), which seems very unnatural. Looking at $f = f_3$, which preserves all non-maximal elements, we see that the order on arguments generated by f satisfies indeed 1.-3. This is immediate by $\alpha \prec \beta \leftrightarrow \alpha \subseteq f_3(\alpha \cup \beta) \leftrightarrow \forall x \in \alpha \exists y \in \alpha \cup \beta(x < y) \leftrightarrow \forall x \in \alpha \exists y \in \beta(x < y)$, the last equivalence by transitivity of the partial order on Σ.

We have now for $\delta \in \Delta$ ($\prec \bullet = \emptyset$ again, M' being the set of minimal inconsistent subsets of Δ, thus by $\prec \bullet = \emptyset$ $M = M'$): $\delta \in \Delta' \leftrightarrow \forall A \in M'(\delta \notin X_A) \leftrightarrow \forall A \in M'(\delta \in A \to \delta \in f(A)) \leftrightarrow \forall A \in M'(\delta \in A \to \{\delta\} \prec A - \{\delta\}) \leftrightarrow \forall \beta \subseteq \Delta(\neg Con(\delta, \beta) \to \{\delta\} \prec \beta)$. In the last equivalence, we use property 2.: "$\leftarrow$" is trivial. "$\to$": Suppose there is $\beta, \neg Con(\delta, \beta), \neg\{\delta\} \prec \beta$. Take $\beta' \subseteq \beta$ s.th. $\{\delta\} \cup \beta'$ is minimal inconsistent, so by prerequisite, $\{\delta\} \prec \beta'$, and by property 2. $\{\delta\} \prec \beta$.

Moreover, for $\alpha \subseteq \Delta'$, we have $\neg Con(\alpha \cup \beta) \to \alpha \prec \beta$: Let $\beta' \subseteq \beta$ s.th. $\alpha \cup \beta'$ is minimal inconsistent. As $\alpha \subseteq \Delta'$, $X_{\alpha \cup \beta'} \cap \alpha = \emptyset$, thus $\alpha \subseteq f(\alpha \cup \beta')$ and $\alpha \prec \beta'$, by 2. $\alpha \prec \beta$.

3.3 Defaults as Generalized Quantifiers

Let us resume the discussion of $\prec \bullet$. What are the meaning and properties of the $\prec \bullet$-relation? Let $A \prec \bullet B$. Thus, the elimination of the inconsistency A by eliminating the elements of X_A is estimated so strong that the inconsistency B need not be considered any more. In other words, the argument $B - X_B$ against X_B looses its importance. Now, we have by f a relation $\prec$ on arguments. Moreover, we have a notion of definiteness, resulting from f as well, as introduced at the beginning of the Section. A natural definition (suggested by S. Lorenz) of $\prec \bullet$ will be e.g.: $A \prec \bullet B :\leftrightarrow$ 1. $\overline{f}(A) \subseteq B$, 2. $A - B \prec B$, 3. $d(f(A)) = 1$, $d(f(B)) \neq 1$. By 3., any such $\prec \bullet$ will be acyclic, by the same reason, we can't have $A \prec \bullet B \prec \bullet C$. To conclude $A \prec \bullet B \subseteq C \to A \prec \bullet C$ for C s.th. $d(f(C)) \neq 1$, we need $A - C \prec C$, which will not always hold. (It will hold for $\prec$ defined by f_3.)

Approach 4

As a last alternative, we shall discuss a more "procedural approach", which tries to directly find a suitable subset $\Delta' \subseteq \Delta$. Again we suppose some subset function f to be given. Let $\phi_1 < \phi_2 < \phi_3$, $\phi_1 < \psi < \rho < \phi_4$ be defaults ordered by their sorts. The ψ, ρ are (logically) unimportant here and might be tautologies. Suppose further $\{\phi_1, \phi_2, \phi_4\}$ and $\{\phi_2, \phi_3\}$ are the minimal inconsistent subsets. Suppose further $f(\{\phi_1, \phi_2, \phi_4\}) = \{\phi_1\}$, $f(\{\phi_2, \phi_3\}) = \{\phi_2\}$, which seems a natural choice. Proceeding now by the rank of the ϕ_i in the $<$-order, at rank 2 ϕ_3 will be eliminated, as the elimination of ϕ_2 and ϕ_4 will be discovered only at rank 3. In other words, we have here a result similar to the one discussed in the first approach. (It corresponds again to choosing the empty order $\prec \bullet$ on minimal inconsistent subsets.)